Lecture Notes in Physics

Springer
Berlin
Heidelberg
New York
Barcelona
Hong Kong
London
Milan
Paris
Singapore
Tokyo

Physics and Astronomy ONLINE LIBRARY

http://www.springer.de/phys/

Editorial Policy

The series *Lecture Notes in Physics* (LNP), founded in 1969, reports new developments in physics research and teaching -- quickly, informally but with a high quality. Manuscripts to be considered for publication are topical volumes consisting of a limited number of contributions, carefully edited and closely related to each other. Each contribution should contain at least partly original and previously unpublished material, be written in a clear, pedagogical style and aimed at a broader readership, especially graduate students and nonspecialist researchers wishing to familiarize themselves with the topic concerned. For this reason, traditional proceedings cannot be considered for this series though volumes to appear in this series are often based on material presented at conferences, workshops and schools (in exceptional cases the original papers and/or those not included in the printed book may be added on an accompanying CD ROM, together with the abstracts of posters and other material suitable for publication, e.g. large tables, colour pictures, program codes, etc.).

Acceptance

A project can only be accepted tentatively for publication, by both the editorial board and the publisher, following thorough examination of the material submitted. The book proposal sent to the publisher should consist at least of a preliminary table of contents outlining the structure of the book together with abstracts of all contributions to be included.
Final acceptance is issued by the series editor in charge, in consultation with the publisher, only after receiving the complete manuscript. Final acceptance, possibly requiring minor corrections, usually follows the tentative acceptance unless the final manuscript differs significantly from expectations (project outline). In particular, the series editors are entitled to reject individual contributions if they do not meet the high quality standards of this series. The final manuscript must be camera-ready, and should include both an informative introduction and a sufficiently detailed subject index.

Contractual Aspects

Publication in LNP is free of charge. There is no formal contract, no royalties are paid, and no bulk orders are required, although special discounts are offered in this case. The volume editors receive jointly 30 free copies for their personal use and are entitled, as are the contributing authors, to purchase Springer books at a reduced rate. The publisher secures the copyright for each volume. As a rule, no reprints of individual contributions can be supplied.

Manuscript Submission

The manuscript in its final and approved version must be submitted in camera-ready form. The corresponding electronic source files are also required for the production process, in particular the online version. Technical assistance in compiling the final manuscript can be provided by the publisher's production editor(s), especially with regard to the publisher's own Latex macro package which has been specially designed for this series.

Online Version/ LNP Homepage

LNP homepage (list of available titles, aims and scope, editorial contacts etc.):
http://www.springer.de/phys/books/lnpp/

LNP online (abstracts, full-texts, subscriptions etc.):
http://link.springer.de/series/lnpp/

H. Latal W. Schweiger (Eds.)

Methods of Quantization

Lectures Held at the 39. Universitätswochen für Kern- und Teilchenphysik, Schladming, Austria

Springer

Editors

H. Latal
W. Schweiger
Institut für Theoretische Physik
Universität Graz
Universitätsplatz 5
8010 Graz, Austria

Supported by the Österreichische Bundesministerium für Wirtschaft, Verkehr und Kunst, Vienna, Austria

Cover picture: see contribution by B. Bakker in this volume.

Library of Congress Cataloging-in-Publication Data applied for.

Die Deutsche Bibliothek - CIP-Einheitsaufnahme

Methods of quantization : lectures held at the 39. Universitätswochen für Kern- und Teilchenphysik, Schladming, Austria / H. Latal ; W. Schweiger (ed.). - Berlin ; Heidelberg ; New York ; Barcelona ; Hong Kong ; London ; Milan ; Paris ; Singapore ; Tokyo : Springer, 2001
(Lecture notes in physics ; 572)
(Physics and astronomy online library)
ISBN 3-540-42100-9

ISSN 0075-8450
ISBN 3-540-42100-9 Springer-Verlag Berlin Heidelberg New York

Typesetting: Camera-ready by the authors/editors
Camera-data conversion by Steingraeber Satztechnik GmbH Heidelberg
Cover design: *design & production*, Heidelberg

Printed on acid-free paper
SPIN: 10836063 55/3141/du - 5 4 3 2 1 0

Preface

This volume contains the written versions of invited lectures presented at the "39. Internationale Universitätswochen für Kern- und Teilchenphysik" in Schladming, Austria, which took place from February 26th to March 4th, 2000. The title of the school was "Methods of Quantization". This is, of course, a very broad field, so only some of the new and interesting developments could be covered within the scope of the school.

About 75 years ago Schrödinger presented his famous wave equation and Heisenberg came up with his algebraic approach to the quantum-theoretical treatment of atoms. Aiming mainly at an appropriate description of atomic systems, these original developments did not take into consideration Einstein's theory of special relativity. With the work of Dirac, Heisenberg, and Pauli it soon became obvious that a unified treatment of relativistic and quantum effects is achieved by means of local quantum field theory, i.e. an intrinsic many-particle theory. Most of our present understanding of the elementary building blocks of matter and the forces between them is based on the quantized version of field theories which are locally symmetric under gauge transformations. Nowadays, the prevailing tools for quantum-field theoretical calculations are covariant perturbation theory and functional-integral methods. Being not manifestly covariant, the Hamiltonian approach to quantum-field theories lags somewhat behind, although it resembles very much the familiar nonrelativistic quantum mechanics of point particles. A particularly interesting Hamiltonian formulation of quantum-field theories is obtained by quantizing the fields on hypersurfaces of the Minkowsi space which are tangential to the light cone. The "time evolution" of the system is then considered in "light-cone time" $x^+ = t + z/c$. The appealing features of "light-cone quantization", which are the reasons for the renewed interest in this formulation of quantum field theories, were highlighted in the lectures of Bernard Bakker and Thomas Heinzl. One of the open problems of light-cone quantization is the issue of spontaneous symmetry breaking. This can be traced back to zero modes which, in general, are subject to complicated constraint equations. A general formalism for the quantization of physical systems with constraints was presented by John Klauder. The perturbative definition of quantum field theories is in general afflicted by singularities which are overcome by a regularization and renormalization procedure. Structural aspects of the renormalization problem in the case of gauge invariant field theories were discussed in the lecture of Klaus Sibold. A review of the mathematics underlying the functional-integral quantization was given by Ludwig Streit.

Apart from the topics included in this volume there were also lectures on the Kaluza–Klein program for supergravity (P. van Nieuwenhuizen), on dynamical r-matrices and quantization (A. Alekseev), and on the quantum Liouville model as an instructive example of quantum integrable models (L. Faddeev). In addition, the school was complemented by many excellent seminars. The list of seminar speakers and the topics addressed by them can be found at the end of this volume. The interested reader is requested to contact the speakers directly for detailed information or pertinent material.

Finally, we would like to express our gratitude to the lecturers for all their efforts and to the main sponsors of the school, the Austrian Ministry of Education, Science, and Culture and the Government of Styria, for providing generous support. We also appreciate the valuable organizational and technical assistance of the town of Schladming, the Steyr-Daimler-Puch Fahrzeugtechnik, Ricoh Austria, Styria Online, and the Hornig company. Furthermore, we thank our secretaries, S. Fuchs and E. Monschein, a number of graduate students from our institute, and, last but not least, our colleagues from the organizing committee for their assistance in preparing and running the school.

Graz,
March 2001

Heimo Latal
Wolfgang Schweiger

Table of Contents

List of Contributors

B.L.G. Bakker
Vrije Universiteit
Dept. of Physics and Astronomy
De Boelelaan 1081
NL-1081 Amsterdam
The Netherlands
blgbkkr@nat.vu.nl

T. Heinzl
Friedrich-Schiller-Universität
Theoretisch-Physikalisches Institut
Max-Wien-Platz 1
D-07743 Jena
Germany
heinzl@tpi.uni-jena.de

J.R. Klauder
University of Florida
Physics Department
P.O. Box 118440
Gainesville, FL 32611-8440
USA
klauder@phys.ufl.edu

K. Sibold
Universität Leipzig
Institut für Theoretische Physik
Augustusplatz 10
D-04109 Leipzig
Germany
sibold@physik.uni-leipzig.de

L. Streit
Universität Bielefeld
BiBoS
D-33615 Bielefeld
Germany
streit@physik.uni-bielefeld.de

Forms of Relativistic Dynamics

Bernard L.G. Bakker

Vrije Universiteit, Department of Physics and Astronomy,
NL-1081 Amsterdam, De Boelelaan 1081,The Netherlands

Abstract. Since Dirac wrote his famous article on forms of relativistic dynamics, it has been realized that the front form, or light-front dynamics, is ideally suited for the solution of the bound state problem in quantum field theory. Still, it is useful to know what the other forms are and what makes the front form so well-adapted to non-perturbative problems.

First, a brief discussion is given of the Poincaré group and its connection to different forms of dynamics as described by Dirac. Next the question of equivalence of the different forms of dynamics is discussed. It is shown that the field-theoretical formulae for the Poincaré generators follow Dirac's classification: kinematic vs. dynamic.

A difficulty that always arises in quantum field theory is the need for regularization to render the results of actual computations finite. In a Hamiltonian framework one cannot immediately apply all methods devised for covariant approaches: e.g. dimensional regularization. Thus new methods must be used and the results compared to calculations carried out in the standard, covariant way. This is done in perturbation theory applied to the case of light-front quantization, where many results are known from the literature, so Hamiltonian methods can be checked explicitly. In this part examples are treated in some detail to illustrate the characteristic features of a light-front calculation.

1 Introduction

The two fundamental revolutions in physics of the twentieth century: relativity theory and quantum mechanics, force us to formulate questions about the smallest building blocks of matter in a language that accounts for the quantum nature of those systems, yet respects the fundamental space-time symmetries. Relativistic quantum field theory provides such a language. After more than a half century of development it is clear that the manifestly covariant formulation, pioneered by Feynman, has many advantages if one deals with problems that may be solved by perturbative methods. The questions concerning the regulation of divergent integrals appearing in the naive application of the Feynman rules have been answered in various ways and the program of renormalization was successfully carried out for almost all interesting field theories. (Gravitation is a well known exception.)

Notwithstanding these achievements, there is room for alternative approaches. A purely theoretical reason for following another path is that the study of alternatives tends to highlight the strengthes and weaknesses of either approach. Secondly, one formulation may be intuitively more appealing than the other. Hamiltonian formulations of field theory, being not *manifestly* covariant, are not immediately recognized as equivalent to the Feynman way. Nevertheless, they are closer to the familiar quantum mechanics of point particles and were historically the first to be used. This is the reason that some authors refer to Hamiltonian methods as "old fashioned". Furthermore, it seems that they lend themselves naturally to the solution of bound-state problems.

As any physical observable, S-matrix element, bound-state mass, magnetic moment, ... must be invariant under proper space-time transformations, the challenge of practical calculations in the framework of Hamiltonian dynamics is to produce invariant results for observables. An important application of Hamiltonian dynamics is to nuclear physics. Traditionally nonrelativistic model Hamiltonians were used in this field, but since the advent of powerful accelerators that can boost hadrons to energies far exceeding their masses, it has become clear that the implementation of a relativistic framework is unavoidable. In addition, the common practice of leaning heavily on field theory to construct the so called realistic nuclear forces, made it clear that also in nuclear physics one needs to take the requirements of special relativity seriously.

The concept of "relativistic Hamiltonian dynamics" needs to be properly defined. This is our first topic. The appropriate symmetry, the Poincaré group, will be briefly discussed. This leads naturally to the different forms of dynamics, introduced in a famous paper by Dirac [5]. Later two more forms of dynamics were described in [16], bringing the total number to five. There is a fivefold ambiguity of relativistic dynamics, as can be seen by analyzing the classification of all subgroups of the Poincaré group.

In view of the challenge to maintain the space-time symmetries, one may wonder why one should consider the Hamiltonian formulation at all. One reason is that nonperturbative problems may be solved by matrix diagonalization, as one is used to in many-body theory. In order to make this program viable, it is necessary to guarantee that the dimensions of the matrices involved, are within the limits that present day computers pose. So it is important to investigate whether any of the five forms of dynamics is more suited to implementation on the computer than the others. One consideration comes to mind immediately: the Fock-state expansion is in principle different for the various forms of dynamics, as its terms are not invariant. Therefore the investigation of the Fock column must be an issue. We shall discuss one example in detail.

Finally, some definite examples in one particular form: the "front form" also known as Light-Cone Quantization or Light-Front Quantization or Dynamics, are discussed in detail. Light-Front Dynamics (LFD) is argued to be most suitable for numerical treatment as the vacuum is particularly simple in this form. The examples are taken from the perturbative domain for several reasons: (i) one cannot hope to solve the problems of maintaining symmetry and regularization/renormalization in a nonperturbative context if they are not solved in perturbation theory, (ii) the techniques used and the results obtained are interesting in themselves and much more easy to illustrate in perturbation theory, and (iii) many more results are available in the perturbative domain.

It will turn out that LFD in the perturbative regime contains additional singularities, so called "longitudinal" ones, that do not occur in the covariant formulation. If these are subtracted, LF perturbation theory reproduces the results of the Feynman approach.

The equivalence of the Hamiltonian methods in perturbation theory being established, one may turn to nonperturbative problems. The scope of this lecture does not include a treatment of bound states. However, here we mention the very promising development called Discretized Light Cone Quantization, that aims at fully solving bound states in field theory, its accuracy limited only by the capacity of available computers. A review of this method can be found in [3].

These lectures do not aim at a comprehensive treatment of the different forms of relativistic dynamics. Some excellent reviews on the subject have been written. We mention [14] and [17] for a discussion of systems with a fixed number of particles. The progress on LFD can be traced in the proceedings of several workshops devoted to that subject e.g. [13,27,6,7,12]. A different approach to LFD is advocated by Carbonell et al. [4]. Pioneering work on nonperturbative QCD in LF quantization was done by Wilson et al. [29].

2 The Poincaré Group

It is the main purpose of this short section to fix the notation. A discussion of the Poincaré group can be found in numerous books on group theory, as well as in the literature devoted to field theory and particle physics.

We denote the generators of space-time transformations by

P^μ space-time translations,
$M^{\mu\nu}$ pure Lorentz transformations.

Their commutation relations determine the Poincaré algebra, which in its turn determines the Poincaré group locally. They are

$$\begin{aligned}
[P^\mu, P^\nu] &= 0\,, \\
[M^{\mu\nu}, P^\sigma] &= i\,(P^\mu g^{\nu\sigma} - P^\nu g^{\mu\sigma})\,, \\
[M^{\mu\nu}, M^{\rho\sigma}] &= i\,(g^{\nu\rho} M^{\mu\sigma} - g^{\mu\rho} M^{\nu\sigma} + g^{\mu\sigma} M^{\nu\rho} - g^{\nu\sigma} M^{\mu\rho})\,. \qquad (1)
\end{aligned}$$

The well known physical interpretation of these operators is

$$\begin{aligned}
J^i &= \tfrac{1}{2}\epsilon_{ijk} M^{jk}\,, \\
K^i &= M^{0i}\,. \qquad (2)
\end{aligned}$$

One can split the angular-momentum tensor $M^{\mu\nu}$ into two pieces: one part $L^{\mu\nu}$, that corresponds to the orbital angular momentum and another, $S^{\mu\nu}$, that corresponds to the intrinsic spin [10,9].

Irreducible representations of the symmetry are characterized by invariants. They are the mass m and the intrinsic spin s. The mass is a constraint on the components of the momentum,

$$P_\mu P^\mu = m^2. \tag{3}$$

The intrinsic spin is determined by the Pauli-Lubanski pseudovector

$$W^\mu = -\tfrac{1}{2}\epsilon^{\mu\nu\rho\sigma} P_\nu S_{\rho\sigma}, \tag{4}$$

its square being an invariant

$$W_\mu W^\mu = -m^2 \boldsymbol{S}^2. \tag{5}$$

Substitution of $M^{\mu\nu}$ for $S^{\mu\nu}$ (4) would make no difference to W.

The components of W have a simple interpretation; the zeroth component is proportional to the helicity

$$W^0 = \boldsymbol{P} \cdot \boldsymbol{S} \tag{6}$$

and

$$\boldsymbol{W} = P^0 \boldsymbol{S} \tag{7}$$

is proportional to the intrinsic spin.

If the mass is determined as the square root of the eigenvalue of P^2, then the spin can be calculated by dividing the eigenvalue of $-W^2$ by m^2.

It is the subject of relativistic dynamics to find representations of these operators in a physical form, e.g. as differential, integral or matrix operators on states. The simplest realization is the one called an "elementary particle" which according to Wigner is a unitary, irreducible representation: a state of definite mass and spin. Next one may consider a collection of noninteracting particles of different masses and spins and construct realizations of the Poincaré algebra for them. This task is almost trivial as the tensor product of representations does the job. Much more difficult is the construction of representations in the case of interacting particles. This is the topic of relativistic dynamics proper. One way of doing it is covariant field theory. The generators are then expressed in terms of integrals of the energy-momentum tensor. Such a construction is not always straightforward, but as a starting point it is very useful. The next section deals with the question of what different forms dynamics may take.

3 Forms of Relativistic Dynamics

In his ground breaking paper, Dirac [5] formulated two requirements on relativistic dynamical systems:

> General relativity requires that physical laws expressed in terms of a system of curvilinear coordinates in space-time, shall be invariant under transformations from one such coordinate system to another.

and

> A second general requirement for dynamical theory has been brought to light through the discovery of quantum mechanics by Heisenberg and Schrödinger, namely the requirement that the equations of motion shall be expressible in the Hamiltonian form.

These conditions do not by themselves define a dynamical system, but rather limit the possible forms it may take. A proper determination of the dynamics involves the specification of the interactions. In nonrelativistic dynamics only one unique way is allowed: the interaction must be included in the Hamiltonian. All other generators–of the Galilei group in this case–are independent of the interaction.

The evolution of a system with nonrelativistic dynamics is governed fully by the Hamiltonian: given the state of the system at some time $t = 0$, one may calculate its state at any other time using the evolution operator $U(t) = \exp(-iHt)$. The state specification at the surface $t = 0$, an *instant in time*, represents the initial conditions. For the Galilei group the instant is the only appropriate initial surface.

For systems that are governed by Einstein relativity, more possibilities are open as the family of world lines is more restricted. Any hypersurface Σ in Minkowski-space that does not contain timelike directions (lightlike directions are allowed) can be used to formulate the initial conditions. If no more limitations are set, the choice is infinite, but it is useful to try and find surfaces with the highest possible symmetries. This leads to the concept of the *stability group* G_Σ, the subgroup of the Poincaré group that maps the surface Σ onto itself. The subset of generators of the full group that generate elements of G_Σ are said to be *kinematical* operators. The other generators map Σ into another surface, $\Sigma \to \Sigma'$. They are said to be *dynamical* operators. (Dirac called them "Hamiltonians" but we shall not follow this terminology.) If Σ has the property

$$\forall x, y \in \Sigma : \exists\, g \in G_\Sigma \to x = gy, \tag{8}$$

then it is said to be *transitive* and all points in Σ are equivalent. Now, if we limit our initial surfaces to transitive ones, there exist just five different – inequivalent – possibilities, corresponding to the five subgroups of the Poincaré group. Dirac himself discussed three forms

$$\begin{aligned} &\text{Instant Form} && x^0 = 0, \\ &\text{Point Form} && x^2 = a^2 > 0,\ x^0 > 0, \\ &\text{Front Form} && x^0 + x^3 = 0. \end{aligned} \tag{9}$$

After a full classification of the subgroups of the Poincaré group was given the remaining ones could be found. They are given by Leutwyler and Stern, viz.

$$(x^0)^2 - (x^1)^2 - (x^2)^2 = a^2 > 0,\ x^0 > 0,$$
$$(x^0)^2 - (x^3)^2 = a^2 > 0,\ x^0 > 0. \tag{10}$$

As the latter have not been used in practice, we shall not discuss them any further.

The instant form is of course the one best known. The other two forms discussed by Dirac have also been applied, the point form less widely than the front form.

3.1 Comparison of Instant Form, Front Form, and Point Form

In Table 3.1 we summarize the classification of the three forms, instant form (IF), front form (LF) and point form (PF) dynamics. We need the following notation for any vector A^μ in the front form

$$A^\pm = \frac{1}{\sqrt{2}}(A^0 \pm A^3), \quad \boldsymbol{A}^\perp = (A^1, A^2), \tag{11}$$

hence

$$A \cdot B = A_\mu B^\mu = A^- B^+ + A^+ B^- - \boldsymbol{A}^\perp \cdot \boldsymbol{B}^\perp. \tag{12}$$

In the instant form we use the four-velocity, denoted by u^μ, and related to the four momentum by

$$p^\mu = m u^\mu, \quad u^\mu u_\mu = 1. \tag{13}$$

From Table 3.1 read off that the instant form has the advantage that the rotations are all kinematical, which means that a classification of states with respect to their spins is immediate. The point form shares this property with the IF, and has in addition kinematical boosts. However, in the point form all components of the four-momentum are dynamical. The front form is in between, as it has kinematical as well as dynamical rotations and boosts. A disadvantage of LFD is that only boosts in the z-direction and rotations around the z-axis are kinematical.

Naturally the question arises how to describe interacting systems. In field theory, to which we shall turn later, this question is solved in a standard way. In the case of a system consisting of a fixed, finite number of particles the answer is complicated.

Dirac [5] identified the "real difficulties" for the three forms. Some simple requirements can be given, related to the commutators of the Poincaré generators that are linear in the interactions. We give them here. In the formulae below the summation runs over the particle labels.

Instant Form

$$\begin{aligned} P^0 &= \sum \sqrt{\boldsymbol{p}^2 + m^2} + V, \\ M^{0r} &= \sum x^r \sqrt{\boldsymbol{p}^2 + m^2} + V^r, \end{aligned} \tag{14}$$

where V is a three-dimensional scalar, independent of the origin of the coordinates $\boldsymbol{x}$, and $\boldsymbol{V}$ is a three-dimensional vector, such that

$$\boldsymbol{V} = \boldsymbol{x} V + \boldsymbol{V}', \tag{15}$$

where $\boldsymbol{V}'$ is again independent of the origin of the coordinates. The real difficulty is to satisfy the commutators $[V, \boldsymbol{V}]$ and $[V^i, V^j]$ that follow from the Poincaré algebra.

Point Form

$$\begin{aligned} P^\mu &= \sum [p^\mu + x^\mu B(p^2 - m^2)] + V^\mu, \\ B(p^2 - m^2) &= \frac{1}{x^2}\left[\sqrt{(p\cdot x)^2 - x^2(p^2 - m^2)} - p\cdot q\right]. \end{aligned} \tag{16}$$

The interaction V^μ must be a four-vector and the real difficulty is to satisfy the commutators $[V^\mu, V^\nu]$ that follow from $[P^\mu, P^\nu] = 0$.

Front Form

$$\begin{aligned} P^- &= \sum \frac{\boldsymbol{p}^{\perp\,2} + m^2}{2p^+} + V, \\ M^{-\,i} &= \sum \left[x^i \frac{\boldsymbol{p}^{\perp\,2} + m^2}{2p^+} - x^- p^i\right] + V^i. \end{aligned} \tag{17}$$

The interaction V must be invariant under all transformations of $\boldsymbol{x}^\perp$ and x^-, except those of the form $x^- \to \lambda x^-$, in which case $V \to \lambda V$. The interactions $\boldsymbol{V}^\perp$ can be written as

$$\boldsymbol{V}^\perp = \boldsymbol{x}^\perp V + \boldsymbol{V}'^\perp, \tag{18}$$

where $\boldsymbol{V}'^\perp$ is subject to the same limitations as V, and in addition transforms as a vector under rotations around the z-axis.

A complete construction of the generators was given by Bakamjian and Thomas [1] starting from an invariant mass operator. Their method is peculiar in that all interaction dependence is introduced solely through this operator. It was proven by Sokolov and Shatny [26] that this leads to *equivalent* forms of dynamics. These authors consider two forms equivalent if their Hamiltonians are related by a unitary similarity transformation and the S-matrix elements calculated in these two forms coincide.

Table 3.1. Comparison of three different forms of dynamics

Instant Form	**Front Form**	**Point Form**
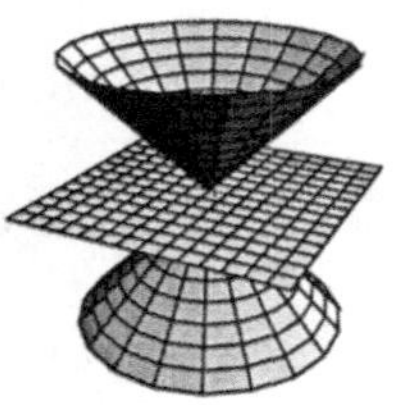	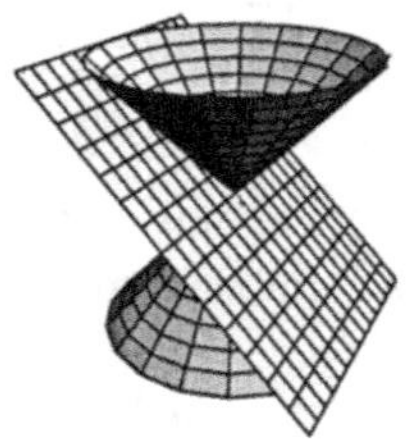	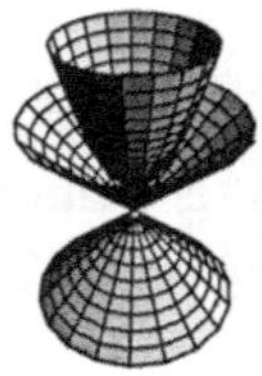
Quantization Surface		
$x^0 = 0$	$x^0 + x^3 = 0$	$x^2 = a^2 > 0$, $x^0 > 0$
Kinematical Generators		
$\boldsymbol{P}$ $\boldsymbol{J}$	P^+, $\boldsymbol{P}^\perp$ $E^1 = M^{+1} = \frac{K_x + J_y}{\sqrt{2}}$ $E^2 = M^{+2} = \frac{K_y - J_x}{\sqrt{2}}$ $J_z = M^{12}$ $K_z = M^{-+}$	$M^{\mu\nu}$
Dynamical Generators		
P^0 $\boldsymbol{K}$	P^- $F^1 = M^{-1} = \frac{K_x - J_y}{\sqrt{2}}$ $F^2 = M^{-2} = \frac{K_y + J_x}{\sqrt{2}}$	P^μ
Plane-wave Representation		
$\vert\boldsymbol{p}\rangle$ $p^0 = \pm\sqrt{\boldsymbol{p}^2 + m^2}$ $p^0 > 0$ and $p^0 < 0$ not kinematically disjoint	$\vert p^+, \boldsymbol{p}^\perp\rangle$ $p^- = \frac{\boldsymbol{p}^{\perp\,2} + m^2}{2p^+}$ $p^- > 0 \leftrightarrow p^+ > 0$ $p^- < 0 \leftrightarrow p^+ < 0$	$\vert\boldsymbol{u}\rangle$ $u^\mu = p^\mu / m$, $u^2 = 1$ $u^0 = \pm\sqrt{\boldsymbol{u}^2 + 1}$ not kinemat. disjoint
Measure		
$\int \frac{\mathrm{d}^3 p}{2p^0}$	$\int \frac{\mathrm{d}^2 p^\perp \mathrm{d}p^+}{2p^+}$	$\int \frac{\mathrm{d}^3 u}{2g^0}$

Using this construction one obtains for instance

$$\begin{aligned} P^0 &= \sqrt{\boldsymbol{p}^2 + M_{\mathrm{IF}}^2}, \quad \text{(instant form)}, \\ P^\mu &= M_{\mathrm{PF}}\, u^\mu, \quad \text{(point form)} \\ P^- &= \frac{\boldsymbol{p}^{\perp\,2} + M_{\mathrm{LF}}^2}{2p^+}. \quad \text{(front form)} \end{aligned} \tag{19}$$

The three mass operators are not identical, but related.

These results express the formal limitations set by relativistic invariance, but do not determine the interactions explicitly. For guidance in interaction choice one may, and frequently does, resort to field theory. In Sect. 5 we discuss some aspects of Hamiltonian dynamics in that context.

4 Light-Front Dynamics

Up till now the discussion has been rather general. Now we turn to some specific problems in order to illustrate some ideas discussed so far. We shall limit ourselves to LFD, because it has some unusual and unexpected features.

In the present section we discuss the entanglement of the Fock-space expansion with space-time symmetries. For pedagogical reasons spin is ignored here, so the specific case considered is not very realistic. After discussing how the generators of the Poincaré group are related to the underlying Lagrangean in Sect. 5, we give a more realistic example in a field-theoretical context.

Light-front dynamics (LFD) is singled out for corresponding to the smallest number of dynamical generators: three. This property by itself is not important enough to warrant a preference for LFD. Much more important is the fact that one can make a useful distinction between over-all and relative variables, in a way quite similar to CM and relative variables in nonrelativistic theories.

Another advantage, already stressed by Dirac, is the *spectrum property*. It is connected to the condition that for massive physical particles both P^2 and P^0 must be positive. Then P^+ and P^- must be positive too. Now in IFD this condition must be implemented separately, but in LFD the positivity of P^- follows from the positivity of P^+: states of positive and negative energies are separated kinematically. This property is of eminent importance for the role the vacuum plays in LFD. As the vacuum has energy zero, only particles with mass zero can be created from the LF vacuum, unlike the IF vacuum that can create particles with nonvanishing energy, if their energies sum up to zero.

4.1 Relative Momentum, Invariant Mass

In this subsection we define a relative momentum such that the invariant mass can be expressed in terms of the relative-momentum components only. We shall discuss first the case where four-momentum conservation can be used and next the case where it cannot.

Conserved Four Momentum Consider the case of two free particles, with masses m_1 and m_2. For free particles the four momenta add up, so we have

$$P = p_1 + p_2 \Leftrightarrow P^{\pm} = p_1^{\pm} + p_2^{\pm}, \quad \boldsymbol{P}^{\perp} = \boldsymbol{p}_1^{\perp} + \boldsymbol{p}_2^{\perp}. \tag{20}$$

The invariant mass is of course given by

$$P^2 = M^2, \tag{21}$$

leading to the energy-momentum dispersion relation

$$P^- = \frac{\boldsymbol{P}^{\perp 2} + M^2}{2P^+}. \tag{22}$$

We define relative variables

$$x = \frac{p_1^+}{P^+}, \quad \boldsymbol{q}^\perp = (1-x)\boldsymbol{p}_1^\perp - x\boldsymbol{p}_2^\perp. \tag{23}$$

The inverse relation is

$$\boldsymbol{p}_1^\perp = x\boldsymbol{P}^\perp + \boldsymbol{q}^\perp, \quad \boldsymbol{p}_2^\perp = (1-x)\boldsymbol{P}^\perp - \boldsymbol{q}^\perp. \tag{24}$$

One can find the invariant mass in terms of x and $\boldsymbol{q}^\perp$,

$$\begin{aligned} M^2 &= 2P^+P^- - \boldsymbol{P}^{\perp 2} \\ &= \frac{\boldsymbol{q}^{\perp 2} + m_1^2}{x} + \frac{\boldsymbol{q}^{\perp 2} + m_2^2}{1-x}. \end{aligned} \tag{25}$$

One can define a vector $\boldsymbol{q}$ by adding to $\boldsymbol{q}^\perp$ the third component q_z, given by

$$q_z = x(1-x)\frac{\partial M}{\partial x} = (x - \tfrac{1}{2})M - \frac{m_1^2 - m_2^2}{2M}. \tag{26}$$

In order to justify the use of the word vector for $\boldsymbol{q}$ we must prove that its length is independent of the reference frame.

From the expression of M in terms of $\boldsymbol{q}^\perp$ we derive

$$\boldsymbol{q}^{\perp 2} = x(1-x)M^2 - (1-x)m_1^2 - xm_2^2. \tag{27}$$

Then we can calculate the square of $\boldsymbol{q}$:

$$\boldsymbol{q}^{\perp 2} + q_z^2 = \frac{(M^2 - m_1^2 - m_2^2)^2 - 4m_1^2m_2^2}{4M^2}. \tag{28}$$

As M is an invariant, $\boldsymbol{q}^2$ is also an invariant. In fact, $\boldsymbol{q}$ is the momentum of particle 1 in the center of momentum frame ($\boldsymbol{P} = 0$). Consequently, we can equate the invariant mass with the energy in the CM frame

$$M = \sqrt{m_1^2 + \boldsymbol{q}^2} + \sqrt{m_2^2 + \boldsymbol{q}^2} \equiv E_1 + E_2. \tag{29}$$

Using the last expression one can relate x to q_z as follows

$$x = \frac{\sqrt{m_1^2 + \boldsymbol{q}^2} + q_z}{\sqrt{m_1^2 + \boldsymbol{q}^2} + \sqrt{m_2^2 + \boldsymbol{q}^2}}. \tag{30}$$

The vector property of $\boldsymbol{q}$ can be related to the orbital angular momentum operator $\boldsymbol{L}$. If one defines the LF helicity L_3 by

$$L_3 = -i\left(q_1\frac{\partial}{\partial q_2} - q_2\frac{\partial}{\partial q_1}\right), \tag{31}$$

and defines the two other components by

$$L_r = i\epsilon_{rs}\left[-\frac{q_s}{M}\frac{\partial}{\partial x} + x(1-x)\frac{\partial M}{\partial x}\frac{\partial}{\partial q_s}\right], \tag{32}$$

then these three components together are related to the momentum $\boldsymbol{q}$ in the usual way

$$\boldsymbol{L} = -i\boldsymbol{q} \times \boldsymbol{\nabla}_q. \tag{33}$$

Off-Energy-Shell The considerations above are relevant for on-energy-shell states, i.e. states that have the same total energy and kinematical momenta. Then they have the same invariant mass too. If one evaluates diagrams beyond tree level, either in perturbation theory or as part of the kernel of an integral equation, one has to deal with intermediate states that are off the energy shell, and which are connected to each other by the action of interactions. Then P^- is not conserved, as is of course to be expected, as the interactions are the dynamical ingredients. So let us lift the condition that $P^- = p_1^- + p_2^-$. The algebra can still be carried out as before, but we shall express everything in terms of the quantity

$$M^2 = 2P^+(p_{1\,\mathrm{on}}^- + p_{2\,\mathrm{on}}^-) - \boldsymbol{P}^{\perp 2}, \quad p_{i\,\mathrm{on}}^- = \frac{\boldsymbol{p}_i^{\perp 2} + m_i^2}{2p^+}. \tag{34}$$

If four-momentum is not conserved, this is *not* an invariant, so the "vector" $\boldsymbol{q}$ is not the CM momentum of particle 1 and its square is not an invariant under rotations. This can be contrasted to IFD, where three-vectors are kinematical, so their squares are invariant under rotations.

Phase Space The IFD phase space $\mathrm{d}^3p/2E$ translates into $\mathrm{d}p^\perp\mathrm{d}^2p^+/2p^+$ in LFD. If we use the relative coordinates, then we find for the two-body phase space

$$\begin{aligned}
\mathrm{d}^2p_1^\perp\frac{\mathrm{d}p_1^+}{2p_1^+}\mathrm{d}^2p_2^\perp\frac{\mathrm{d}p_2^+}{2p_2^+} &= \mathrm{d}^2P^\perp\mathrm{d}^2q^\perp\mathrm{d}P^+\mathrm{d}x\frac{P^+}{2p_1^+2p_2^+} \\
&= \mathrm{d}^2P^\perp\mathrm{d}P^+\frac{\mathrm{d}x}{2x(1-x)}\mathrm{d}^2q^\perp.
\end{aligned} \tag{35}$$

The usefulness of $\boldsymbol{q}$ becomes apparent again if we express the phase space in terms of $\boldsymbol{q}$ too. The relevant Jacobian is

$$\frac{\partial x}{\partial q_z} = \frac{x(1-x)M}{E_1E_2}, \tag{36}$$

so using (35, 36) we find the internal phase space expressed in terms of $\boldsymbol{q}$:

$$\frac{\mathrm{d}x}{2x(1-x)}\mathrm{d}^2q^{\perp} = \mathrm{d}^3\boldsymbol{q}\frac{M}{2E_1E_2}. \tag{37}$$

Exchange Diagrams As an example of the difference between covariant and LF diagrams we calculate the one-boson exchange diagram. The kinematics is defined in Fig. 1. The solid lines denote "nucleons" with mass m_1 and m_2, the dashed lines "mesons" with mass μ. As we shall see later, in LF perturbation theory amplitudes are expressed in terms of energy denominators and phase-space factors, instead of Feynman propagators. The energy denominators D_a and D_b are

$$\begin{aligned} D_a &= P^- - p^-_{1\,\mathrm{on}} - p^-_{4\,\mathrm{on}} - q^-_{\mathrm{on}} \\ &= P^- - \frac{m_1^2 + \boldsymbol{p}_1^{\perp\,2}}{2p_1^+} - \frac{m_4^2 + \boldsymbol{p}_4^{\perp\,2}}{2p_4^+} - \frac{\mu^2 + \boldsymbol{q}^{\perp\,2}}{2(p_3^+ - p_1^+)}, \end{aligned} \tag{38}$$

$$\begin{aligned} D_b &= P^- - p^-_{2\,\mathrm{on}} - p^-_{3\,\mathrm{on}} - q^-_{\mathrm{on}} \\ &= P^- - \frac{m_2^2 + \boldsymbol{p}_2^{\perp\,2}}{2p_2^+} - \frac{m_3^2 + \boldsymbol{p}_3^{\perp\,2}}{2p_3^+} - \frac{\mu^2 + \boldsymbol{q}^{\perp\,2}}{2(p_1^+ - p_3^+)}. \end{aligned} \tag{39}$$

The two time-ordered diagrams are equal to the same covariant amplitude, but in two different kinematical domains: (a) $p_3^+ - p_1^+ = p_2^+ - p_4^+ > 0$ and (b) $p_3^+ - p_1^+ = p_2^+ - p_4^+ < 0$, respectively. The perpendicular momentum transfer is

$$\boldsymbol{q}^{\perp} = \boldsymbol{p}_3^{\perp} - \boldsymbol{p}_1^{\perp} = \boldsymbol{p}_2^{\perp} - \boldsymbol{p}_4^{\perp}. \tag{40}$$

We can write this in invariant form in terms of the relative coordinates

$$\boldsymbol{q}_{12}^{\perp} = x_2\boldsymbol{p}_1^{\perp} - x_1\boldsymbol{p}_2^{\perp}, \quad \boldsymbol{q}_{34}^{\perp} = x_4\boldsymbol{p}_3^{\perp} - x_3\boldsymbol{p}_4^{\perp} \tag{41}$$

and

$$x_i = p_i^+/P^+. \tag{42}$$

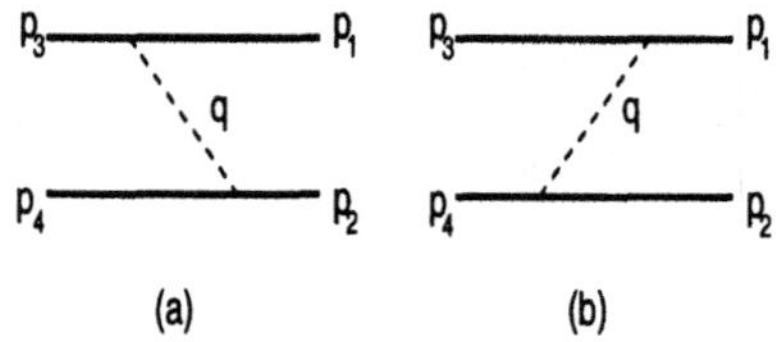

Fig. 1. Kinematics of LF time ordered exchange diagrams. (a) $q^+ = p_3^+ - p_1^+ > 0$, (b) $q^+ = p_1^+ - p_3^+ > 0$.

The formulae are

$$2P^+ D_a = s - \left[\frac{m_1^2 + \boldsymbol{q}_{12}^{\perp\,2}}{x_1} + \frac{m_4^2 + \boldsymbol{q}_{34}^{\perp\,2}}{x_4} + \frac{\mu^2 + (\boldsymbol{q}_{12}^\perp - \boldsymbol{q}_{34}^\perp)^2}{x_3 - x_1}\right] \equiv s - M_a^2\,, \tag{43}$$

$$2P^+ D_b = s - \left[\frac{m_2^2 + \boldsymbol{q}_{12}^{\perp\,2}}{x_2} + \frac{m_3^2 + \boldsymbol{q}_{34}^{\perp\,2}}{x_3} + \frac{\mu^2 + (\boldsymbol{q}_{12}^\perp - \boldsymbol{q}_{34}^\perp)^2}{x_1 - x_3}\right] \equiv s - M_b^2\,. \tag{44}$$

If the states with the momenta p_i^+ and $\boldsymbol{p}_i^\perp$ are *not* on the energy shell, this is our final formula. Such is the case if the exchange diagrams are parts of the kernel of an integral equation to be used in a nonperturbative calculation. However, in the other case we can write

$$P^2 = s = M_{12}^2 = M_{34}^2. \tag{45}$$

Then we can rewrite $s - M_a^2$ and $s - M_b^2$ in terms of a four-momentum transfer squared. We shall do this for $M_{12}^2 - M_a^2$.

$$\begin{aligned} M_{12}^2 - M_a^2 = {} & \frac{m_1^2 + \boldsymbol{q}_{12}^{\perp\,2}}{x_1} + \frac{m_2^2 + \boldsymbol{q}_{12}^{\perp\,2}}{x_2} \\ & - \left[\frac{m_1^2 + \boldsymbol{q}_{12}^{\perp\,2}}{x_1} + \frac{m_4^2 + \boldsymbol{q}_{34}^{\perp\,2}}{x_4} + \frac{\mu^2 + (\boldsymbol{q}_{12}^\perp - \boldsymbol{q}_{34}^\perp)^2}{x_3 - x_1}\right] \\ = {} & \frac{m_2^2 + \boldsymbol{q}_{12}^{\perp\,2}}{x_2} - \frac{m_4^2 + \boldsymbol{q}_{34}^{\perp\,2}}{x_4} - \frac{\mu^2 + (\boldsymbol{q}_{12}^\perp - \boldsymbol{q}_{34}^\perp)^2}{x_3 - x_1} \end{aligned} \tag{46}$$

As we are now in the covariant case: external particles on shell *and* states on energy shell, we can calculate the square of the four-momentum transfer in the ordinary way. We find

$$(p_2 - p_4)^2 = 2(p_2^+ - p_4^+)(p_2^- - p_4^-) - (\boldsymbol{p}_2^\perp - \boldsymbol{p}_4^\perp)^2. \tag{47}$$

We can simplify this expression

$$(p_2 - p_4)^2 = (x_2 - x_4)\left[\frac{m_2^2 + \boldsymbol{q}_{12}^{\perp\,2}}{x_2} - \frac{m_4^2 + \boldsymbol{q}_{34}^{\perp\,2}}{x_4}\right] - (\boldsymbol{q}_{12}^\perp - \boldsymbol{q}_{34}^\perp)^2\,. \tag{48}$$

It is straightforward to derive from this equation the formulae for case (a), $x_3 - x_1 = x_2 - x_4 > 0$

$$(p_2 - p_4)^2 - \mu^2 = (x_2 - x_4)(M_{12}^2 - M_a^2), \tag{49}$$

$$(p_3 - p_1)^2 - \mu^2 = (x_3 - x_1)(M_{34}^2 - M_a^2), \tag{50}$$

and for case (b); $x_1 - x_3 = x_4 - x_2 > 0$

$$(p_1 - p_3)^2 - \mu^2 = (x_1 - x_3)(M_{12}^2 - M_b^2), \tag{51}$$

$$(p_4 - p_2)^2 - \mu^2 = (x_4 - x_2)(M_{34}^2 - M_b^2). \tag{52}$$

If the states are not on the energy shell, M_{12} and M_{34} are not the same and may both differ from $\sqrt{s}$, so $(p_2 - p_4)^2$ and $(p_1 - p_3)^2$ are in general different. One can derive the amusing identity

$$M_{12}^2 + M_{34}^2 = M_a^2 + M_b^2. \tag{53}$$

4.2 The Box Diagram

The tree-level diagrams discussed previously, when used to calculate S-matrix elements, are invariant. However, if they are embedded in a larger diagram, e.g. as kernels in a Lippmann-Schwinger type approach to non-perturbative dynamics, one needs off-energy-shell amplitudes. Then it appears that space-time symmetries are violated if Fock space is truncated improperly.

The simplest place to illustrate this feature is the box diagram in the same purely scalar theory: heavy scalar particles with mass m ("nucleons") interacting with light scalars with mass μ ("mesons"). We look at the process of two nucleons with momenta p and q respectively, coming in and exchanging two mesons of mass μ. The outgoing nucleons have momenta p' and q'. The kinematics is given in Fig. 2. The internal momenta are related to the external ones by four-momentum conservation, which hold for those components of the momenta that are conserved.

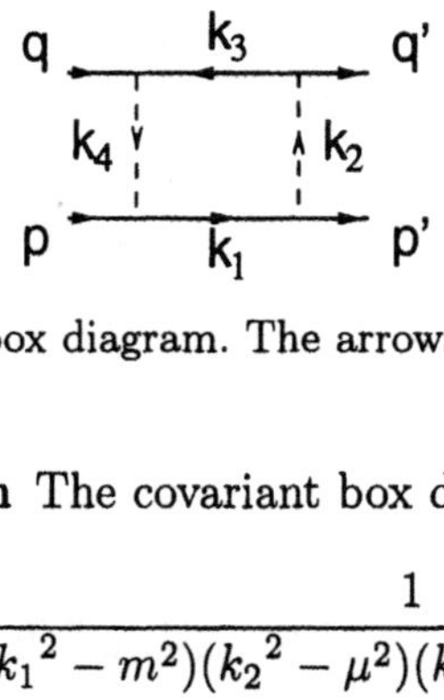

Fig. 2. Kinematics for the box diagram. The arrows denote the momentum flow.

Covariant Box Diagram The covariant box diagram is given by

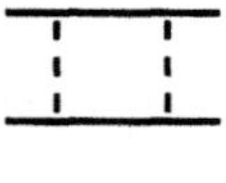

$$= -i \int \mathrm{d}^4 k \frac{1}{({k_1}^2 - m^2)({k_2}^2 - \mu^2)({k_3}^2 - m^2)({k_4}^2 - \mu^2)}, \tag{54}$$

where the imaginary parts $i\epsilon$ of the masses are not written explicitly and the dependence on the coupling constant and factors of 2π are all left out. The amplitude can be evaluated in the usual way by using Feynmans trick to write the integrand in terms of a single denominator and performing a Wick rotation.

LF Time-Ordered Diagrams It is well-known [15] how to construct the LF time-ordered diagrams. We shall illustrate the construction explicitly in the more complicated case of a Yukawa model with nucleons of spin-1/2 later. Here we just mention that LF time ordered diagrams are obtained by integrating over the minus component of the free loop momentum. As a result one obtains several LF time ordered diagrams corresponding to one covariant amplitude. For the box we find

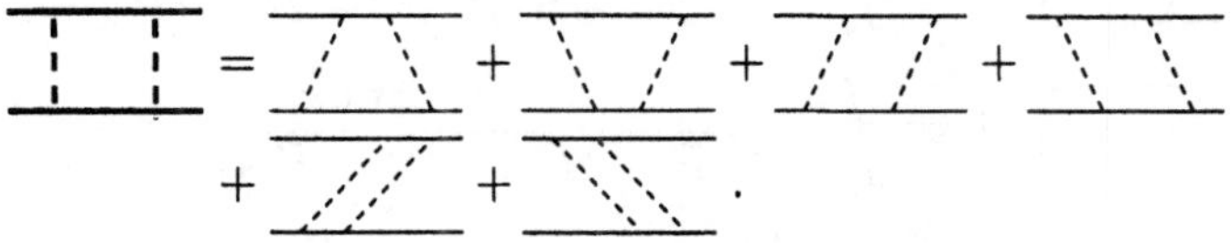

(55)

The first four diagrams contain two- and three-particle Fock states, the last two – the so called "stretched boxes" – contain also four-particle Fock states. Their contribution measures the importance of four-particle states for the calculation of the box diagram.

The time-ordered amplitudes are expressed in terms of energy denominators and phase-space factors. The phase space factor is

$$\Phi = 16|k_1^+ k_2^+ k_3^+ k_4^+|. \tag{56}$$

Without loss of generality we can take $p^+ \geq p'^+$. The internal particles are on mass-shell, however, the intermediate states are off energy-shell. A number of intermediate states occur. We label the corresponding kinetic energies according to which of the internal particles, labeled by $k_1 \dots k_4$ in Fig. 2, are in this state.

$$H_{14} = q^- + \frac{{k_1^\perp}^2 + m^2}{2k_1^+} - \frac{{k_4^\perp}^2 + \mu^2}{2k_4^+}, \tag{57}$$

$$H_{13} = \frac{{k_1^\perp}^2 + m^2}{2k_1^+} - \frac{{k_3^\perp}^2 + m^2}{2k_3^+}, \tag{58}$$

$$H_{12} = q'^- + \frac{{k_1^\perp}^2 + m^2}{2k_1^+} - \frac{{k_2^\perp}^2 + \mu^2}{2k_2^+}, \tag{59}$$

$$H_{34} = p^- - \frac{{k_3^\perp}^2 + m^2}{2k_3^+} + \frac{{k_4^\perp}^2 + \mu^2}{2k_4^+}, \tag{60}$$

$$H_{24} = q'^- + p^- + \frac{{k_2^\perp}^2 + \mu^2}{2k_2^+} - \frac{{k_4^\perp}^2 + \mu^2}{2k_4^+}, \tag{61}$$

$$H_{23} = p'^- + \frac{{k_2^\perp}^2 + \mu^2}{2k_2^+} - \frac{{k_3^\perp}^2 + m^2}{2k_3^+}. \tag{62}$$

A minus sign occurs if the particle goes in the direction opposite to the direction defined in Fig. 2. All particles are on mass-shell, including the external ones:

$$\begin{aligned} q^- &= \frac{{q^\perp}^2 + m^2}{2q^+}, & q'^- &= \frac{{q'^\perp}^2 + m^2}{2q'^+}, \\ p^- &= \frac{{p^\perp}^2 + m^2}{2p^+}, & p'^- &= \frac{{p'^\perp}^2 + m^2}{2p'^+}. \end{aligned} \tag{63}$$

We can now construct the LF time-ordered diagrams.

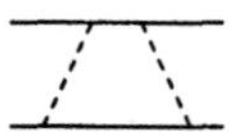

$$= \int \mathrm{d}^2 k^{\perp} \int_0^{p'^+} \frac{-2\pi \, \mathrm{d}k^+}{\Phi\,(P^- - H_{14})\,(P^- - H_{13})\,(P^- - H_{12})}, \tag{64}$$

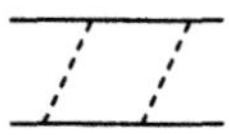

$$= \int \mathrm{d}^2 k^{\perp} \int_{p'^+}^{p^+} \frac{-2\pi \, \mathrm{d}k^+}{\Phi\,(P^- - H_{14})\,(P^- - H_{13})\,(P^- - H_{23})}, \tag{65}$$

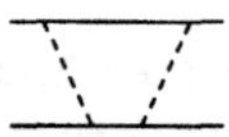

$$= \int \mathrm{d}^2 k^{\perp} \int_{p^+}^{p^+ + q^+} \frac{-2\pi \, \mathrm{d}k^+}{\Phi\,(P^- - H_{34})\,(P^- - H_{13})\,(P^- - H_{23})}, \tag{66}$$

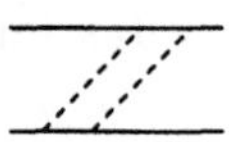

$$= \int \mathrm{d}^2 k^{\perp} \int_{p'^+}^{p^+} \frac{-2\pi \, \mathrm{d}k^+}{\Phi\,(P^- - H_{14})\,(P^- - H_{24})\,(P^- - H_{23})}, \tag{67}$$

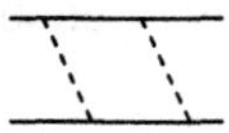 $=$ 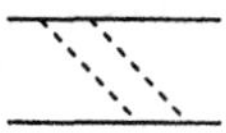 $= 0. \qquad (68)$

The factor 2π is the product of $-2\pi i$ from the k^--integration and the factor $-i$ in (54). The last two diagrams are zero because we have taken $p^+ \geq p'^+$ and therefore these diagrams have an empty k^+-range. If we take $p^+ \leq p'^+$, the diagrams in (68) have nonvanishing contributions.

Numerical Experiment In order to estimate how important the higher Fock states can be in practice, and to illustrate the dependence of the different LF diagrams on the orientation of the reference frame, we give the results of a "numerical experiment". We look at the scattering of two particles over an angle of $\pi/2$. In Fig. 3 the process is viewed in two different ways.

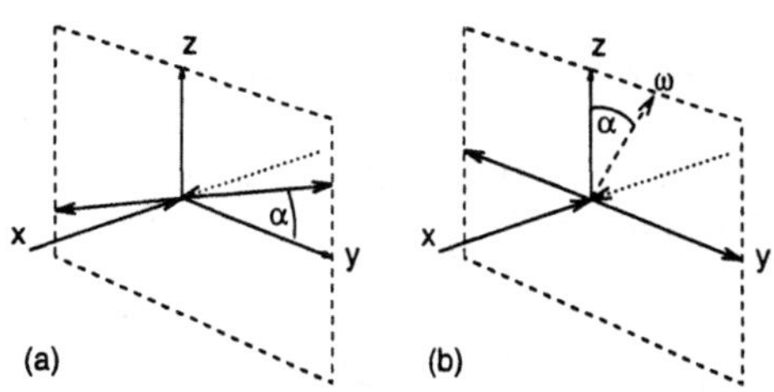

Fig. 3. (a) Two particles come in along the x-axis. They scatter into the $y-z$ plane over an angle of $\pi/2$. The azimuthal angle is given by α. (b) Another viewpoint. The outgoing particles move along the y-axis. The normal on the light-front ω makes an angle α with respect to the z-axis.

Fig. 3a pictures the situation where the scattering plane is rotated around the x-axis. The viewpoint in Fig. 3b concentrates on the influence of the orientation of the quantization plane. Both viewpoints should render identical results, since all angles between the five relevant directions (the quantization axis and the four external particles) are the same. We choose for the momenta

$$p^\mu = (v^0, +v^x, \quad 0, \quad 0), \tag{69}$$

$$q^\mu = (v^0, -v^x, \quad 0, \quad 0), \tag{70}$$

$$p'^\mu = (v^0, \quad 0, -v^y, -v^z), \tag{71}$$

$$q'^\mu = (v^0, \quad 0, +v^y, +v^z). \tag{72}$$

indicating that we have chosen the fixed quantization plane $x^+ = 0$ (Fig. 3a). The incoming and outgoing particles are required to have the same absolute values of the momenta in the CM system. Therefore

$$|\boldsymbol{v}|^2 = (v^x)^2 = (v^y)^2 + (v^z)^2 = |\boldsymbol{v}'|^2. \tag{73}$$

If v^0 and $|\boldsymbol{v}| = |\boldsymbol{v}'|$ are kept constant, while the azimuthal angle α given by

$$\alpha = \arctan \frac{v^z}{v^y}, \tag{74}$$

varies, the Mandelstam variables s, t, and u are constant too and so must be the invariant amplitude.

We are now ready to perform a numerical experiment. Two parameters are focused on. We vary the azimuthal angle α in the y-z-plane, and the incoming CM momentum $v = v^x$. In the remainder we will omit the units for the masses (MeV/c^2).

Numerical Results Two nucleons of mass $m = 940$ scatter via the exchange of scalar mesons of mass $\mu = 140$. First we varied the direction of $\boldsymbol{v}'$, given by the azimuthal angle α, but kept its length fixed. Therefore the Mandelstam variables are independent of α, and the full amplitude must be invariant. We tested this numerically for a number of values of v. In the region $0 \leq \alpha \leq \pi$ we used the formulas (64-67). In the region $\pi \leq \alpha \leq 2\pi$ the diagrams (65) and (67) vanish. However, then there are contributions from the diagrams in (68). The results are shown in Fig. 4. The results are normalized to the value of the covariant amplitude. The contributions from the different diagrams vary strongly with the angle α. Since the imaginary parts are always positive, they are necessarily in the range $[0, 1]$ when divided by the imaginary part of the covariant amplitude. The real parts can behave much more wildly, especially for higher values of the incoming CM momentum v. Clearly the LF time-ordered diagrams add up to the covariant amplitude, so we see that in all cases we obtain covariant (in particular rotationally invariant) results for both the real and the imaginary part.

After this numerical investigation of the dependence of the LF-time ordered diagrams on the kinematics, we also investigated the energy dependence of the different contributions. As the stretched boxes are maximal for $\alpha = \pi/2$, we give the results for that case, i.e. scattering in the x-z-plane.

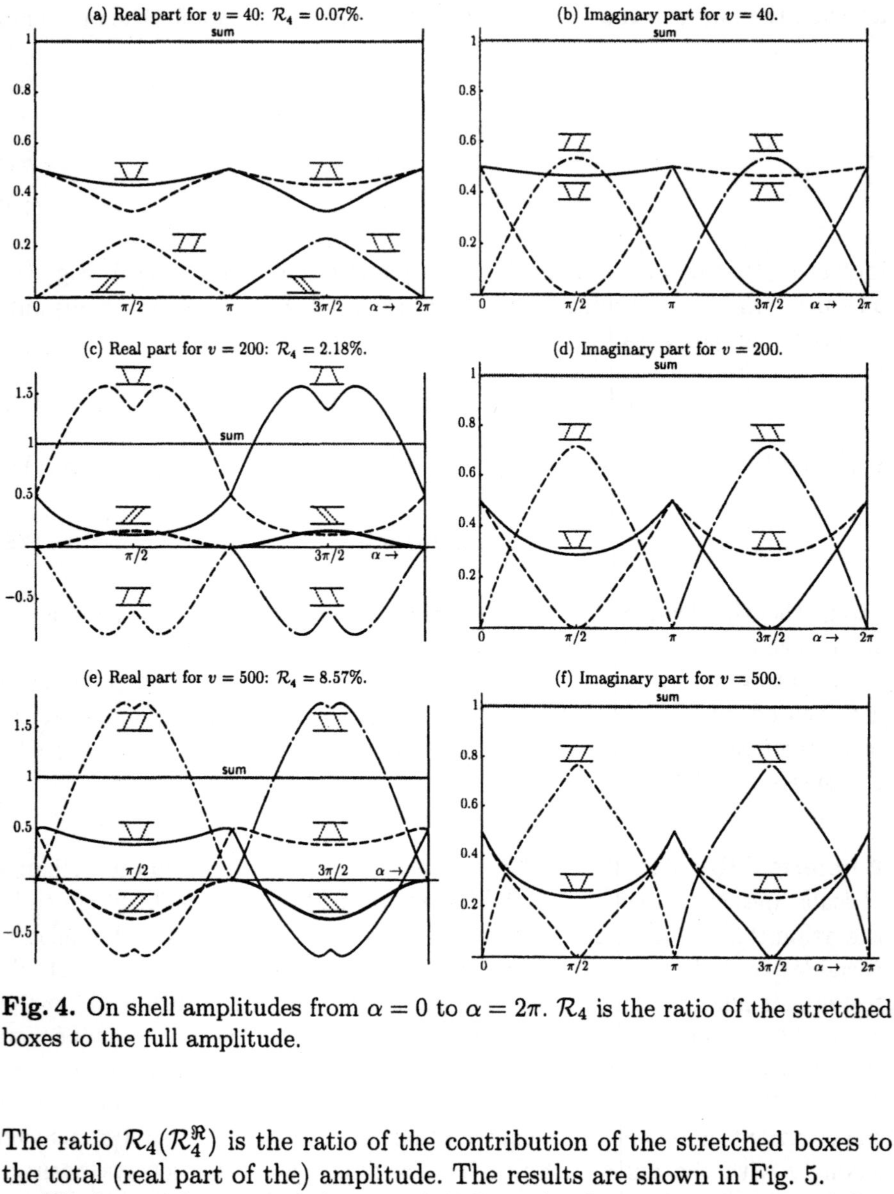

Fig. 4. On shell amplitudes from $\alpha = 0$ to $\alpha = 2\pi$. $\mathcal{R}_4$ is the ratio of the stretched boxes to the full amplitude.

The ratio $\mathcal{R}_4(\mathcal{R}_4^{\Re})$ is the ratio of the contribution of the stretched boxes to the total (real part of the) amplitude. The results are shown in Fig. 5.

We depict the ratio of the stretched box, the diagram with two simultaneously exchanged mesons, to both the real part and to the magnitude of the total amplitude. Since the real part has a zero near $v = 280$, the ratio $\mathcal{R}_4^{\Re}$ becomes infinite at that value of the incoming momentum. Therefore $\mathcal{R}_4$ gives a better impression of the contribution of the stretched box. We conclude from our numerical results that the stretched box is relatively small at low energies, but becomes rather important at higher energies.

If the angular momentum operator would have been kinematical, as in IFD, then each of the six diagrams would have been independent of the

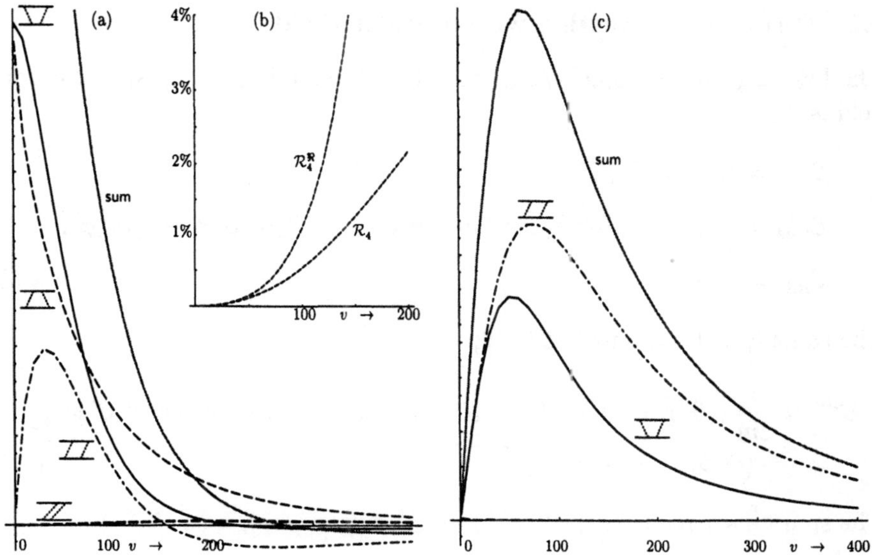

Fig. 5. Real (a) and imaginary part (c) of the LF time-ordered boxes for $\alpha = \pi/2$ as a function of the momentum of the incoming particles v. The inset (b) shows the ratio of the stretched box to the real part of the amplitude ($\mathcal{R}_4^{\Re}$) and to the absolute value ($\mathcal{R}_4$).

orientation of the reference frame (but in IFD many more diagrams would occur). The results illustrate the fact that the angular momentum is a dynamical operator, only its third component is kinematical. So the separate diagrams depend on the orientation of the reference frame, or, equivalently, the orientation of the light front.

5 Poincaré Generators in Field Theory

Covariant field theory is defined in terms of a Lagrangean density, which can be integrated over space-time to yield the action, a true Lorentz scalar:

$$A = \int \mathrm{d}^4 x \mathcal{L}(x). \tag{75}$$

If the action is an invariant, the Euler-Lagrange equations are covariant.

The object that is most useful when constructing the generators of space-time translations and Lorentz transformations, is the energy-momentum tensor $\Theta^{\mu\nu}$. It can be used to determine the generators of the Poincaré group. We shall derive them for two forms, IFD and LFD, in a specific case: the interaction of fermions and scalar bosons.

5.1 Fermions Interacting with a Scalar Field

The Lagrangean for a spin-1/2 fermion field interacting with a spin-0 boson field is

$$\begin{aligned}\mathcal{L} &= \mathcal{L}_{\text{free}} + \mathcal{L}_{\text{int}}\,,\\ \mathcal{L}_{\text{free}} &= \frac{i}{2}(\bar{\psi}\gamma^\mu\partial_\mu\psi - (\partial_\mu\bar{\psi})\gamma^\mu\psi) - m\bar{\psi}\psi + \tfrac{1}{2}(\partial_\mu\phi)(\partial^\mu\phi) - \tfrac{1}{2}m_b^2\phi^2,\\ \mathcal{L}_{\text{int}} &= -g\bar{\psi}\psi\phi. \end{aligned} \tag{76}$$

The canonical stress tensor is [2,10]

$$\begin{aligned}\Theta^{\mu\nu} = \frac{1}{2m}&\left[(\partial^\mu\bar{\psi})(\partial^\nu\psi) + (\partial^\nu\bar{\psi})(\partial^\mu\psi) + ig\phi\left\{(\partial^\nu\bar{\psi})\gamma^\mu\psi - \bar{\psi}\gamma^\mu(\partial^\nu\psi)\right\}\right]\\ &+(\partial^\mu\phi)(\partial^\nu\phi) - g^{\mu\nu}\mathcal{L}. \end{aligned} \tag{77}$$

The spin tensor reads

$$M^{\rho\mu\nu} = \frac{i}{4m}\left[\bar{\psi}\sigma^{\mu\nu}\partial^\rho\psi - (\partial^\rho\bar{\psi})\sigma^{\mu\nu}\psi + ig\phi\bar{\psi}\left\{\sigma^{\mu\nu}\gamma^\rho + \gamma^\rho\sigma^{\mu\nu}\right\}\psi\right]. \tag{78}$$

5.2 Instant Form

We calculate the components of the energy-momentum tensor, the generators of the space-time translations and the Lorentz transformations first in instant form. In the following section they will be calculated in front form.

Free Boson Poincaré Generators The boson part of the stress tensor reads:

$$\Theta^{\mu\nu} = (\partial^\mu\phi)(\partial^\nu\phi) - \tfrac{1}{2}g^{\mu\nu}\left[(\partial_\rho\phi)(\partial^\rho\phi) - m_b^2\phi^2\right]. \tag{79}$$

Hence

$$\Theta^{00} = \tfrac{1}{2}\left[(\partial^0\phi)^2 + (\nabla\phi)^2 + m_b^2\right], \quad \Theta^{0i} = \partial^0\phi\partial^i\phi. \tag{80}$$

The space-time translation operators are (the subscript "free" denotes the case where no interaction is included in the Lagrangean)

$$\begin{aligned}P^0_{\text{free}}(t) &= \int \mathrm{d}^3\boldsymbol{x}\tfrac{1}{2}\left[(\partial^0\phi)^2 + (\nabla\phi)^2 + m_b^2\right],\\ P^i_{\text{free}}(t) &= \int \mathrm{d}^3\boldsymbol{x}\,(\partial^0\phi)(\partial^i\phi). \end{aligned} \tag{81}$$

The Lorentz generators are

$$\begin{aligned}M^{ij}_{\text{free}}(t) &= \int \mathrm{d}^3\boldsymbol{x}\left[x^i(\partial^0\phi)(\partial^j\phi) - x^j(\partial^0\phi)(\partial^i\phi)\right],\\ M^{0i}_{\text{free}}(t) &= x^0P^i_{\text{free}} - \int \mathrm{d}^3\boldsymbol{x}\,x^i\left[(\partial^0\phi)^2 + (\nabla\phi)^2 + m_b^2\right]. \end{aligned} \tag{82}$$

Free Fermion Poincaré Generators The free fermion stress tensor reads

$$\Theta^{\mu\nu} = \frac{i}{2}[\bar{\psi}\gamma^{\mu}\partial^{\nu}\psi - (\partial^{\nu}\bar{\psi})\gamma^{\mu}\psi] - g^{\mu\nu}\mathcal{L}. \tag{83}$$

Hence (making use of the Dirac equation we see that the part $g^{\mu\nu}\mathcal{L}$ does not contribute to $\Theta^{\mu\nu}$) we obtain for the momentum operators

$$P^{\mu} = \int \mathrm{d}^3\boldsymbol{x}\, \frac{i}{2}\left[\bar{\psi}\gamma^0\partial^{\mu}\psi - (\partial^{\mu}\bar{\psi})\gamma^0\psi\right]. \tag{84}$$

For the construction of the Lorentz generators we make use of the covariant splitting of orbital (O) and spin (S) angular momentum by Hilgevoord and Wouthuysen [10]. One derives with the aid of the Dirac equation that

$$\begin{aligned}
M^{\mu\nu}_{\text{free}}(t) &= L^{\mu\nu}_{\text{free}}(t) + S^{\mu\nu}_{\text{free}}(t),\\
L^{\mu\nu}_{\text{free}}(t) &= \int \mathrm{d}^3\boldsymbol{x}\, M_O^{0\mu\nu} = \int \mathrm{d}^3\boldsymbol{x}\,(x^{\mu}\Theta^{0\nu} - x^{\nu}\Theta^{0\mu}),\\
S^{\mu\nu}_{\text{free}}(t) &= \int \mathrm{d}^3\boldsymbol{x}\, M_S^{0\mu\nu} = \frac{i}{4m}\int \mathrm{d}^3\boldsymbol{x}\,\left\{\bar{\psi}\sigma^{\mu\nu}\partial_0\psi - (\partial_0\bar{\psi})\sigma^{\mu\nu}\psi\right\}.
\end{aligned} \tag{85}$$

5.3 Front Form (LF)

In this section we repeat the calculation in the front form.

Free Boson Poincaré Generators The boson stress tensor is of course the same as before. For the front-form components we find

$$\begin{aligned}
\Theta^{-+} &= \tfrac{1}{2}\left[(\partial^{\perp}\phi)^2 + m_b^2\phi^2\right],\\
\Theta^{+\mu} &= (\partial^{+}\phi)(\partial^{\mu}\phi), \quad \mu = 1,2,+.
\end{aligned} \tag{86}$$

The space-time translation operators are

$$\begin{aligned}
P^{-}_{\text{free}}(x^+) &= \int \mathrm{d}^2\boldsymbol{x}^{\perp}\mathrm{d}x^{-}\,\tfrac{1}{2}\left[(\partial^{\perp}\phi)^2 + m_b^2\phi^2\right],\\
P^{\mu}_{\text{free}}(x^+) &= \int \mathrm{d}^2\boldsymbol{x}^{\perp}\mathrm{d}x^{-}\,\partial^{+}\phi\partial^{\mu}\phi, \quad \mu = 1,2,+.
\end{aligned} \tag{87}$$

The Lorentz generators are

$$\begin{aligned}
M^{+\mu}_{\text{free}}(x^+) &= x^{+}P^{\mu}_{\text{free}} - \int \mathrm{d}^2\boldsymbol{x}^{\perp}\mathrm{d}x^{-}\,x^{\mu}(\partial^{+}\phi)^2, \quad \mu = 1,2,-\\
M^{-r}_{\text{free}}(x^+) &= \int \mathrm{d}^2\boldsymbol{x}^{\perp}\mathrm{d}x^{-}\,\left[x^{-}(\partial^{+}\phi)(\partial^{r}\phi) - x^{r}\tfrac{1}{2}\left[(\partial^{\perp}\phi)^2 + m_b^2\phi^2\right]\right], \; r = 1,2\\
M^{12}_{\text{free}}(x^+) &= \int \mathrm{d}^2\boldsymbol{x}^{\perp}\mathrm{d}x^{-}\,\left[x^{1}(\partial^{+}\phi)(\partial^{2}\phi) - x^{2}(\partial^{+}\phi)(\partial^{1}\phi)\right].
\end{aligned} \tag{88}$$

Free Fermion Poincaré Generators The free fermion stress tensor also remains unchanged. Hence we obtain for the momentum operators

$$P^\mu(x^+) = \int \mathrm{d}^2\boldsymbol{x}^\perp \mathrm{d}x^- \frac{i}{2}\left[\bar{\psi}\gamma^+\partial^\mu\psi - (\partial^\mu\bar{\psi})\gamma^+\psi\right]. \tag{89}$$

For the construction of the Lorentz generators we make again use of the covariant splitting of orbital and spin angular momentum. One derives with the aid of the Dirac equation that

$$\begin{aligned} L^{\mu\nu}(x^+) &= \int \mathrm{d}^2\boldsymbol{x}^\perp \mathrm{d}x^- \, M_O^{+\mu\nu} = \int \mathrm{d}^2\boldsymbol{x}^\perp \mathrm{d}x^- \, (x^\rho\Theta^{+\sigma} - x^\sigma\Theta^{+\rho}), \\ S^{\mu\nu}(x^+) &= \int \mathrm{d}^2\boldsymbol{x}^\perp \mathrm{d}x^- \, M_S^{+\mu\nu} = \frac{i}{4m}\int \mathrm{d}^2\boldsymbol{x}^\perp \mathrm{d}x^- \, \left[\bar{\psi}\sigma^{\mu\nu}\partial^+\psi - \partial^+\bar{\psi}\sigma^{\mu\nu}\psi\right]. \end{aligned} \tag{90}$$

Note that the spin and orbital angular momenta are separately conserved.

The kinematical LFD Lorentz generators now read

$$\begin{aligned} E^i_{\text{free}} &= L^{+i}_{\text{free}} + S^{+i}_{\text{free}}, \\ K^3_{\text{free}} &= L^{-+}_{\text{free}} + S^{-+}_{\text{free}}, \\ J^3_{\text{free}} &= L^{12}_{\text{free}} + S^{12}_{\text{free}}. \end{aligned} \tag{91}$$

The dynamical generators are P^- and

$$F^i_{\text{free}} = L^{-i}_{\text{free}} + S^{-i}_{\text{free}}. \tag{92}$$

5.4 Interacting and Non-interacting Generators on an Instant and on the Light Front

In order to investigate which Poincaré generators will contain interactions, we need to determine which Noether charges will contain interacting terms, in the IFD as well as in the LFD.

Poincaré Generators on an Instant In the instant form, we find for the translation operators

$$P^0_{\text{int}} = -\int d^3\boldsymbol{x}\mathcal{L}_{\text{int}}, \quad P^i_{\text{int}} = 0. \tag{93}$$

For the Lorentz generators, we have

$$M^{\mu\nu}_{\text{int}} = \int d^3\boldsymbol{x}(g^{0\mu}x^\nu - g^{0\nu}x^\mu)\mathcal{L}_{\text{int}} \tag{94}$$

from which it follows immediately that

$$J^k_{\text{int}} = \tfrac{1}{2}\epsilon_{ijk}M^{ij}_{\text{int}} = 0, \quad K^i_{\text{int}} = M^{0i}_{\text{int}} = \int d^3\boldsymbol{x}\, x^i\mathcal{L}_{\text{int}}. \tag{95}$$

So in a field theory without a derivative coupling the interacting operators are

$$P^0, \quad K^i, \tag{96}$$

whereas the kinematical operators read

$$\boldsymbol{P}, \quad \boldsymbol{J}. \tag{97}$$

Poincaré Generators on a Light Front We now have for the translation generators

$$P^-_{\text{int}} = -\int \mathrm{d}^2\boldsymbol{x}^\perp \mathrm{d}x^- \mathcal{L}_{\text{int}}, \quad P^+_{\text{int}} = \boldsymbol{P}^\perp_{\text{int}} = 0. \tag{98}$$

The expressions for the Lorentz generators is:

$$M^{\mu\nu}_{\text{int}} = \int \mathrm{d}^2\boldsymbol{x}^\perp \mathrm{d}x^- (g^{+\mu}x^\nu - g^{+\nu}x^\mu)\mathcal{L}_{\text{int}}, \tag{99}$$

whence we derive

$$\begin{aligned} F^j_{\text{int}} &= M^{-j}_{\text{int}} = \int \mathrm{d}^2\boldsymbol{x}^\perp \mathrm{d}x^- x^j \mathcal{L}_{\text{int}} \\ K^3_{\text{int}} &= M^{-+}_{\text{int}} = \int \mathrm{d}^2\boldsymbol{x}^\perp \mathrm{d}x^- x^+ \mathcal{L}_{\text{int}} = -x^+ P^-_{\text{int}} \\ E^j_{\text{int}} &= M^{+j}_{\text{int}} = 0. \end{aligned} \tag{100}$$

So in the case of no derivative coupling the interacting operators are

$$P^-, \quad F^j \, (j = 1, 2). \tag{101}$$

The kinematical operators read for $x^+ = 0$

$$P^+, \quad \boldsymbol{P}^\perp, \quad E^j \, (j = 1, 2), \quad K^3. \tag{102}$$

Summarizing, we conclude that in the case that no derivative coupling is present in the interaction $\mathcal{L}_{\text{int}}$, the dynamical generators contain the interaction, whereas kinematical operators are interaction-free, both in IFD and LFD . This is precisely the 'intuitive' case, which was also discussed by Dirac [5].

6 Light-Front Perturbation Theory

A sophisticated way to discuss Hamiltonian dynamics is to implement all constraints present and write down the equations of motion for the dynamical degrees of freedom. This method has been recommended by Dirac and already applied to the electromagnetic field in [5]. For illustration we consider the free Dirac equation

$$(i\gamma^\mu \partial_\mu - m)\psi = 0, \tag{103}$$

which can be rewritten in LF variables as

$$[i(\gamma^+\partial^- + \gamma^-\partial^+ - \gamma^\perp\partial^\perp) - m]\psi = 0. \tag{104}$$

If one defines projection operators and projections as follows

$$\Lambda^\pm = \tfrac{1}{2}\gamma^\mp\gamma^\pm, \quad \psi^\pm = \Lambda^\pm\psi, \tag{105}$$

then one finds that the two components ψ^+ and ψ^- are not independent but are related through a *constraint equation*

$$\psi^- = \frac{1}{i\partial^+}(i\alpha^\perp \cdot \partial^\perp + \gamma^0 m)\psi^+. \tag{106}$$

One considers ψ^+ to be the dynamical part. By elimination of ψ^- from the Dirac equation one finds for ψ^+ the dynamical equation

$$i\partial^-\psi^+ = \frac{-(\partial^\perp)^2 + m^2}{i\partial^+}\psi^+. \tag{107}$$

The operator ∂^- is differentiation with respect to LF time x^+; the r.h.s. of (107) contains differentiation with respect to the LF coordinates x^-, $\boldsymbol{x}^\perp$ only. It is the Hamiltonian operator for the free Dirac particle in LFD.

This simple example illustrates the role of constraints and makes it clear that the Hamiltonian formulation may become quite involved. First the dynamical degrees of freedom must be identified and next the quantization must be formulated in a consistent way. We shall not follow this path here, but rather take a short cut, using the method of Kogut and Soper [15].

6.1 Connection of Covariant Amplitudes to Light-Front Amplitudes

We consider a covariant theory to be defined by its Feynman diagrams. This definition is at least consistent in perturbation theory. For nonperturbative problems it may not be fully adequate. In Discretized Light-Cone Quantization the Hamiltonian is written down in terms of tree-level amplitudes; the kernel of the Lippmann-Schwinger equation is also written in terms of perturbative diagrams, so there are several cases where the use of diagrams in perturbation theory may be sufficient.

For any covariant diagram, which is a tensor integrated over the free momenta occurring in the loops, the associated LF-time ordered diagrams are derived by first performing the integration over the minus-components of the loop momenta. We shall later illustrate this procedure in two specific cases. Before doing so we first discuss the LF propagator.

Propagators Consider first the propagator of a scalar particle

$$D(x) = \int \mathrm{d}^4 p \frac{e^{-ipx}}{p^2 - m^2 + i\epsilon} = \int \mathrm{d}^4 p \frac{e^{-i(p^- x^+ + p^+ x^- - p^\perp x^\perp)}}{2p^+ (p^- - \frac{m^2 + p^{\perp 2} - i\epsilon}{2p^+})} . \tag{108}$$

If one performs the integration over p^- first, one may choose to evaluate the integral by closing the contour either in the upper or the lower half of the complex p^--plane. At this point the fact that positive and negative LF energies are kinematically disjoint is crucial. Positive energy is associated with positive p^+. If $p^+ > 0$, then the pole of the integrand is located in the lower halfplane. In order to interpret the propagator as a physical one we impose the *spectrum condition*, i.e. all plus momenta must be positive.

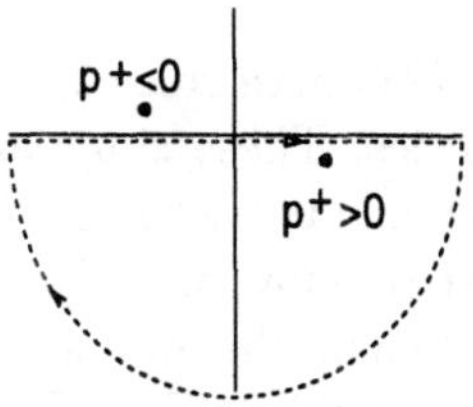

Fig. 6. Contour in the complex p^- plane

As p^+ is a kinematical quantity, it is conserved, so for any state the plus-components of the particles involved must be positive, as well as their sum. Therefore, the vacuum, which has zero plus momentum, cannot create states containing particles. There actually exists a loophole here: we cannot exclude $p^+ = 0$. Massless particles may have vanishing plus momentum and, moreover, for ultrarelativistic particles their plus momenta may tend to zero. We shall see later that such states are indeed occurring, but they may not be too dangerous. A full discussion of these so called "zero modes" is outside the scope of the present lectures.

Next we consider the free fermion propagator. The quantity $p^-_{\rm on}$ is defined as in (34).

$$\frac{i(\not p + m)}{p^2 - m^2 + i\epsilon} = \frac{i(\not p_{\rm on} + m)}{p^2 - m^2 + i\epsilon} + \frac{i(\not p - \not p_{\rm on})}{p^2 - m^2 + i\epsilon} \tag{109}$$

$$= \frac{i(\not p_{\rm on} + m)}{p^2 - m^2 + i\epsilon} + \frac{i(p^- - p^-_{\rm on})\gamma^+}{(p^- - p^-_{\rm on})2p^+ + i\epsilon} \tag{110}$$

$$= \frac{i\sum |u><u|}{p^2 - m^2 + i\epsilon} \qquad (\text{on} - \text{shell}) \tag{111}$$

$$= \frac{i\sum |u><u|}{p^2 - m^2 + i\epsilon} + \frac{i\gamma^+}{2p^+} \qquad (\text{off} - \text{shell}) \tag{112}$$

where

$$p_{\text{on}} = (p^-_{\text{on}}, p^+, \boldsymbol{p}^\perp) . \tag{113}$$

The part $\gamma^+/2p^+$, when transformed to coordinate space, is proportional to $\delta(x^+)$, so it is called the *instantaneous part* of the fermion propagator. The other part we will denote as the LF propagating part.

One may wonder what the corresponding formulae will be in the instant form. In fact, a derivation along the same line as given in (112) will show that upon integration of the covariant expression over the energy p^0 first, gives always two contributions. There are again instantaneous terms, but they cancel exactly. This is the reason why one never considers them in "old fashioned perturbation theory" in the instant form.

6.2 Regularization

The naive application of the Feynman rules to construct covariant diagrams leads in many cases to infinities. These singularities must be regulated. In the covariant calculations one may use several different methods, dimensional regularization being the most popular one. This recipe, as most of the well-known methods, relies very much on the manifest symmetry of the amplitudes with respect to Lorentz tranformations. In LFD this symmetry is no longer manifest and one needs methods that can handle time-ordered diagrams. One of these methods is Pauli-Villars regularization (see, e.g., [11]). It works for time-ordered as well as covariant amplitudes. We shall not discuss it here, as it is well known.

A method that was specifically devised for LFD is the so called "minus regularization", introduced by Ligterink [19]. The main idea is to expand the amplitude in terms of the independent invariants that can be built from the external momenta, in a Taylor series. The terms with infinite coefficients are subtracted, leaving a regulated amplitude. Of course, this method is already known in the literature: it was introduced by Hepp and Zimmermann (see [8,30,31]). The novelty of Ligterink's method, later extended in [23], is the way it is adapted to LFD amplitudes.

6.3 Minus Regularization

First consider a covariant amplitude, say $M(p_i^2, p_i \cdot p_j)$, depending on four-momenta $p_1, \dots p_n$. It is obtained by integration in the usual way

$$M(p_i^2, p_i \cdot p_j) = \int \mathrm{d}^4 k \mathcal{M}(p_i^2, p_i \cdot p_j; k), \tag{114}$$

where the integration runs over the independent loops, collectively denoted by k. The result depends on the invariants p_i^2 and $p_i \cdot p_j$, not all of them being necessarily independent. A subtraction point is chosen, say $p_i^2 = 0$, $p_i \cdot p_j = 0$.

The amplitude is then written in terms of the parameter λ in the following way

$$M(\lambda p_i^2, \lambda p_i \cdot p_j) = \int \mathrm{d}^4 k \mathcal{M}(\lambda p_i^2, \lambda p_i \cdot p_j; k). \tag{115}$$

The regularized amplitude is obtained by first differentiating $\mathcal{M}(\lambda p_i^2, \lambda p_i \cdot p_j; k)$ as many times with respect to λ as is necessary to obtain a convergent integral and, after performing the integrals over the momenta $k_1, \ldots k_l$, integrate the result over λ as many times as the integrand was differentiated. Schematically

$$M(\lambda p_i^2, \lambda p_i \cdot p_j) = \left(\int_0^1 \mathrm{d}\lambda \right)^n \int \mathrm{d}^4 k \frac{\partial^n}{\partial \lambda^n} \mathcal{M}(\lambda p_i^2, \lambda p_i \cdot p_j; k). \tag{116}$$

At this point it is important to realize that both differentiation and integration are linear operations, so if a covariant amplitude is split in the way described above by differentiation over k^- into LF time ordered diagrams, it is easy to see that one can use the same procedure for the latter ones. The only change that needs to be made is to accommodate the fact that the LF time ordered diagrams cannot usually be expressed in terms of invariants. To adapt the algorithm to that circumstance write for the LF diagram

$$M_{\mathrm{LF}}(\lambda p_i^-, p_i^+, \lambda \boldsymbol{p}_i^\perp) = \left(\int_0^1 \mathrm{d}\lambda \right)^n \int \mathrm{d}^3 k \frac{\partial^n}{\partial \lambda^n} \mathcal{M}_{\mathrm{LF}}(\lambda p_i^-, p_i^+, \lambda \boldsymbol{p}_i^\perp; k) . \tag{117}$$

The integration variables are now k^+ and $\boldsymbol{k}^\perp$.

The path from $\lambda = 0$ to $\lambda = 1$ in p space has the same begin and end points as the path followed in the case of the covariant amplitude. The only concern is to avoid singularities in the integrals. Note that we have left the plus-components alone. Because of the form the scalar product takes in LFD, a parameterization of the minus and perpendicular components suffices to trace out a path in $p_i \cdot p_j$ space. This has the advantage that the limits of the k^+ integrals, which are determined by the values of p_i^+, remain unchanged.

This method was successfully applied to QED by Ligterink and Bakker [19], where it could be shown to respect the Ward-Takahasi identities and yield the covariant result. In [23] the same method was applied to the two- and three-point functions in Yukawa theory. In the next section we shall give a detailed example: the electromagnetic current of a scalar particle consisting of either bosonic or fermionic constituents. There we shall encounter a peculiarity of LFD: additional singularities occur, that are not present in the associated covariant amplitude. We shall see that they can be regulated by minus regularization too.

7 Triangle Diagram in Yukawa Theory

We consider the electromagnetic current matrix element of a composite system composed of two charged fermions where the light-cone wavefunctions are known explicitly from perturbation theory. To construct the model, we consider a 3+1 dimensional system represented by the Lagrangean:

$$\mathcal{L} = \bar{\psi}_a(i(\not\partial + ie_a\not A) - m_a)\psi_a + \bar{\psi}_b(i(\not\partial + ie_b\not A) - m_b)\psi_b$$
$$+\frac{1}{2}\partial_\mu\phi\partial^\mu\phi - \frac{1}{2}m^2\phi\phi + g\phi(\bar{\psi}_a\psi_b + \bar{\psi}_b\psi_a). \tag{118}$$

The covariant diagram is shown in Fig. 7. In general, there are two contributions: the photon may couple to particle a or particle b. The corresponding currents are denoted by $\mathcal{M}_a^\mu$ and $\mathcal{M}_b^\mu$ respectively. They are related by the interchange $a \leftrightarrow b$ in the formulas written in the rest of this section.

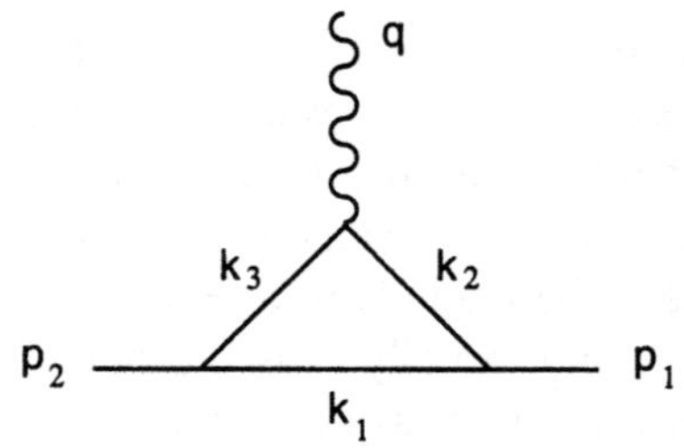

Fig. 7. Covariant triangle diagram

The momenta are chosen as follows: the external hadrons have mass m and momenta p_1 and p_2 resp., the photon has momentum q. The constituents have masses m_a and m_b resp., their momenta are k_1, k_2, and k_3 (see Fig. 7). Then we have the kinematical relations

$$k_2 = p_1 + k_1, \quad k_3 = p_2 + k_1, q = p_2 - p_1. \tag{119}$$

The amplitude we are going to evaluate contains one integration. We take momentum $-k_1$ as the integration variable and denote it by k. Then the momenta become

$$k_1 = -k, \quad k_2 = p_1 - k, \quad k_3 = p_2 - k. \tag{120}$$

7.1 Covariant Calculation

In this section we shall calculate the amplitude using the common covariant techniques.

The covariant amplitude follows from the Feynman rules. It is

$$\mathcal{M}_a^\mu = -e_a g^2 \int \frac{\mathrm{d}^4k}{(2\pi)^4} \frac{\mathrm{Tr}[(\not k_1 + m_b)(\not k_2 + m_a)\gamma^\mu(\not k_3 + m_a)]}{(k_1^2 - m_b^2 + i\epsilon)\,(k_2^2 - m_a^2 + i\epsilon)\,(k_3^2 - m_a^2 + i\epsilon)}\,. \tag{121}$$

Lorentz invariance requires the matrix element to be of the form

$$\mathcal{M}_a^\mu = -i\, e_a\, F_a\, (p_1^\mu + p_2^\mu), \tag{122}$$

if the external legs with momenta p_1 and p_2 are on shell and $p_1^2 = p_2^2 = m^2$.

The numerator of (121) is

$$\begin{aligned} T^\mu &= \mathrm{Tr}[(\not{k}_1 + m_b)(\not{k}_2 + m_a)\gamma^\mu(\not{k}_3 + m_a)] \\ &= 4[(m_a^2 - k_2 \cdot k_3)k_1^\mu + (m_a m_b + k_1 \cdot k_3)k_2^\mu + (m_a m_b + k_1 \cdot k_2)k_3^\mu], \end{aligned} \tag{123}$$

or, in terms of the integration variable k and the external momenta $p_{1,2}$

$$T^\mu = 4(Ak^\mu + B_1 p_1^\mu + B_2 p_2^\mu). \tag{124}$$

The factors A, B_1, and B_2 are easily found.

Following the usual procedure, Feynman parameterization and Wick rotation, one ends up with two types of integrals. First, space-time integrals of the form

$$\int \frac{\mathrm{d}^D k}{(2\pi)^D} f(k^2), \quad \int \frac{\mathrm{d}^D k}{(2\pi)^D} k^\mu k^\nu h(k^2) = \frac{1}{D} g^{\mu\nu} \int \frac{\mathrm{d}^D k}{(2\pi)^D} k^2 h(k^2). \tag{125}$$

Secondly, integrals over the Feynman parameters x and y, that occur because we obtain, using Feynman's trick, an integrand containing the factor

$$D(k; x, y) = x(k_2^2 - m_a^2) + y(k_3^2 - m_a^2) + (1 - x - y)(k_1^2 - m_b^2). \tag{126}$$

Then we can express $\mathcal{M}_a^\mu$ in terms of two integrals

$$I_0 = \int \frac{\mathrm{d}^D k}{(2\pi)^D} \frac{1}{D(k; x, y)^3}, \quad I_2 = \int \frac{\mathrm{d}^D k}{(2\pi)^D} \frac{k^2}{D(k; x, y)^3}. \tag{127}$$

If the relations (120) are substituted one finds for D

$$D(k; x, y) = k^2 - M^2(x, y). \tag{128}$$

After Wick rotation one obtains Euclidean integrals. The standard rules of dimensional regularization give in D dimensions

$$I_0^{\mathrm{E}} = \frac{\Gamma(3 - D/2)}{(4\pi)^{D/2}\Gamma(3)} M^{D-6}, \quad I_2^{\mathrm{E}} = \frac{D}{2} \frac{\Gamma(2 - D/2)}{(4\pi)^{D/2}\Gamma(3)} M^{D-4}. \tag{129}$$

Clearly there occurs a singularity in I_2 in 3+1 dimensions. It can be regularized with the method we just described.

The quantity M^2 can be written as

$$\begin{aligned} M^2 &= [(x + y)(x + y - 1)m^2 + xyQ^2] + [(1 - x - y)m_b^2 + (x + y)m_a^2] \\ &\equiv M_e^2 + M_i^2 \end{aligned} \tag{130}$$

with $Q^2 = -q^2 = -(p_2 - p_1)^2$. The regularization now consists in multiplying $m^2 = p_1^2 = p_2^2$ and Q^2 with λ and perform the differentiation and integration with respect to this parameter. The result is

$$I_2^{\mathrm{E\,reg}} = \int_0^1 d\lambda \int_{\mathrm{E}} \frac{d^4k}{(2\pi)^2} \frac{\partial}{\partial\lambda} \frac{k^2}{(k^2 + \lambda M_e^2 + M_i^2)^3} . \tag{131}$$

The differentiation and integration over λ are trivial. They give

$$\begin{aligned} I_2^{\mathrm{E\,reg}} &= -3M_e^2 \int_0^1 d\lambda \int_{\mathrm{E}} \frac{d^4k}{(2\pi)^2} \frac{k^2}{(k^2 + \lambda M_e^2 + M_i^2)^4} \\ &= \frac{-M_e^2}{(4\pi)^2} \int_0^1 d\lambda \frac{1}{\lambda M_e^2 + M_i^2} \\ &= \frac{-1}{(4\pi)^2} \log \left(\frac{M_e^2 + M_i^2}{M_i^2} \right) . \end{aligned} \tag{132}$$

It is amusing that these formulae coincide exactly with what one would obtain using dimensional regularization. This need not be so, as two different forms of regularization may lead to results that differ by a finite constant.

In 1+1 dimensions the integrals are

$$I_0^{\mathrm{E}} = \frac{1}{8\pi} \frac{1}{M^4(x,y;Q^2)} \quad I_2^{\mathrm{E}} = \frac{1}{8\pi} \frac{1}{M^2(x,y;Q^2)} \tag{133}$$

which are both finite.

The final results are given by integration over x and y.

7.2 Construction of the Current in LFD

In order to derive the LF time dependent amplitudes, we take the covariant expression and integrate over k^-. As the numerator contains for $\mu = -$ the integration variable too, for a proper identification of the poles and the residues, one needs to separate the Feynman propagators into the LF propagating part and the instantaneous part, as described in Sect. 6. After this split has been performed, the poles and residues can be identified properly.

To facilitate the discussion we introduce the notations

$$\Lambda_i = \not{k}_{i\,\mathrm{on}} + m_i, \quad \Gamma_i = (k_i^- - k_{i\,\mathrm{on}}^-)\gamma^+, \quad \Omega = \not{k}_i + m_i = \Lambda_i + \Gamma_i . \tag{134}$$

Then the product $\Omega_1 \Omega_2 \gamma^\mu \Omega_3$, occurring in T^μ can be expanded into eight terms

$$\begin{array}{llllll} \Lambda_1\Lambda_2\gamma^\mu\Lambda_3 & (a) & \Gamma_1\Lambda_2\gamma^\mu\Lambda_3 & (b) & \Lambda_1\Gamma_2\gamma^\mu\Lambda_3 & (c) \\ \Lambda_1\Lambda_2\gamma^\mu\Gamma_3 & (d) & \Gamma_1\Gamma_2\gamma^\mu\Lambda_3 & (e) & \Gamma_1\Lambda_2\gamma^\mu\Gamma_3 & (f) \\ \Lambda_1\Gamma_2\gamma^\mu\Gamma_3 & (g) & \Gamma_1\Gamma_2\gamma^\mu\Gamma_3 & (h) & & \end{array} \tag{135}$$

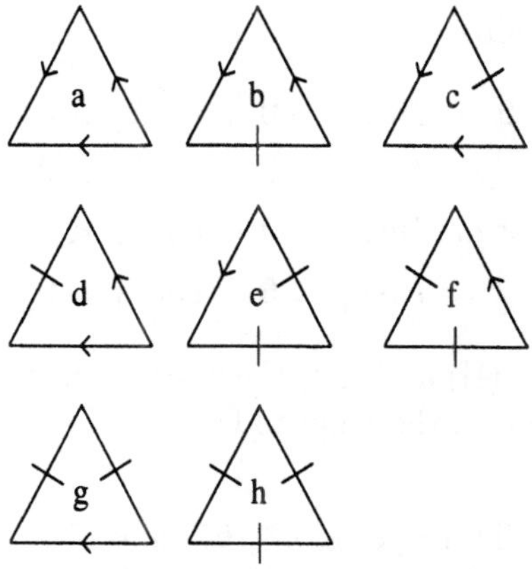

Fig. 8. Skeleton graphs (a) – (h) for the fermion triangle. A tag on the line indicates an instantaneous propagator.

These terms are drawn in Fig. 8. In this figure the lines with arrows symbolize the LF propagating parts and those with tags the instantaneous parts. Here we mention that a calculation of the same diagram, but with the spinor constituents replaced by scalar bosons, would be much simplified. Everywhere the trace T^μ must be replaced by $p_1^\mu + p_2^\mu - 2k^\mu$.Then the integrands are free of longitudinal singularities, i.e. singularities of the form $1/(k^+ - p_i^+)$.

After the split described above is done, the poles and residues can be properly determined. We find the following poles

$$\begin{aligned} k_1^- &= \frac{k^{\perp 2} + m_b^2 - i\epsilon}{2k^+} \equiv H_1 - \frac{i\epsilon}{2k^+}, \\ k_2^- &= p_1^- - \frac{(p_1^\perp - k^\perp)^2 + m_a^2 - i\epsilon}{2(p_1^+ - k^+)} \equiv H_2 + \frac{i\epsilon}{2(p_1^+ - k^+)}, \\ k_3^- &= p_2^- - \frac{(p_2^\perp - k^\perp)^2 + m_a^2 - i\epsilon}{2(p_2^+ - k^+)} \equiv H_3 + \frac{i\epsilon}{2(p_2^+ - k^+)}. \end{aligned} \tag{136}$$

The denominator in the expression for the covariant amplitude can be written as

$$8k^+(p_1^+ - k^+)(p_2^+ - k^+)(k^- - H_1)(k^- - H_2)(k^- - H_3). \tag{137}$$

The factor in front of the three poles is the phase-space factor

$$\Phi = 8k^+(p_1^+ - k^+)(p_2^+ - k^+). \tag{138}$$

The positions of these poles depend on the values of p_1^+ and p_2^+. Without loss of generality we shall consider the case $q^+ = p_2^+ - p_1^+ > 0$ only. Then we find the following domains

$$\begin{array}{lllll}
(\mathrm{I}) & k^+ < 0 & Im\, k_1^- > 0, & Im\, k_2^- > 0, & Im\, k_3^- > 0; \\
(\mathrm{II}) & 0 < k^+ < p_1^+ & Im\, k_1^- < 0, & Im\, k_2^- > 0, & Im\, k_3^- > 0; \\
(\mathrm{III}) & p_1^+ < k^+ < p_2^+ & Im\, k_1^- < 0, & Im\, k_2^- < 0, & Im\, k_3^- > 0; \\
(\mathrm{IV}) & k^+ > p_2^+ & Im\, k_1^- < 0, & Im\, k_2^- < 0, & Im\, k_3^- < 0.
\end{array} \tag{139}$$

So, only in the domains (II) and (III) there are poles at either side of the real k^--axis; they contribute to the amplitude.

Skeleton Graph (a) Three poles: k_1^-, k_2^-, and k_3^-.
The two domains (II) and (III) contribute to the amplitude. In domain (II) we take the pole $k_1^- = H_1 - i\delta$ and find

$$\mathcal{M}_a^\mu(\mathrm{II.a}) = \frac{-e_a g^2}{(2\pi)^4} \int \mathrm{d}^2 k_1^\perp \int_0^{p_1^+} \mathrm{d}k_1^+ \frac{-2\pi i}{\Phi} \frac{\mathrm{Tr}[\Lambda_1 \Lambda_2 \gamma^\mu \Lambda_3]}{(H_1 - H_2)(H_1 - H_3)}. \tag{140}$$

The energy denominators are

$$D_1 = p_1^- - \frac{{k^\perp}^2 + m_b^2}{2k^+} - \frac{(p_1^\perp - k^\perp)^2 + m_a^2}{2(p_1^+ - k^+)} \tag{141}$$

and

$$D_2 = p_2^- - \frac{{k^\perp}^2 + m_b^2}{2k^+} - \frac{(p_2^\perp - k^\perp)^2 + m_a^2}{2(p_2^+ - k^+)} \tag{142}$$

Then straightforward algebra gives $H_2 - H_1 = D_1$ and $H_3 - H_1 = D_2$, so we can write

$$\mathcal{M}_a^\mu(\mathrm{II.a}) = \frac{i e_a g^2}{(2\pi)^3} \int \mathrm{d}^2 k^\perp \int_0^{p_1^+} \mathrm{d}k^+ \frac{1}{\Phi} \frac{\mathrm{Tr}[\Lambda_1 \Lambda_2 \gamma^\mu \Lambda_3]}{D_2 D_1}. \tag{143}$$

In domain (III) we take the pole $k^- = k_3^- = -H_3 + i\delta$. We find for the amplitude

$$\mathcal{M}_a^\mu(\mathrm{III.a}) = \frac{-e_a g^2}{(2\pi)^4} \int \mathrm{d}^2 k_1^\perp \int_{p_1^+}^{p_2^+} \mathrm{d}k^+ \frac{-2\pi i}{\Phi'} \frac{\mathrm{Tr}[\Lambda_1 \Lambda_2 \gamma^\mu \Lambda_3]}{(k_3^- - k_1^-)(k_3^- - k_2^-)}, \tag{144}$$

where Φ' is given by

$$\Phi' = 2k^+ 2(p_2^+ - k^+) 2(k^+ - p_1^+). \tag{145}$$

The energy denominators are now

$$D_1' = p_2^- - p_1^- - \frac{(k^\perp - p_1^\perp)^2 + m_a^2}{2(k^+ - p_1^+)} - \frac{(p_2^\perp - k^\perp)^2 + m_a^2}{2(p_2^+ - k^+)} \tag{146}$$

and D_2 is not changed. The domain of integration for k^+ is now $p_1^+ < k^+ < p_2^+$. Again we find in terms of energy denominators

$$\mathcal{M}_a^\mu(\text{III.a}) = \frac{ie_a g^2}{(2\pi)^3}\int \mathrm{d}^2k^\perp \int_{p_1^+}^{p_2^+} \mathrm{d}k^+ \frac{1}{\Phi'}\frac{\mathrm{Tr}[\Lambda_1\Lambda_2\gamma^\mu\Lambda_3]}{D_2 D_1'}. \tag{147}$$

Skeleton Graph (b) Two poles: k_2^- and k_3^-. The factor $k_1^- - k_{1\,\text{on}}^-$ in the denominator is canceled by the same factor in the numerator. The two remaining poles in domain (II) have positive imaginary parts. In domain (III) the imaginary parts of the two poles have different signs. So only one time-ordered diagram remains here. The corresponding amplitude is

$$\mathcal{M}_a^\mu(\text{III.b}) = \frac{ie_a g^2}{(2\pi)^3}\int \mathrm{d}^2k^\perp \int_{p_1^+}^{p_2^+} \mathrm{d}k^+ \frac{1}{\Phi'}\frac{\mathrm{Tr}[\gamma^+\Lambda_2\gamma^\mu\Lambda_3]}{D_1'}. \tag{148}$$

Skeleton Graph (c) Two poles: k_1^- and k_3^-. Here it is the factor $k_2^- - k_{2\,\text{on}}^-$ that cancels. The remaining poles are k_1^- and k_3^-, which lie at different sides of the real axis in both the domains (II) and (III). The corresponding amplitudes are

$$\mathcal{M}_a^\mu(\text{II.c}) = \frac{ie_a g^2}{(2\pi)^3}\int \mathrm{d}^2k^\perp \int_{0}^{p_1^+} \mathrm{d}k^+ \frac{1}{\Phi}\frac{\mathrm{Tr}[\Lambda_1\gamma^+\gamma^\mu\Lambda_3]}{D_2} \tag{149}$$

and

$$\mathcal{M}_a^\mu(\text{III.c}) = \frac{ie_a g^2}{(2\pi)^3}\int \mathrm{d}^2k^\perp \int_{p_1^+}^{p_2^+} \mathrm{d}k^+ \frac{1}{\Phi'}\frac{\mathrm{Tr}[\Lambda_1\gamma^+\gamma^\mu\Lambda_3]}{D_2}. \tag{150}$$

Skeleton Graph (d) Two poles: k_1^- and k_2^-. Now the factor $k_3^- - k_{3\,\text{on}}^-$ cancels. Only in domain (II) there are two poles at different sides of the real axis. The amplitude is

$$\mathcal{M}_a^\mu(\text{II.d}) = \frac{ie_a g^2}{(2\pi)^3}\int \mathrm{d}^2k^\perp \int_{0}^{p_1^+} \mathrm{d}k^+ \frac{1}{\Phi}\frac{\mathrm{Tr}[\Lambda_1\Lambda_2\gamma^\mu\gamma^+]}{D_1}. \tag{151}$$

Skeleton Graph (g) One pole: k_1^-. Only one pole remains: k_1^-. So one can close the contour in the half-plane not containing this pole. Consequently, this graph does not contribute.

We summarize the situation graphically in Fig. 9. These diagrams contain separately longitudinal singularities. They appear because the LF propagator has $\not{k}_{\text{on}}$ in the numerator, which has a piece $k_{\text{on}}^- = (\boldsymbol{k}^{\perp 2} + m_i^2)\gamma^+/2k_i^+$. The $1/k^+$ singularity that occurs here has exactly the same form as the one coming from the instantaneous part. So, after we have correctly identified the poles and residues in the k^--integration, the parts with identical denominators and defined in the same domain, can be added. This procedure produces the so called "blinks" [20]. The blink construction removes the $1/k^+$-singularities connected to the internal point $k^+ = p_1^+$. Of the other two, $k^+ = 0, p_2^+$, the first one is harmless, but the second one remains.

In order to keep matters as simple as possible, we shall continue our discussion for the 1+1 dimensional case. The diagrams in the first column at the r.h.s. of Fig. 9 can be added. This produces blinks for the lines k_2 and k_3. The result is called the *valence* amplitude, denoted by $\mathcal{M}^\mu(\text{val})$. Adding the other diagrams produces an amplitude with blinks on lines k_1 and k_2. It is called the *nonvalence* amplitude $\mathcal{M}^\mu(\text{nv})$.

Now we are ready to calculate the *good* and the *terrible* current. (The terminology originates from the "infinite momentum" interpretation of LFD. It is explained in [16].)

The peculiar behaviour of the amplitude is best seen in 1+1 dimensions. The reason is that in 3+1 dimensions the covariant amplitude is divergent, so regularization is needed. In lower dimensions, however, one expects the amplitude to be regular. In LFD this turns out not to be the case. There occurs an additional singularity in the integration over k^+. As it is also present in the 3+1 dimensional case, but is removed there by minus regularization, this peculiar phenomenon can best be seen in 1+1 dimensions. Therefore we give our results for that case only.

Good Current $\mathcal{M}^+$ The trace for the valence contribution is:

$$T^+(\text{val}) = 4\left[(m_a m_b + m_b^2)(p_1^+ + p_2^+) - (m_a + m_B)^2 k^+ - m_b^2 \frac{p_1^+ p_2^+}{k^+}\right]. \tag{152}$$

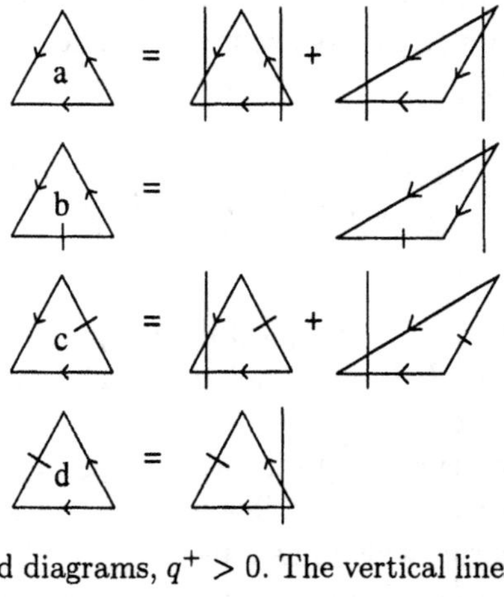

Fig. 9. LF-time ordered diagrams, $q^+ > 0$. The vertical lines denote energy denominators.

So we can write for the amplitude in 1+1 dimensions

$$\mathcal{M}_a^\mu(\mathrm{val}) = \frac{ie_a g^2}{2\pi}\int_0^{p_1^+} \mathrm{d}k^+ \frac{1}{\Phi}\frac{T^+(\mathrm{val})}{D_2 D_1}. \tag{153}$$

The end-point singularity $k^+ = 0$ is not real, as it cancels in the numerator and the denominator.

In domain (III) the trace is

$$\begin{aligned} T^+(\mathrm{nv}) = 4\,&\big[(m_a m_b + m_a^2)p_1^+ + m_a m_b p_2^+ \\ &- 2(m_a m_b + m_a^2)k^+ - 2p_2^-(k^+ - p_1^+)(k^+ - p_2^+)\big]\,. \end{aligned} \tag{154}$$

The amplitude is

$$\mathcal{M}_a^\mu(\mathrm{nv}) = \frac{ie_a g^2}{2\pi}\int_{p_1^+}^{p_2^+} \mathrm{d}k^+ \frac{1}{\Phi'}\frac{T^+(\mathrm{nv})}{D_2 D_1}. \tag{155}$$

Terrible Current $\mathcal{M}^-$ As the denominators and the phase-space factors do not change, we give the traces only.

For the valence part we find

$$T^-(\mathrm{val}) = 4[(m_a m_b + m_b^2)(p_1^- + p_2^-) - (m_a + m_b)^2 k^- - 2k^+ p_1^- p_2^-]. \tag{156}$$

The non-valence part is

$$\begin{aligned} T^-(\mathrm{nv}) = 4\Bigg[&-m_a^2 p_2^- + m_a m_b(p_1^- - p_2^-) + (m_a^2 + 2m_a m_b)\frac{m_a^2}{2(p_2^+ - k^+)} \\ &-2k^+(p_1^- - p_2^-)\frac{m_a^2}{2(p_2^+ - k^+)} - 2k^+\left(\frac{m_a^2}{2(p_2^+ - k^+)}\right)^2\Bigg]\,. \end{aligned} \tag{157}$$

In the next subsection we give the explicit formulas.

Explicit Formulae Here we gather the explicit expressions.

$$\begin{aligned} \mathcal{M}_a^+(\mathrm{val}) &= \frac{ie_a g^2}{2\pi}\int_0^{p_1^+} \mathrm{d}k^+ \frac{1}{8k^+(p_1^+ - k^+)(p_2^+ - k^+)} \\ &\times \frac{4[(m_a m_b + m_b^2)(p_1^+ + p_2^+) - (m_a + m_b)^2 k^+ - m_b^2\frac{p_1^+ p_2^+}{k^+}]}{\left(p_2^- - \frac{m_b^2}{2k^+} - \frac{m_a^2}{2(p_2^+ - k^+)}\right)\left(p_1^- - \frac{m_b^2}{2k^+} - \frac{m_a^2}{2(p_1^+ - k^+)}\right)} \\ &= \frac{-ie_a g^2}{2\pi}\int_0^{p_1^+} \mathrm{d}k^+ 2[(m_a m_b + m_b^2)k^{+\,2} \\ &\qquad\qquad - (m_a m_b + m_b^2)(p_1^+ + p_2^+)k^+ + m_b^2 p_1^+ p_2^+] \\ &\times \frac{1}{(2p_2^- k^+(p_2^+ - k^+) - m_b^2(p_2^+ - k^+) - m_a^2 k^+)} \\ &\times \frac{1}{(2p_1^- k^+(p_1^+ - k^+) - m_b^2(p_2^+ - k^+) - m_a^2 k^+)}\,, \end{aligned} \tag{158}$$

$$\begin{aligned}\mathcal{M}_a^+(\mathrm{nv}) &= \frac{ie_a g^2}{2\pi}\int_{p_1^+}^{p_2^+} \mathrm{d}k^+ \frac{1}{8k^+(k^+-p_1^+)(p_2^+-k^+)}\\ &\quad\times 4[m_a m_b(p_1^+ + p_2^+) - (2m_a m_b + m_a^2)k^+ \\ &\qquad\qquad -(m^2 - m_a^2 - 2k^+ p_2^-)(p_1^+ - k^+)]\\ &\quad\times \frac{1}{\left(p_2^- - \frac{m_b^2}{2k^+} - \frac{m_a^2}{2(p_2^+-k^+)}\right)\left(p_2^- - p_1^- - \frac{m_a^2}{2(k^+-p_1^+)} - \frac{m_a^2}{2(p_2^+-k^+)}\right)}\\ &= \frac{ie_a g^2}{2\pi}\int_{p_1^+}^{p_2^+} \mathrm{d}k^+ 2[m_a m_b(p_1^+ + p_2^+) - (2m_a m_b + m_a^2)k^+\\ &\qquad\qquad -(m^2 - m_a^2 - 2k^+ p_2^-)(p_1^+ - k^+)](p_2^+ - k^+)\\ &\quad\times \frac{1}{(2p_2^- k^+(p_2^+ - k^+) - m_b^2(p_2^+ - k^+) - m_a^2 k^+)}\\ &\quad\times \frac{1}{(2(p_2^- - p_1^-)(k^+ - p_1^+)(p_2^+ - k^+) - m_a^2(p_2^+ - p_1^+))}\,.\end{aligned} \tag{159}$$

Terrible Current

$$\begin{aligned}\mathcal{M}_a^-(\mathrm{val}) &= \frac{ie_a g^2}{2\pi}\int_0^{p_1^+} \mathrm{d}k^+ 2k^+\\ &\quad\times\left[(m_a m_b + m_b^2)(p_1^- + p_2^-) - (m_a + m_b)\frac{m_b^2}{2k^+} - 2k^+ p_1^- p_2^-\right]\\ &\quad\times \frac{1}{(2p_2^- k^+(p_2^+ - k^+) - m_b^2(p_2^+ - k^+) - m_a^2 k^+)}\\ &\quad\times \frac{1}{(2p_1^- k^+(p_1^+ - k^+) - m_b^2(p_1^+ - k^+) - m_a^2 k^+)}\,,\end{aligned} \tag{160}$$

$$\begin{aligned}\mathcal{M}_a^-(\mathrm{nv}) &= \frac{ie_a g^2}{2\pi}\int_{p_1^+}^{p_2^+} \mathrm{d}k^+ \left[2(m_a m_b p_1^- - (m_a^2 + m_a m_b)p_2^-)(p_2^+ - k^+)\right.\\ &\quad\left. +(m_a^2 + 2m_a m_b - 2k^+(p_1^- - p_2^-))m_a^2 - \frac{k^+ m_a^4}{p_2^+ - k^+}\right]\\ &\quad\times \frac{1}{(2p_2^- k^+(p_2^+ - k^+) - m_b^2(p_2^+ - k^+) - m_a^2 k^+)}\\ &\quad\times \frac{1}{(2(p_2^- - p_1^-)(k^+ - p_1^+)(p_2^+ - k^+) - m_a^2(p_2^+ - p_1^+))}\,.\end{aligned} \tag{161}$$

This amplitude contains a singularity at $k^+ = p_2^+$. It has the form

$$\mathcal{M}_a^-(\mathrm{nv}) = \frac{ie_a g^2}{2\pi}\int_{p_1^+}^{p_2^+} \mathrm{d}k^+ \left(-\frac{1}{p_2^+ - p_1^+}\frac{1}{p_2^+ - k^+}\right). \tag{162}$$

As this singularity does not depend on p_i^-, $i = 1, 2$, it can be removed by minus-regularization. If we scale k^+ as follows: $k^+ = (1+x)p_1^+$ and write $p_2^+ = (1+y)p_1^+$, then we find it to be of the form

$$\frac{-1}{p_2^+ - p_1^+} \int_0^y \mathrm{d}x \frac{1}{y-x}. \tag{163}$$

So the singularity does not depend on any physical parameter: it is a pure (infinite) number. We shall simply subtract it and see whether this leads to the desired covariant result.

7.3 Numerical Results

In order to see the connection between the spin of the constituents and the occurrence of the longitudinal singularity at $k^+ = p_2^+$ we considered spin-1/2 constituents as well as spin-0 constituents and found dramatic differences between the two cases. Comparing with the covariant Feynman calculations, we notice that the common belief of equivalence between the manifestly covariant calculation and the LF calculation linked by the LF energy integration of the Feynman amplitude is not always realized. The minus component of the LF current generated by the fermion loop has a persistent end-point singularity that must be removed to assure covariance and current conservation. A similar singularity was observed in the calculation of the fermion selfenergy in [22]. The plus component of the LF current, however, is immune to this disorder and provides a form factor identical to the one obtained doing the covariant Feynman calculation. This phenomenon is also associated with the spin-effect of the constituents because the calculation with scalar (spin-0) constituents does not have the same symptom.

Decomposing the LF amplitude into the valence and nonvalence parts, it is interesting to note that the end-point singularity exists only in the nonvalence vertex contribution. We have numerically estimated the importance of the nonvalence vertices in both cases. In all cases, our results show that if the meson is weakly bound then the contributions from the valence and the nonvalence vertices to the plus current are separately almost the same as those for the minus current. Of course, their sums add up to the same number in both the plus and minus cases.

8 Four Variations on a Theme in ϕ^3 Theory

The purpose of this chapter is to illustrate in the simplest possible example the different techniques.

Consider the simplest diagram with one loop in ϕ^3 theory. The particle mass is m, the coupling constant is g. The Feynman rules then give

$$A(p) = \frac{g^2}{2(2\pi)^4} \int \mathrm{d}^4k \frac{1}{(k^2 - m^2 + i\epsilon)((k-p)^2 - m^2 + i\epsilon)} . \tag{164}$$

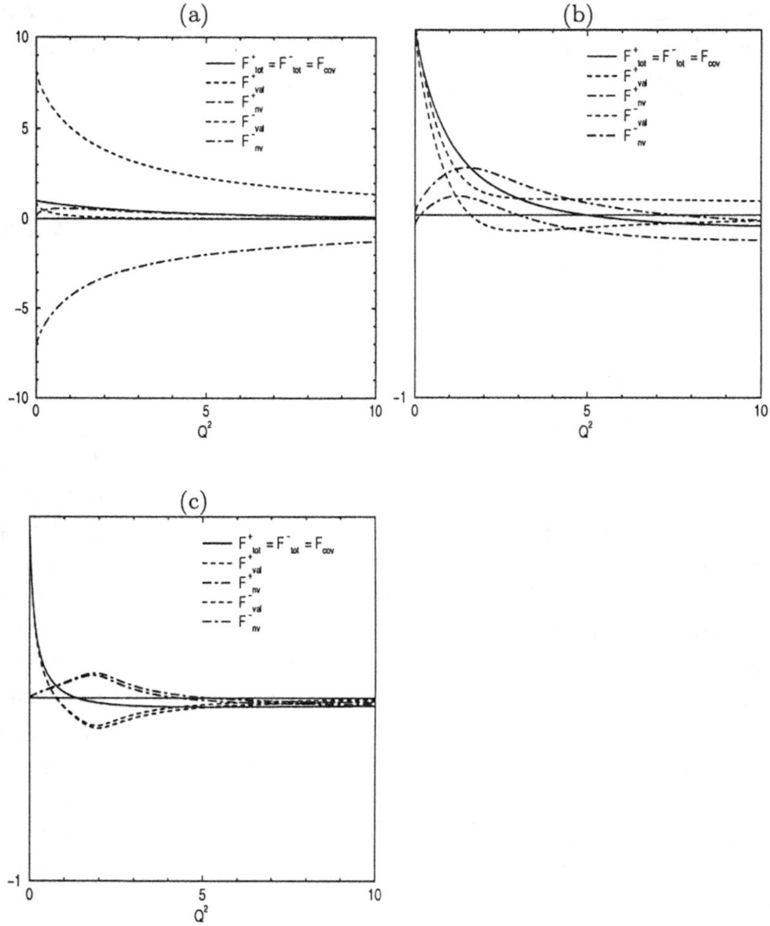

Fig. 10. Form factor of scalar meson with spinor constituents. Constituent masses are $m_a = m_b = 1.00$. Meson masses (a) $m = 1.00$, (b) $m = 1.90$, (c) $m = 1.99$. Fat lines correspond to the plus-current, thin lines to the minus-current. The solid line is the full form factor. It is the sum of the valence and the non-valence contributions. The separate contributions differ but the sums coincide. The form factor determined from the covariant amplitude is identical with the full form factor determined in the LF calculation.

In D dimensions we deal with the integral

$$I_D(p^2) = \int \mathrm{d}^D k \frac{1}{(k^2 - m^2 + i\epsilon)((k-p)^2 - m^2 + i\epsilon)} \,. \tag{165}$$

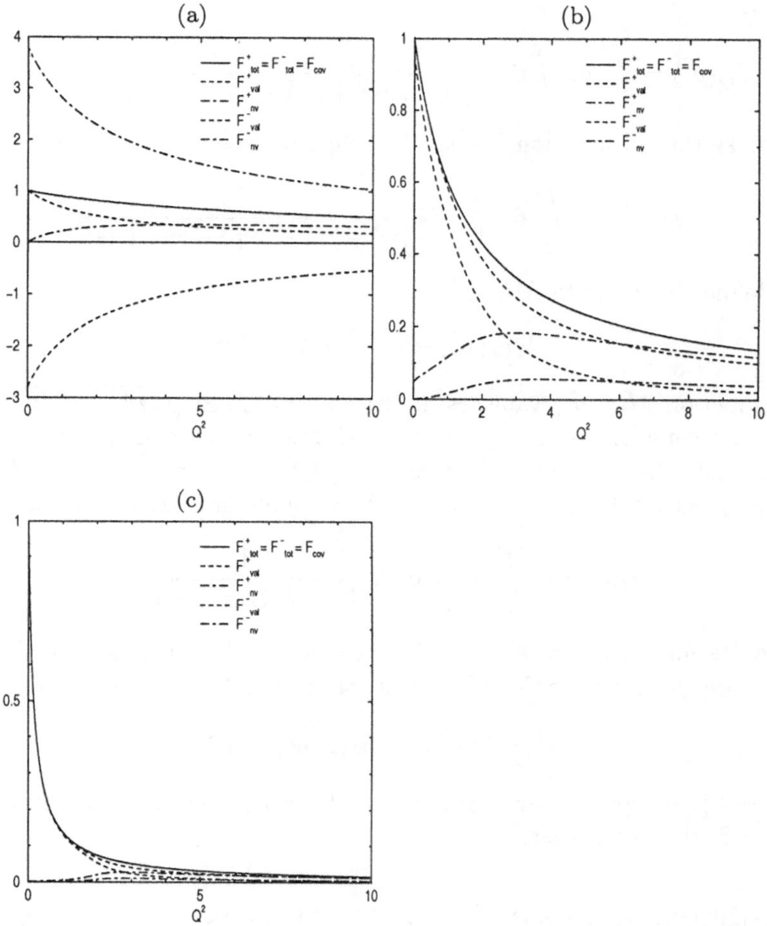

Fig. 11. Form factor of scalar meson with boson constituents. Constituent masses are $m_a = m_b = 1.00$. Meson masses (a) $m = 1.00$, (b) $m = 1.90$, (c) $m = 1.99$. The lines have the same meaning as in Fig. 10

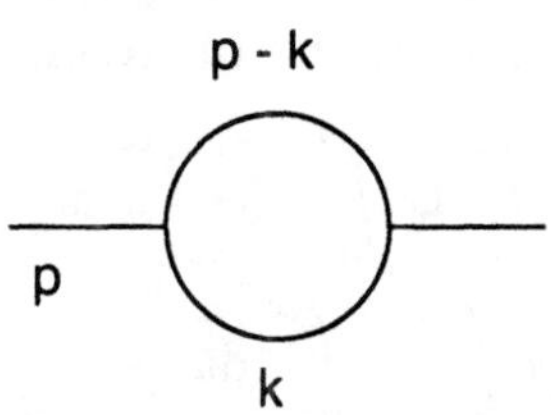

Fig. 12. Boson loop in ϕ^3 theory

8.1 Covariant Calculation

Upon using Feynman's formula

$$\frac{1}{AB} = \int_0^1 \mathrm{d}x \frac{1}{(xA + (1-x)B)^2} \tag{166}$$

we obtain

$$I_D(p^2) = \int_0^1 \mathrm{d}x \int \mathrm{d}^D k \frac{1}{((1-x)k^2 + x(k-p)^2 - m^2 + i\epsilon)^2}. \quad (167)$$

We make the substitution $k - xp \to k$, then we find

$$I_D(p^2) = \int_0^1 \mathrm{d}x \int \mathrm{d}^D k \frac{1}{(k^2 + x(1-x)p^2 - m^2 + i\epsilon)^2}. \quad (168)$$

We define the quantity $M(x;p^2)$:

$$M(x;p^2) = m^2 - x(1-x)p^2. \quad (169)$$

The function $M(x;p^2)$ vanishes for $x = x_\pm = 1/2 \pm \sqrt{1/4 - m^2/p^2}$. These zeros are complex for $0 < p^2 < 4m^2$. Outside that p^2-interval $M(x;p^2)$ is zero, inside the integration interval $x \in [0,1]$ only for $p^2 \geq 4m^2$. We shall discuss the case $0 < p^2 < 4m^2$ only. Then the integral (168) becomes

$$I_D(p^2) = \int_0^1 \mathrm{d}x \int \mathrm{d}^D k \frac{1}{(k^2 - M(x;p^2) + i\epsilon)^2}. \quad (170)$$

We write for k^2: $k^2 = (k^0)^2 - \boldsymbol{k}^2$, where $\boldsymbol{k}$ is a $D-1$-dimensional vector. Therefore, the singularities of the integrand in (170) are located at

$$k^0 = (\pm\sqrt{\boldsymbol{k}^2 + M(x;p^2)} - i\epsilon')). \quad (171)$$

In case the integrand converges, we are allowed to perform a Wick-rotation. If $D = 2$, this is the case.

1+1-Dimensional Case, $D = 2$ We write for $(k^0, k^1) \to (iy_1, y_2)$. Then $\mathrm{d}^D k = i\mathrm{d}y_1 \mathrm{d}y_2$. So we find

$$I_2(p^2) = i \int_0^1 \mathrm{d}x \int_{-\infty}^{\infty} \mathrm{d}y_1 \int_{-\infty}^{\infty} \mathrm{d}y_2 \frac{1}{(y_1^2 + y_2^2 + M(x;p^2) - i\epsilon)^2}. \quad (172)$$

Because $M(x;p^2) > 0$ for $p^2 > 0$, $0 \leq x \leq 1$, this integral is regular. Upon using plane polar coordinates, we find the result

$$\begin{aligned} I_2(p^2) &= i \int_0^1 \mathrm{d}x \int_0^{2\pi} \mathrm{d}\phi \int_0^{\infty} \mathrm{d}r\, r \frac{1}{(r^2 + M)^2} \\ &= i\pi \int_0^1 \mathrm{d}x \int_0^{\infty} \mathrm{d}u \frac{1}{(u + M)^2} \\ &= i\pi \int_0^1 \mathrm{d}x \frac{1}{m^2 - x(1-x)p^2} \\ &= \frac{4\pi i}{\sqrt{p^2(4m^2 - p^2)}} \arctan \sqrt{\frac{p^2}{4m^2 - p^2}}. \end{aligned} \quad (173)$$

3+1-Dimensional Case, $D = 4$ In 3+1-dimensions divergences occur. In order to regularize them, we use dimensional regularization (see [21]). We take the dimension D to $D = 4 - 2\epsilon$ and the limit $\epsilon \to 0$ is performed eventually. The correct dimensions are maintained if one multiplies the coupling constant with $\mu^{2-D/2}$, where μ has the dimension of a momentum. Consequently we calculate the integral:

$$I_D(p^2) = \mu^{4-D} \int_0^1 \mathrm{d}x \int \mathrm{d}^D k \frac{1}{(k^2 - M + i\epsilon)^2}. \tag{174}$$

Upon Wick rotation we obtain

$$I_D(p^2) = i\mu^{4-D} \int_0^1 \mathrm{d}x \int \mathrm{d}^D y \frac{1}{(y^2 + M)^2}. \tag{175}$$

Dimensional Regularization Using the formulae of Appendix A, we find

$$I_D(p^2) = i\mu^{4-D} \pi^{D/2} \Gamma(2 - D/2) \int_0^1 \mathrm{d}x M^{D/2-2}. \tag{176}$$

As $2 - D/2 = \epsilon$, we find for the limit $\epsilon \to 0$ of this expression

$$I_4(p^2) = i \int_0^1 \mathrm{d}x \left(\frac{\mu^2}{M}\right)^\epsilon \pi^{2-\epsilon} \Gamma(\epsilon). \tag{177}$$

The limit $\epsilon \to 0$ gives

$$I_4(p^2) = i\pi^2 \int_0^1 \mathrm{d}x \left(1 + \epsilon \log\left(\frac{\mu^2}{M}\right)\right) \left(\frac{1}{\epsilon} - \gamma + \mathcal{O}(\epsilon)\right). \tag{178}$$

If we substitute the expression for M in this equation, we find

$$I_4(p^2) = i\pi^2 \left(\frac{1}{\epsilon} - \gamma + \log \pi\right) - i\pi^2 \int_0^1 \mathrm{d}x \log\left(\frac{m^2 - x(1-x)p^2}{\mu^2}\right). \tag{179}$$

We write the final formula for two renormalizations:
on-shell renormalization, $I_4^{\mathrm{ren}}(m^2) = 0$:

$$I_4^{\mathrm{ren}}(p^2) = I_4(p^2) - I_4(m^2) = -i\pi^2 \int_0^1 \mathrm{d}x \log\left(\frac{m^2 - x(1-x)p^2}{m^2(1 - x(1-x))}\right). \tag{180}$$

and
$p^2 = 0$ *renormalization,* $I_4^{\mathrm{ren}}(0) = 0$:

$$I_4^{\mathrm{ren}}(p^2) = I_4(p^2) - I_4(0) = -i\pi^2 \int_0^1 \mathrm{d}x \log\left(\frac{m^2 - x(1-x)p^2}{m^2}\right). \tag{181}$$

In both cases we have to keep in mind that the argument of the logarithm is positive for $p^2 < 4m^2$, which means that $I_4(p^2)$ is real below the two-body threshold.

8.2 Instant-Form Calculation

Our point of departure is (165). We write for the integration element $\mathrm{d}^D k = \mathrm{d}k^0 \mathrm{d}^{D-1}\boldsymbol{k}$. Then it is natural to define for a momentum $\boldsymbol{q}$: $\omega(\boldsymbol{q}) = \sqrt{\boldsymbol{q}^2 + m^2}$. The factors in the denominator in (165) becomes:

$$k^2 - m^2 + i\epsilon = (k^0)^2 - \omega^2(\boldsymbol{k}) + i\epsilon = (k^0 - \omega(\boldsymbol{k}) + i\epsilon')(k^0 + \omega(\boldsymbol{k}) - i\epsilon') \quad (182)$$

and

$$\begin{aligned}(k-p)^2 - m^2 + i\epsilon &= (k^0 - p^0)^2 - \omega^2(\boldsymbol{k}-\boldsymbol{p}) + i\epsilon \\ &= [(k^0 - (p^0 + \omega(\boldsymbol{k}-\boldsymbol{p}) - i\epsilon')][(k^0 - (p^0 - \omega(\boldsymbol{k}-\boldsymbol{p}) + i\epsilon')].\end{aligned} \quad (183)$$

So, there are two poles on either side of the real k^0-axis. For $D = 2$ the integral converges and the circle at infinity does not contribute. Therefore, we now consider $D = 2$ and calculate $\int \mathrm{d}k^0$ by contour integration.

1+1-Dimensional Case In order to expose the symmetry between the two legs of the bubble we change the variables to $k = \frac{1}{2}(p+q)$, so $p - k = \frac{1}{2}(p-q)$ and $\mathrm{d}^D k = 2^{-D}\mathrm{d}^D q$. Then we obtain

$$\begin{aligned}I_2(p^2) &= \int \frac{\mathrm{d}^2 q}{4} \frac{1}{(\frac{p+q}{2})^2 - m^2 + i\epsilon} \frac{1}{(\frac{p-q}{2})^2 - m^2 + i\epsilon} \\ &= \int \frac{\mathrm{d}^2 q}{4} \frac{4}{(p+q)^2 - 4m^2 + i\epsilon} \frac{4}{(p-q)^2 - 4m^2 + i\epsilon}\end{aligned} \quad (184)$$

There are four poles, $P_1, \ldots, P_4$ with residues $R_1, \ldots, R_4$

$$\begin{aligned}P_1 &: q^0 = p^0 - \omega_- + i\epsilon; \qquad P_2 : q^0 = p^0 + \omega_- - i\epsilon; \\ P_3 &: q^0 = -p^0 + \omega_+ - i\epsilon; \quad P_4 : q^0 = -p^0 - \omega_+ + i\epsilon.\end{aligned} \quad (185)$$

The quantities $\omega_\pm$ are defined as $\omega_\pm = \sqrt{(\boldsymbol{p} \pm \boldsymbol{q})^2 + 4m^2}$. We will close the contour in the upper q^0-plane, so we need the residues R_1 and R_4:

$$\begin{aligned}R_1 &= \frac{-4}{2\omega_-(2p^0 - \omega_- + \omega_+)(2p^0 - \omega_- - \omega_+)}, \\ R_4 &= \frac{-4}{2\omega_+(2p^0 - \omega_- + \omega_+)(-2p^0 - \omega_- - \omega_+)}.\end{aligned} \quad (186)$$

In order to facilitate the interpretation, we reshuffle R_1 and R_4 as follows:

$$\frac{1}{2p^0 - \omega_- + \omega_+} \frac{1}{2p^0 - \omega_- - \omega_+ + i\epsilon} = \frac{1}{2\omega_+}\left[\frac{1}{2p^0 - \omega_- - \omega_+ + i\epsilon} - \frac{1}{2p^0 - \omega_- + \omega_+}\right], \quad (187)$$

$$\frac{1}{2p^0 - \omega_- + \omega_+} \frac{1}{2p^0 + \omega_- + \omega_+ - i\epsilon} = \frac{-1}{2\omega_-} \left[\frac{1}{2p^0 + \omega_- + \omega_+ - i\epsilon} - \frac{1}{2p^0 - \omega_- + \omega_+} \right]. \tag{188}$$

Consequently, the sum $R_1 + R_4$ can be written as

$$R_1 + R_4 = \frac{1}{\omega_-} \frac{1}{\omega_+} \left[\frac{-1}{2p^0 - \omega_- - \omega_+ + i\epsilon} + \frac{-1}{-2p^0 - \omega_- - \omega_+ + i\epsilon} \right]. \tag{189}$$

The complex denominators can be interpreted as energy denominators of time-ordered diagrams.

$$\begin{aligned} I_2^+(p^2) &= 2\pi i \int_{-\infty}^{\infty} \mathrm{d}q \frac{1}{\omega_- \omega_+} \frac{-1}{2p^0 - \omega_- - \omega_+ + i\epsilon}, \\ I_2^-(p^2) &= 2\pi i \int_{-\infty}^{\infty} \mathrm{d}q \frac{1}{\omega_- \omega_+} \frac{-1}{-2p^0 - \omega_- - \omega_+ + i\epsilon}. \end{aligned} \tag{190}$$

Because we know already that I_2 is a function of the variable p^2, we may

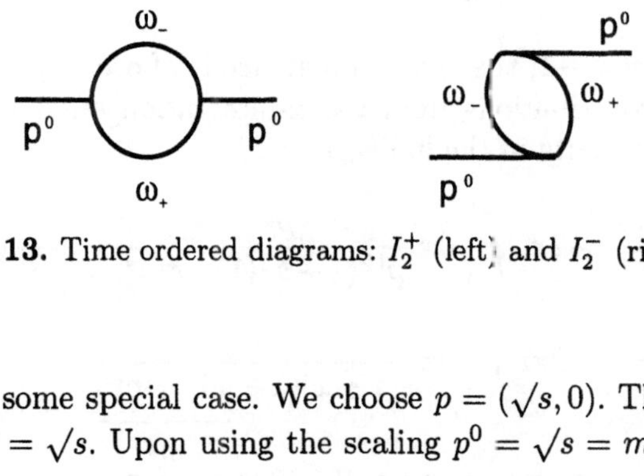

Fig. 13. Time ordered diagrams: I_2^+ (left) and I_2^- (right).

calculate it for some special case. We choose $p = (\sqrt{s}, 0)$. Then $\omega_+ = \omega_- = \sqrt{\boldsymbol{q}^2 + 4m^2}, p^0 = \sqrt{s}$. Upon using the scaling $p^0 = \sqrt{s} = mz$, $|\boldsymbol{q}| = mx$ we find

$$\begin{aligned} I_2^{\pm}(p^2) &= \mp \frac{i\pi}{m^2} \int_{-\infty}^{\infty} \mathrm{d}x \frac{1}{x^2 + 4} \frac{1}{z \mp \sqrt{x^2 + 4}} \\ &= \frac{i\pi}{m^2} \left\{ \mp \frac{\pi}{2\sqrt{z}} + \frac{\pi}{\sqrt{z(4 - z)}} \left[\pm 1 + \frac{2}{\pi} \arctan \sqrt{\frac{z}{4 - z}} \right] \right\}, \end{aligned} \tag{191}$$

hence

$$I_2(p^2) = I_2^+(p^2) + I_2^-(p^2) = \frac{4\pi i}{\sqrt{s(4m^2 - s)}} \arctan \sqrt{\frac{s}{4m^2 - s}}. \tag{192}$$

Infinite-Momentum Frame Now consider the amplitude in the *infinite momentum frame* (IMF). This is the limit where $|\boldsymbol{p}|$ is taken to infinity. It is clear that in this limit I_2^- vanishes, as all denominators in its integrand are of order $|\boldsymbol{p}|$ for all $|\boldsymbol{q}|$. For I_2^+ this is not so, ω_+ and ω_- are of order $|\boldsymbol{p}|$, but $2p^0 - \omega_+ - \omega_-$ is not.

Before taking the limit $|\boldsymbol{p}| \to \infty$, we change the integration variable from $|\boldsymbol{q}|$ to $x \equiv |\boldsymbol{q}|/|\boldsymbol{p}|$. In this way we ensure that the limits $|\boldsymbol{p}| \to \infty$ and $|\boldsymbol{q}| \to \pm\infty$ (the boundaries of the integration region) are taken properly. We find:

$$\begin{aligned} \omega_\pm(\boldsymbol{p},\boldsymbol{q};m) &= \sqrt{(\boldsymbol{p}\pm\boldsymbol{q})^2+4m^2} = |\boldsymbol{p}|\sqrt{(1\pm x)^2+4m^2/\boldsymbol{p}^2} \\ &\stackrel{|\boldsymbol{p}|\to\infty}{\sim} |\boldsymbol{p}||1\pm x| + \frac{2m}{|1\pm x||\boldsymbol{p}|}, \end{aligned} \tag{193}$$

and

$$p^0(p^2=s) = \sqrt{\boldsymbol{p}^2+s} \stackrel{|\boldsymbol{p}|\to\infty}{\sim} |\boldsymbol{p}| + \frac{s}{2|\boldsymbol{p}|}. \tag{194}$$

The absolute value signs are essential. We see this when we examine the energy difference

$$2p^0 - \omega_+ - \omega_- \stackrel{|\boldsymbol{p}|\to\infty}{\sim} 2|\boldsymbol{p}| + \frac{s}{|\boldsymbol{p}|} - |\boldsymbol{p}||1+x| - \frac{2m^2}{|1+x||\boldsymbol{p}|} - |\boldsymbol{p}||1-x| - \frac{2m^2}{|1-x||\boldsymbol{p}|}. \tag{195}$$

When $x > 1$ or $x < -1$, the energy difference is of order $|\boldsymbol{p}|$. So, in the limit $|\boldsymbol{p}| \to \infty$ the contributions from the x-integration outside $[-1,1]$ vanish. Consequently, we write in the limit $|\boldsymbol{p}| \to \infty$

$$\begin{aligned} I_2(p^2=s) &\stackrel{|\boldsymbol{p}|\to\infty}{\sim} -2\pi i \int_{-1}^{1} \mathrm{d}x \frac{|\boldsymbol{p}|}{|\boldsymbol{p}|^2(1+x)(1-x)} \frac{1}{\frac{s}{|\boldsymbol{p}|} - \frac{2m^2}{(1+x)|\boldsymbol{p}|} - \frac{2m^2}{(1-x)|\boldsymbol{p}|}} \\ &= -2\pi i \int_{-1}^{1} \mathrm{d}x \frac{1}{s(1+x)(1-x) - 4m^2} \\ &= \frac{4\pi i}{\sqrt{p^2(4m^2-p^2)}} \arctan\sqrt{\frac{p^2}{4m^2-p^2}}. \end{aligned} \tag{196}$$

The "success" of the IMF limit is somewhat misleading. It leans heavily on the fact that the spin-0 propagator does not contain p^- in the numerator. If spin-1/2 constituents were considered, the instantaneous terms would spoil the naive IMF limit.

3+1-Dimensional Case In the case of three spatial dimensions we obtain

$$\begin{aligned} I_4(p^2) &= \int \frac{\mathrm{d}^4q}{2^4} \frac{1}{(\frac{p+q}{2})^2 - m^2 + i\epsilon} \frac{1}{(\frac{p-q}{2})^2 - m^2 + i\epsilon} \\ &= \int \frac{\mathrm{d}^4q}{2^4} \frac{4}{(p+q)^2 - 4m^2 + i\epsilon} \frac{4}{(p-q)^2 - 4m^2 + i\epsilon}. \end{aligned} \tag{197}$$

There are again four poles, $P_1, \ldots, P_4$, corresponding to the residues $R_1, \cdots, R_4$, that are treated in the same way as before. The final integrals for the time-ordered diagrams become:

$$I_4^+(p^2) = 2\pi i \int \mathrm{d}^3 q \frac{1}{2\omega_- 2\omega_+} \frac{-1}{2p^0 - \omega_- - \omega_+ + i\epsilon},$$
$$I_4^-(p^2) = 2\pi i \int \mathrm{d}^3 q \frac{1}{2\omega_- 2\omega_+} \frac{-1}{-2p^0 - \omega_- - \omega_+ + i\epsilon}. \tag{198}$$

We calculate these integrals for the case $\boldsymbol{p} = 0$. Then $\omega_+ = \omega_- = \sqrt{\boldsymbol{q}^2 + 4m^2}, (p^0)^2 = p^2$;

$$I_4(p^2) = I_4^+ + I_4^- = 2\pi i \int \mathrm{d}^3 q \frac{1}{4\sqrt{q^2 + 4m^2}} \frac{1}{q^2 + 4m^2 - p^2}. \tag{199}$$

Here $q = |\boldsymbol{q}|$. This integral is clearly logarithmically divergent. To see most clearly the divergence we write it as

$$I_4(p^2) = 2\pi i \Omega_3 \int_0^\infty \mathrm{d}q \frac{1}{4\sqrt{q^2 + 4m^2}} \frac{q^2}{q^2 + 4m^2 - p^2}. \tag{200}$$

The fraction $q^2/(q^2+4m^2-p^2)$ can be split into two parts as follows: $q^2/(q^2 + 4m^2 - p^2) = 1 - (4m^2 - p^2)/(q^2 + 4m^2 - p^2)$. The first term leads to a divergent integral that has a pole in dimension space at dimension $D = 3$ and is independent of p. So we can regularize the integral by splitting off this part. The regularized amplitude becomes

$$I_4^{\mathrm{reg}}(p^2) = -i8\pi^2(4m^2 - p^2) \int_0^\infty \mathrm{d}q \frac{1}{4\sqrt{q^2 + 4m^2}(q^2 + 4m^2 - p^2)}. \tag{201}$$

The integral is known in closed form, and we get:

$$I_4^{\mathrm{reg}}(p^2) = -i2\pi^2 \sqrt{\frac{4m^2 - p^2}{p^2}} \arctan \sqrt{\frac{p^2}{4m^2 - p^2}}. \tag{202}$$

The renormalized integral is then:
on-shell renormalization, $p^2 = m^2$

$$I_4^{\mathrm{ren}}(p^2) = i2\pi^2(p^2 - m^2) \int_0^\infty \mathrm{d}q \frac{q^2}{\sqrt{q^2 + 4m^2}(q^2 + 3m^2)(q^2 + 4m^2 - p^2)}$$
$$= i2\pi^2 \left[\frac{\pi}{2\sqrt{3}} - \sqrt{\frac{4m^2 - p^2}{p^2}} \arctan \sqrt{\frac{p^2}{4m^2 - p^2}} \right]; \tag{203}$$

on-shell renormalization, $p^2 = 0$

$$I_4^{\mathrm{ren}}(p^2) = i2\pi^2 p^2 \int_0^\infty \mathrm{d}q \frac{q^2}{\sqrt{q^2 + 4m^2}(q^2 + 4m^2)(q^2 + 4m^2 - p^2)}$$
$$= i2\pi^2 \left[1 - \sqrt{\frac{4m^2 - p^2}{p^2}} \arctan \sqrt{\frac{p^2}{4m^2 - p^2}} \right]. \tag{204}$$

Infinite-Momentum Limit Again, we consider the infinite momentum limit. Let the restframe momentum be $p^\mu = (\sqrt{s}, \mathbf{0})$ and boost in the z-direction with the boost parameter χ, *i.e.*, $p^\mu \to (\cosh\chi\sqrt{s}, \mathbf{0}^\perp, \sinh\chi\sqrt{s})$. The infinite momentum limit amounts to $\chi \to \infty$. As we are working in the instant form, we will translate this limit into $p^3 \to \infty$. Then we have $p^0 = p^3 \coth\chi$. In the same way as in the 1+1-dimensional case we can prove that only I^+ survives in the limit $p^3 \to \infty$. The energies $\omega_\pm$ and the energy denominator $2p^0 - \omega_+ - \omega_-$ are

$$\omega_\pm = p^3|1 \pm x| + \frac{\boldsymbol{q}^{\perp 2} + 4m^2}{p^3|1 \pm x|} + \mathcal{O}((p^3)^2), \tag{205}$$

$$2p^0 - \omega_+ - \omega_- = 2p^3 \coth\chi - p^3|1+x| - \frac{\boldsymbol{q}^{\perp 2} + 4m^2}{2p^3|1+x|} - p^3|1-x| - \frac{\boldsymbol{q}^{\perp 2} + 4m^2}{2p^3|1-x|}. \tag{206}$$

The variable q^3 scales with p^3 as $q^3 = p^3 x$, $-1 \le x \le 1$. Therefore we find

$$\begin{aligned} I_4^+(p^2) &= i\frac{\pi}{2}\int_{-1}^{1} \mathrm{d}x \int \mathrm{d}^2 q^\perp \frac{1}{1-x^2} \frac{-1}{2(p^3)^2(\coth\chi - 1) - \frac{\boldsymbol{q}^{\perp 2}+4m^2}{1-x^2}} \\ &= i\frac{\pi}{2}\int_{-1}^{1} \mathrm{d}x \int \mathrm{d}^2 q^\perp \frac{1}{\boldsymbol{q}^{\perp 2} + 4m^2 - p^2(1-x^2)}, \end{aligned} \tag{207}$$

where we used that in the limit $\chi \to \infty$ we get

$$2\,(p^3)^2\,(\coth\chi - 1) \sim 2\,\frac{s}{4}\,\mathrm{e}^{2\chi}\,2\mathrm{e}^{-2\chi} = s = p^2. \tag{208}$$

Of course, this amplitude is also logarithmically divergent. The renormalized amplitudes are obtained by subtracting the value at a chosen renormalization point. Thus we get for the two cases considered before:
on-shell renormalization $p^2 = m^2$

$$\begin{aligned} I_4^{\mathrm{ren}}(p^2) &= i\frac{\pi}{2}\int_{-1}^{1} \mathrm{d}x \int \mathrm{d}^2 q^\perp \\ &\quad \times \left[\frac{1}{\boldsymbol{q}^{\perp 2} + 4m^2 - p^2(1-x^2)} - \frac{1}{\boldsymbol{q}^{\perp 2} + 4m^2 - m^2(1-x^2)}\right] \\ &= i\frac{\pi^2}{2}\int_{-1}^{1} \mathrm{d}x \log\left(\frac{z + 4m^2 - p^2(1-x^2)}{z + 4m^2 - m^2(1-x^2)}\right)_{z=0}^{\infty}, \quad (\text{using } z = \boldsymbol{q}^{\perp 2}) \\ &= -i\frac{\pi^2}{2}\int_{-1}^{1} \mathrm{d}x \log\left(\frac{4m^2 - p^2(1-x^2)}{m^2(3+x^2)}\right) \\ &= i2\pi^2\left[\frac{\pi}{2\sqrt{3}} - \sqrt{\frac{4m^2 - p^2}{p^2}} \arctan\sqrt{\frac{p^2}{4m^2 - p^2}}\right]; \end{aligned} \tag{209}$$

$p^2 = 0$ *renormalization*

$$I_4^{\rm ren}(p^2) = -i\frac{\pi^2}{2}\int_{-1}^{1} \mathrm{d}x \log\left(\frac{4m^2 - p^2(1-x^2)}{4m^2}\right)$$
$$= i2\pi^2\left[1 - \sqrt{\frac{4m^2-p^2}{p^2}}\arctan\sqrt{\frac{p^2}{4m^2-p^2}}\right]. \tag{210}$$

We have recovered, as we should, the results of the previous section, (180, 181), if (252) is taken into account, as well as (203, 204).

8.3 Calculation in Light-Front Coordinates

Before we carry out the front form calculation proper, i.e., compute the integral $I_2(p^2)$ in two steps, we first do the covariant case in light-front coordinates. Those are defined as follows:

$$k^+ = \frac{k^0 + k^3}{\sqrt{2}},\ k^- = \frac{k^0 - k^3}{\sqrt{2}},\ k^1,\ k^2. \tag{211}$$

The scalar product is given by

$$k^2 = 2k^+k^- - (k^\perp)^2, \tag{212}$$

and the integration measure is

$$\mathrm{d}^4k = \mathrm{d}k^+\mathrm{d}k^-\mathrm{d}^2k^\perp. \tag{213}$$

In the 1+1-dimensions the perpendicular degrees of freedom are dropped.

We use the original convention for the momenta: k and $p-k$ in the loop. Then we find

$$I_2(p^2) = \int \mathrm{d}^2k \frac{1}{k^2 - m^2 + i\epsilon}\,\frac{1}{(p-k)^2 - m^2 + i\epsilon}$$
$$= \int_0^1 \mathrm{d}x \int \mathrm{d}^2k \frac{1}{(k^2 - M(x;p^2) + i\epsilon)^2}$$
$$= \int_0^1 \mathrm{d}x \int_{-\infty}^{\infty} \mathrm{d}k^+ \int_{-\infty}^{\infty} \mathrm{d}k^- \frac{1}{(2k^+k^- - M(x;p^2) + i\epsilon)^2}. \tag{214}$$

Naively we have the following situation. For $k^+ \neq 0$ the integral over k^- is convergent and the integral over a semi-circle in the upper half of the complex k^--plane vanishes upon taking the limit of its radius going to infinity. Therefore, the double pole in k^- at $k_P^- = (M(x;p^2) - i\epsilon)/k^+$ gives no contribution to the integral. However, if $k^+ = 0$, the k^--integral diverges. Apparently, this integral defines a distribution with support at $k^+ = 0$.

Our next step illustrates this in another way. We write $z = 2k^+k^-$, $\mathrm{d}k^- = \mathrm{d}z/(2k^+)$. Then we have:

$$I_2(p^2) = \int_0^1 \mathrm{d}x \int_{-\infty}^{\infty} \mathrm{d}k^+ \frac{1}{2|k^+|} \int_{-\infty}^{\infty} \mathrm{d}z \frac{1}{(z - M + i\epsilon)^2}. \tag{215}$$

The modulus of k^+ occurs because the absolute value of the Jacobian is involved. We see that superficially $I_2 = \int_0^1 \mathrm{d}x(\infty \cdot 0)$. This is easy to illustrate by employing a cut-off as follows:

$$\begin{aligned} J_\Lambda(x) &= \frac{1}{2|x|} \int_{-2\Lambda|x|}^{2\Lambda|x|} \mathrm{d}z \frac{1}{(z - M + i\epsilon)^2} \\ &= \frac{1}{2|x|} \left[\frac{-1}{z - M + i\epsilon} \right]_{-2\Lambda|x|}^{2\Lambda|x|} \\ &= \frac{1}{2|x|} \left[\frac{-1}{2\Lambda|x| - M + i\epsilon} - \frac{-1}{-2\Lambda|x| - M + i\epsilon} \right]. \end{aligned} \tag{216}$$

Clearly, $J_\Lambda(x) \to 0$ for $\Lambda \to \infty$ if $x \neq 0$, but $J_\Lambda(0) \to \infty$. In order to complete our calculation we regularize the distribution $(z + i\epsilon)^{-2}$ *à la* [28]

$$\frac{1}{(z + i\epsilon)^2} = \frac{1}{z^2} + i\pi\delta'(z); \tag{217}$$

where the distribution z^{-2} is defined as follows:

$$\left(\frac{1}{z^2}, f(z) \right) \equiv \int_0^\infty \mathrm{d}z \frac{f(z) + f(-z) - 2f(0)}{z^2} \tag{218}$$

Our integral has a piece connected with the real part of the distribution, z^{-2}, that vanishes upon integration using the regularized form with $f(z) \equiv 1$. The imaginary part needs some additional care. We use again a cut-off Λ and find

$$\begin{aligned} &\int_{-\infty}^{\infty} \mathrm{d}k^+ \int_{-\infty}^{\infty} \mathrm{d}k^- \frac{1}{(2k^+k^- - M + i\epsilon)^2} = \\ &\qquad \int_{-\infty}^{\infty} \mathrm{d}k^+ \lim_{\Lambda \to \infty} \int_{-\Lambda}^{\Lambda} \mathrm{d}k^- \left[\frac{1}{(2k^+k^- - M)^2} + i\pi\delta'(2k^+k^- - M) \right]. \end{aligned} \tag{219}$$

The first part vanishes if interpreted *à la* Vilenkin. The second part contains the integral

$$\begin{aligned} \int_{-\Lambda}^{\Lambda} \mathrm{d}k^- \delta'(2k^+k^- - M) &= \frac{1}{|2k^+|} \int_{-2\Lambda|k^+|}^{2\Lambda|k^+|} \mathrm{d}z \delta'(z - M) \\ &= \frac{1}{2|k^+|} \left[\delta(2\Lambda|k^+| - M) - \delta(2\Lambda|k^+| + M) \right] \\ &= \frac{1}{2M} \left[\delta\left(|k^+| - \frac{M}{2\Lambda} \right) - \delta\left(|k^+| + \frac{M}{2\Lambda} \right) \right] \end{aligned} \tag{220}$$

The second δ-function does not contribute. The full integral is then:

$$\begin{aligned} I_2(p^2) &= \int_0^1 \mathrm{d}x \frac{i\pi}{2M(x;p^2)} \int_{-\infty}^{\infty} \mathrm{d}k^+ \delta\left(|k^+| - \frac{M}{2\Lambda}\right) \\ &= i\pi \int_0^1 \mathrm{d}x \frac{1}{M(x;p^2)} \\ &= \frac{4\pi i}{\sqrt{p^2(4m^2-p^2)}} \arctan \sqrt{\frac{p^2}{4m^2-p^2}}. \end{aligned} \tag{221}$$

In 3+1 dimensions we can rewrite the original formula (170) as follows

$$I_4(p^2) = \int_0^1 \mathrm{d}x \int_{\infty}^{\infty} \mathrm{d}k^+ \int_{\infty}^{\infty} \mathrm{d}k^- \int \mathrm{d}^2 k^\perp \frac{1}{(2k^+k^- - \boldsymbol{k}^{\perp 2} - M(x;p^2) + i\epsilon)^2}. \tag{222}$$

If we make the substitution $M \to M + \boldsymbol{k}^{\perp 2}$ in (221) and realize that we need to integrate in addition over $\boldsymbol{k}^\perp$, we obtain the expression

$$I_4(p^2) = i\pi \int_0^1 \mathrm{d}x \int \mathrm{d}^2 k^\perp \frac{1}{\boldsymbol{k}^{\perp 2} + M(x;p^2)}. \tag{223}$$

Next we substitute the expression for $M(x;p^2)$, (169) and make the substitutions $x = (1+x)/2$ and $\boldsymbol{q}^\perp = \boldsymbol{k}^\perp/2$. Then we find the formula

$$I_4(p^2) = i\frac{\pi}{2} \int_{-1}^1 \mathrm{d}x \int \mathrm{d}^2 q^\perp \frac{1}{\boldsymbol{q}^{\perp 2} + 4m^2 - p^2(1-x^2)}. \tag{224}$$

which is equal to (207).

8.4 Front-Form Calculation

We write I_2 in terms of the momenta $\frac{1}{2}(p \pm q)$ in the loop. It is understood that everywhere $m^2 \to m^2 - i\epsilon$.

$$\begin{aligned} I_2(p^2) &= 4 \int \mathrm{d}^2 q \frac{1}{(p+q)^2 - 4m^2} \frac{1}{(p-q)^2 - 4m^2} \\ &= 4 \int_{-\infty}^{\infty} \mathrm{d}q^+ \int_{-\infty}^{\infty} \mathrm{d}q^- \frac{1}{2(p^+ + q^+)(p^- + q^-) - 4m^2} \\ &\qquad \times \frac{1}{2(p^+ - q^+)(p^- - q^-) - 4m^2}. \end{aligned} \tag{225}$$

We must distinguish five cases ($p^+ > 0$):

$$\begin{aligned} &\text{(i) } q^+ < -p^+, \quad \text{(ii) } -p^+ < q^+ < p^+, \quad \text{(iii) } q^+ > p^+, \\ &\text{(iv) } q^+ = p^+, \qquad \text{(v) } q^+ = -p^+. \end{aligned} \tag{226}$$

Accordingly, we consider the poles in the variable q^-:

$$q_1^- = -p^- + \frac{4m^2 - i\epsilon}{2(p^+ + q^+)}, \quad q_2^- = p^- - \frac{4m^2 - i\epsilon}{2(p^+ - q^+)}. \tag{227}$$

Then we find that $q_1^- \to \infty$ for $q^+ \to -p^+$ (v) and $q_2^- \to \infty$ for $q^+ \to p^+$ (iv). The domains (iv) and (v) have measure zero in q^+, so, unless the integral over q^- produces in these cases a distribution with support in (iv) and (v), the contributions from these intervals vanish. We calculate these first: domain (iv) gives

$$J(q^+; m) = 4 \int_{-\infty}^{\infty} \mathrm{d}q^- \frac{1}{4p^+(p^- + q^-) - 4m^2 + i\epsilon} \frac{1}{-4m^2 + i\epsilon} \tag{228}$$

which reduces to

$$\begin{aligned} J(q^+; m) &= \frac{-1}{4p^+m^2} \int_{-\infty}^{\infty} \mathrm{d}q^- \frac{1}{(p^- + q^- - m^2/p^+ + i\epsilon} \\ &= \left[\int_{-\infty}^{\infty} \mathrm{d}q^- \frac{1}{q^- - q_0^-} - i\pi \right], \end{aligned} \tag{229}$$

where $q_0^- = m^2/p^+ - p^-$ and the integral is understood as a principal value integral. The latter vanishes, so $J(q^+; m)$ is finite and independent of q^+. Consequently, its contribution to I_2 vanishes. In the case (v) we find the same result.

Domain (i). Here $Im\ q_1^-$ and $Im\ q_2^-$ are both positive. As the integral is convergent and the semi circle at infinity does not contribute, we can evaluate the q^--integral using the residue-theorem. The result is zero. In domain (iii) we obtain the same answer.

Domain (ii). The integral is

$$\begin{aligned} I_2(p^2) &= \int_{-p^+}^{p^+} \mathrm{d}q^+ \frac{1}{(p^+ + q^+)(p^+ - q^+)} \int_{-\infty}^{\infty} \mathrm{d}q^- \frac{1}{(q^- - q_1^-)(q^- - q_2^-)} \\ &= \int_{-p^+}^{p^+} \mathrm{d}q^+ \frac{1}{(p^+ + q^+)(p^+ - q^+)} 2\pi i \frac{-1}{q_2^- - q_1^-} \\ &= -i\pi \int_{-p^+}^{p^+} \mathrm{d}q^+ \frac{1}{p^-(p^+ + q^+)(p^+ - q^+) - 2m^2 p^+}. \end{aligned} \tag{230}$$

By the substitutions $q^+ = (2x - 1)p^+$ and $2p^+p^- = p^2$, this integral is transformed into

$$I_2(p^2 = m^2) = -\pi i \int_0^1 \mathrm{d}x \frac{1}{p^2 x(1-x) - m^2}. \tag{231}$$

This result is identical with (173), the covariant result.

The 3+1-dimensional case follows of course the same lines. We write then

$$I_4(p^2) = \int \mathrm{d}q^+ \int \mathrm{d}q^- \int \mathrm{d}^2q^\perp \frac{1}{2(p^+ + q^+)2(p^- + q^-) - (4m^2 + \boldsymbol{q}^{\perp 2}) + i\epsilon}$$
$$\times \frac{1}{2(p^+ - q^+)2(p^- - q^-) - (4m^2 + \boldsymbol{q}^{\perp 2}) + i\epsilon}. \tag{232}$$

The poles are now

$$q_1^- = -p^- + \frac{(4m^2 + \boldsymbol{q}^{\perp 2}) - i\epsilon}{2(p^+ + q^+)}, \quad q_2^- = p^- - \frac{(4m^2 + \boldsymbol{q}^{\perp 2}) - i\epsilon}{2(p^+ - q^+)}. \tag{233}$$

Upon closing the contour in the upper q^--plane one finds again

$$I_4(p^2) = \int_{p^-}^{p^+} \mathrm{d}q^+ \int \mathrm{d}^2q^\perp 2\pi i \frac{-1}{q_2^- - q_1^-}. \tag{234}$$

Next, substitute the values of $q_{1,2}^-$, change variables $q^+ = xp^+$, and use the condition $2p^+p^- = p^2$ for the case where $\boldsymbol{p}^\perp = 0$. Then we find

$$I_4(p^2) = \frac{i\pi}{2} \int_{-1}^{1} \mathrm{d}x \int \mathrm{d}^2q^\perp \frac{1}{\boldsymbol{q}^{\perp 2} + 4m^2 - p^2(1 - x^2)}. \tag{235}$$

This is exactly (207) so we again obtain the same result as before.

9 Dimensional Regularization: Basic Formulae

The dimension of space-time will be denoted by D. The surface area of the D-dimensional sphere in Euclidean space is

$$\mathrm{d}\Omega_D = \prod_{l=1}^{D-1} \mathrm{d}\theta_l (\sin\theta_l)^{D-1-l}; \quad \Omega_D = \frac{2\pi^{D/2}}{\Gamma(D/2)}. \tag{236}$$

The following integral is fundamental to dimensional regularization:

$$\int_0^\infty \mathrm{d}t \frac{t^n}{(1+t)^m} = \int_0^\infty \mathrm{d}t \frac{t^{(n+1)-1}}{(1+t)^{n+1+(m-n-1)}} = \frac{\Gamma(n+1)\Gamma(m-n-1)}{\Gamma(m)}. \tag{237}$$

As an application of these two formulae we derive

$$\int \mathrm{d}^D x \frac{1}{(x^2 + \mu^2)^n} = \Omega_D \int_0^\infty \mathrm{d}r \frac{r^{D-1}}{(r^2 + \mu^2)^n}. \tag{238}$$

Upon the substitution

$$r = \mu t^{1/2}, \quad \mathrm{d}r = \frac{\mu}{2} t^{-1/2} \mathrm{d}t \tag{239}$$

one finds

$$\int \mathrm{d}^D x \frac{1}{(x^2+\mu^2)^n} = \Omega_D \int_0^\infty \mathrm{d}t \frac{t^{D/2-1}}{(1+t)^n}. \tag{240}$$

If one substitutes the expressions for Ω_D and the t-integral one finds

$$\int \mathrm{d}^D x \frac{1}{(x^2+\mu^2)^n} = \pi^{D/2} \mu^{D-2n} \frac{\Gamma(n-D/2)}{\Gamma(n)}. \tag{241}$$

Another relation that is used frequently is

$$\Gamma(\epsilon) = \frac{1}{\epsilon} - \gamma + \mathcal{O}(\epsilon). \tag{242}$$

10 Four-Dimensional Integration

In this appendix we discuss the four-dimensional formulae in Minkowski and Euclidean space. We use the following polar coordinates in Euclidean space:

$$(r^0, \mathbf{r}) = (r \sin\theta_1 \sin\theta_2 \sin\theta_3, r \sin\theta_1 \sin\theta_2 \cos\theta_3, r \sin\theta_1 \cos\theta_2, r \cos\theta_1). \tag{243}$$

The Jacobian is then

$$\left| \frac{\partial(r^0, r^1, r^2, r^3)}{\partial(r, \theta_1, \theta_2, \theta_3)} \right| = r^3 (\sin\theta_1)^2 \sin\theta_2. \tag{244}$$

So the integration measure in Euclidean space is

$$\mathrm{d}^4 r = \mathrm{d}r r^3 \mathrm{d}\theta_1 (\sin\theta_1)^2 \mathrm{d}\theta_2 \sin\theta_2 \mathrm{d}\theta_3. \tag{245}$$

The surface area of the unit sphere in four dimensions is

$$\Omega_4 = \int_0^\pi \mathrm{d}\theta_1 (\sin\theta_1)^2 \int_0^\pi \mathrm{d}\theta_2 \sin\theta_2 \int_0^{2\pi} \mathrm{d}\theta_3 = 2\pi^2. \tag{246}$$

A Wick rotation transform integrals in Minkowski space into Euclidean integrals. We use the transformation

$$k^0 \to i r^0, \mathbf{k} \to \mathbf{r}, \tag{247}$$

so the integration measure becomes

$$\int_{-\infty}^\infty \mathrm{d}k^0 \int \mathrm{d}^3 k \to i \int_{-\infty}^\infty \mathrm{d}r^0 \int \mathrm{d}^3 r. \tag{248}$$

11 Some Useful Integrals

Here we collect some useful integrals. For d defined as $b^2 - 4ac$ we have

$$\mathcal{I}_1 = \int \mathrm{d}x \frac{1}{ax^2 + bx + c} = \frac{1}{\sqrt{b^2 - 4ac}} \log \left| \frac{2ax + b - \sqrt{b^2 - 4ac}}{2ax + b + \sqrt{b^2 - 4ac}} \right|, \quad d > 0,$$
$$= \frac{2}{\sqrt{4ac - b^2}} \arctan \left(\frac{2ax + b}{\sqrt{4ac - b^2}} \right), \quad d < 0. \quad (249)$$

The case where $d > 0$ the integral must be understood as a principal value if the denominator has zeros in the integration interval.

For $0 < p^2 < 4m^2$ we have the special case

$$\mathcal{I}_2 = \int_0^1 \mathrm{d}x \frac{1}{m^2 - x(1-x)p^2} = \frac{4}{p\sqrt{4m^2 - p^2}} \arctan \sqrt{\frac{p^2}{4m^2 - p^2}}. \quad (250)$$

For definite integrals containing an arctan one may use the identity

$$\arctan z_1 \pm \arctan z_2 = \arctan \left(\frac{z_1 \pm z_2}{1 \mp z_1 z_2} \right). \quad (251)$$

A second type of integral is, again for $0 < p^2 < 4m^2$

$$\mathcal{I}_3 = \int_0^1 \mathrm{d}x \log(m^2 - x(1-x)p^2)$$
$$= -2 + \log m^2 + 2\sqrt{\frac{4m^2 - p^2}{p^2}} \arctan \sqrt{\frac{p^2}{4m^2 - p^2}}. \quad (252)$$

A third type of integral we encounter is

$$\mathcal{I}_4 = \int_0^\infty \mathrm{d}q \frac{1}{\sqrt{q^2 + 4m^2}(q^2 + 4m^2 + p^2)} = \frac{1}{p\sqrt{4m^2 - p^2}} \arctan \sqrt{\frac{p^2}{4m^2 - p^2}}. \quad (253)$$

References

1. B. Bakamjian, L.H. Thomas: Phys. Rev. **292**, 1300 (1953)
2. F.J. Belinfante: Physica **6**, 887 (1939)
3. S.J. Brodsky, H.-C. Pauli, S.S. Pinsky: Phys. Rep. **301**, 299 (1998)
4. J. Carbonell, B. Desplanques, V.A. Karmanov, J.F. Mathiot: Phys. Rep. **300**, 215 (1998)
5. P.A.M. Dirac: Rev. Mod. Phys. **21**, 392 (1949)
6. S. Głazek (ed.): *Theory of Hadrons and Light-Front QCD* (World Scientific Publishing Co., Singapore 1995)

7. P. Grangé, A. Neveu, H.C. Pauli, S. Pinsky, E. Werner (eds.): *New Nonperturbative Methods and Quantization on the Light Cone, Proceedings of the Les Houches Winter School, Les Houches, France, February 24-March 7, 1997* (Springer, Berlin; EDP Sciences, Les Ulis, France 1998)
8. K. Hepp: Comm. Math. Phys. **2**, 301 (1966)
9. J. Hilgevoord, E.A. De Kerf: Physica **31**, 1002 (1965)
10. J. Hilgevoord, S.A. Wouthuysen: Nucl. Phys. **40**, 1 (1963)
11. C. Itzykson, J.B. Zuber: *Quantum Field Theory* (McGraw-Hill, New York 1980)
12. C.-R. Ji, D.P. Min (eds.): *New Directions in Quantum Chromodynamics, APCTP Topical Meeting and 11th International Light-Cone School, May 26 - June 18, 1999, Seoul and Kyungju, Korea* (AIP Conference Proceedings 494, Melville NY 1999)
13. M.B. Johnson, L.S. Kisslinger (eds.): *Nuclear and Particle Physics on the Light Cone, LAMPF Workshop, Los Alamos, New Mexico July 18-22, 1988* (World Scientific Publishing Co., Singapore 1989)
14. B.D. Keister, W.N. Polyzou: (1991): Adv. Nucl. Phys. **20**, 225, ed. by J.W. Negele and E.W. Vogt (Plenum Press, New York 1991)
15. J.B. Kogut, D.E. Soper: Phys. Rev. D **1**, 2901 (1970)
16. H. Leutwyler, J. Stern: Ann.Phys. (N.Y.) **112**, 94 (1978)
17. F.M. Lev: Rivista del Nuovo Cim. **16**, 1 (1993)
18. N.E. Ligterink: Light-Front Hamiltonian Field Theory, Covariance and Renormalization. Ph.D. Thesis, Vrije Universiteit, Amsterdam (1996)
19. N.E. Ligterink, B.L.G. Bakker: Phys. Rev. D **52**, 5917 (1995)
20. N.E. Ligterink, B.L.G. Bakker: Phys. Rev. D **52**, 5954 (1995)
21. T. Muta: *Foundations of Quantum Chromodynamics* (World Scientific Publishing Co., Singapore 1987)
22. N.C.J. Schoonderwoerd, B.L.G. Bakker: Phys. Rev. D **57**, 4965 (1998)
23. N.C.J. Schoonderwoerd, B.L.G. Bakker: Phys. Rev. D **58**, 025013 (1998)
24. N.C.J. Schoonderwoerd, B.L.G. Bakker, V.A. Karmanov: Phys. Rev. C **58**, 3093 (1998)
25. N.C.J. Schoonderwoerd: Light-Front Hamiltonian Field Theory, Towards a Relativistic Description of Bound States. Ph.D. Thesis, Vrije Universiteit, Amsterdam (1999)
26. S.N. Sokolov, A.N. Shatny: Theor. Math. Phys. **37**, 291 (1978)
27. J.P. Vary, F. Wolz (eds.): *Light-Front Quantization and Non-Perturbative QCD* (International Institute of Theoretical and Applied Physics, Ames, Iowa 1997)
28. N.Ya. Vilenkin, G.F. Votruba, L.F. Boron: *Functional Analysis* (Wolters-Noordhoff, Groningen 1972)
29. K.G. Wilson, T.S. Walhout, A. Harindranath, W.-M. Zhang, R.J. Perry: Phys. Rev. D **49**, 6720 (1994)
30. W. Zimmerman: Comm. Math. Phys. **11**, 1 (1968)
31. W. Zimmerman: Comm. Math. Phys. **15**, 208 (1969)

Light-Cone Quantization: Foundations and Applications

Thomas Heinzl

Friedrich-Schiller-Universität Jena, Theoretisch-Physikalisches Institut,
D-07743 Jena, Max-Wien-Platz 1, Germany

Abstract. These lecture notes review the foundations and some applications of light-cone quantization. First I explain how to choose a time in special relativity. Inclusion of Poincaré invariance naturally leads to Dirac's forms of relativistic dynamics. Among these, the front form, being the basis for light-cone quantization, is my main focus. I explain a few of its peculiar features such as boost and Galilei invariance or separation of relative and center-of-mass motion. Combining light-cone dynamics and field quantization results in light-cone quantum field theory. As the latter represents a first-order system, quantization is somewhat nonstandard. I address this issue using Schwinger's quantum action principle, the method of Faddeev and Jackiw, and the functional Schrödinger picture. A finite-volume formulation, discretized light-cone quantization, is analysed in detail. I point out some problems with causality, which are absent in infinite volume. Finally, the triviality of the light-cone vacuum is established. Coming to applications, I introduce the notion of light-cone wave functions as the solutions of the light-cone Schrödinger equation. I discuss some examples, among them nonrelativistic Coulomb systems and model field theories in two dimensions. Vacuum properties (like chiral condensates) are reconstructed from the particle spectrum obtained by solving the light-cone Schrödinger equation. In a last application, I make contact with phenomenology by calculating the pion wave function within the Nambu and Jona-Lasinio model. I am thus able to predict a number of observables like the pion charge and core radius, the r.m.s. transverse momentum, the pion structure function and the pion distribution amplitude. The latter turns out to be the asymptotic one.

1 Introduction

The nature of elementary particles calls for a synthesis of relativity and quantum mechanics. The necessity of a quantum treatment is quite evident in view of the microscopic scales involved which are several orders of magnitude smaller than in atomic physics. These very scales, however, also require a relativistic formulation. A typical hadronic scale of 1 fm, for instance, corresponds to momenta of the order of $p \sim \hbar c/1\text{fm} \simeq 200$ MeV. For particles with masses $M \lesssim 1$ GeV, this implies sizable velocities $v \simeq p/M \gtrsim 0.2\ c$.

It turns out that the task of unifying the principles of quantum mechanics and relativity is not a straightforward one. One can neither simply extend ordinary quantum mechanics to include relativistic physics nor quantize relativistic mechanics using the ordinary correspondence rules. Nevertheless,

Dirac and others have succeeded in formulating what is called "relativistic quantum mechanics", which has become a subject of text books since – see e.g. [15,157]. It should, however, be pointed out that this formulation, which is based on the concept of single-particle wave-functions and equations, is not really consistent. It does not correctly account for relativistic causality (retardation effects etc.) and the existence of antiparticles. As a result, one has to struggle with issues like the Klein paradox[1], the definition of position operators [115] and the like.

The well-known solution to these problems is provided by quantum *field* theory, with an inherently correct description of antiparticles that entails relativistic causality. In contrast to single-particle wave mechanics, quantum field theory is a (relativistic) many body formulation that necessarily involves (anti-)particle creation and annihilation. Physical particle states are typically a superposition of an infinite number of 'bare' states, as any particle has a finite probability to emit or absorb other particles at any moment of time. A pion, for example, would be represented in terms of the following Fock expansion,

$$|\pi\rangle = \psi_2|q\bar{q}\rangle + \psi_3|q\bar{q}g\rangle + \psi_4|q\bar{q}q\bar{q}\rangle + \dots , \tag{1}$$

where the ψ_n are the probability amplitudes to find n particles (quarks q, antiquarks $\bar{q}$ or gluons g) in the pion. With the advent of QCD, however, a conceptual difficulty concerning this many-particle picture has appeared. At low energy or momentum transfer, hadrons, the bound states of QCD, are reasonably described in terms of two or three constituent quarks and thus as *few-body* systems. These 'effective' quarks Q are dressed so that they gain an effective mass of the order of 300–400 MeV. They are used as the basic degrees of freedom in the 'constituent quark model'. This model yields a reasonable mass spectroscopy of hadrons [101,109], but its foundations are not very well established theoretically. First, a nonrelativistic treatment of light hadrons is not justified (see above). Second, the model violates many symmetries of QCD (in particular *chiral symmetry*). Third, it is rather unclear how a constituent picture can arise in a quantum field theory such as QCD.

In principle, in order to confirm the constituent quark model, one would have to solve the 'QCD Schrödinger equation' for hadron states $|\text{hadron}\rangle$ of mass M_h,

$$H_\text{QCD}|\text{hadron}\rangle = M_\text{h}|\text{hadron}\rangle , \tag{2}$$

and check whether the eigenstates are reasonably well described in terms of the constituent valence states $|Q\bar{Q}\rangle$ or $|QQQ\rangle$. This is a very hard problem. A more moderate goal would be to 'relativize' the constituent quark model, ideally in such a way that it respects the symmetries of QCD. I will discuss this attempt in detail at the end of these lectures.

To arrive at this point, there is, of course, some way to go. Let me start with the following claim. A particularly useful approach for our purposes is

[1] For a nice recent discussion, see [74].

based on a somewhat unorthodox choice of the 'time arrow' within special relativity: instead of the ordinary 'Galileian time' t, I choose 'light-cone time', $x^+ \equiv t - z/c$. In the course of these lectures, this claim will be substantiated step by step.

I will begin with some general remarks on relativistic dynamics (Sect. 2). As a paradigm example I discuss the free relativistic particle which is the prototype of a reparametrization invariant system. I show that the choice of the time parameter is not unique as it corresponds to a gauge fixing, the purpose of which is to get rid of the reparametrization redundancies. By considering the stability subgroups of the Poincaré group, one finds that there are essentially three reasonable choices of 'time' for a relativistic system, corresponding to Dirac's 'instant', 'point' and 'front' form, respectively. The latter choice is the basis of light-cone dynamics, the main features of which will be discussed in the last part of Sect. 2.

Section 3 is devoted to light-cone *field* quantization. I show how the Poincaré generators are defined in this case and utilize Schwinger's quantum action principle to derive the canonical commutators. This is the first method of quantization to be discussed. The relation between equal-time commutators, the field equations and their solutions for different initial and/or boundary conditions is clarified.

It turns out that light-cone field theories, being of first order in the velocity, generally are constrained systems which require a special treatment. I rederive the canonical light-cone commutators using a second method of quantization (based on phase space reduction) due to Faddeev and Jackiw. I extend this discussion to light-cone quantization in finite volume and point out possible problems with causality in this approach. Going back to infinite volume, I introduce a third method of quantization, the functional Schrödinger picture, and combine it with the light-cone formalism. I close this section with a discussion of the presumably most spectacular feature of light-cone quantum field theory, the triviality of the vacuum.

As a prelude to the applications I introduce the notion of light-cone wave functions in Sect. 4. I show how light-cone wave functions can be obtained by solving the light-cone Schrödinger equation. As examples, I discuss nonrelativistic wave functions as they occur in hydrogen-like systems, some model field theory in 1+1 dimensions and a simple Gaussian model.

In Sect. 5, I finally make contact with phenomenology. I calculate the light-cone wave function of the pion within the Nambu and Jona-Lasinio model. This model is known to provide a good description of spontaneous chiral symmetry breaking, as it is governed by the same symmetry group as low-energy QCD. With the pion wave function at hand I derive a number of observables like the pion charge and core radius, the electromagnetic form factor, the r.m.s. transverse momentum and the pion structure function. I conclude with a calculation of the pion distribution amplitude.

2 Relativistic Particle Dynamics

The physically motivated desire to describe hadrons as bound states of a small, fixed number of constituents is our rationale to go back and reanalyze the relation between Hamiltonian quantum mechanics and relativistic quantum field theory.

Quite generally, bound states are obtained by solving the Schrödinger equation,

$$i\hbar\frac{\partial}{\partial\tau}|\psi(\tau)\rangle = H|\psi(\tau)\rangle \ , \tag{3}$$

for *normalized, stationary states,*

$$|\psi(\tau)\rangle = e^{-iE\tau}|\psi(0)\rangle. \tag{4}$$

This leads to the bound-state equation

$$H|\psi(0)\rangle = E|\psi(0)\rangle \ , \tag{5}$$

where E is the bound state energy. We would like to make this Hamiltonian formalism consistent with the requirements of relativity. It is, however, obvious from the outset that this procedure will not be manifestly covariant as it singles out a time τ (and an energy E, respectively). Furthermore, it is not even clear what the time τ really is as it does not have an invariant meaning.

2.1 The Free Relativistic Point Particle

To see what is involved it is sufficient to consider the classical dynamics of a free relativistic particle. We want to find the associated canonical formulation as a basis for subsequent quantization. We will proceed by analogy with the treatment of classical free strings which is described in a number of textbooks [58,123]. Accordingly, the relativistic point particle may be viewed as an infinitely short string.

Recall that the action for a relativistic particle is essentially given by the arc length of its trajectory

$$S = -ms_{12} \equiv -m\int_1^2 ds \ . \tag{6}$$

This action[2] is a Lorentz scalar as

$$ds = \sqrt{g_{\mu\nu}x^\mu x^\nu} \tag{7}$$

is the (infinitesimal) invariant distance. We can rewrite the action (6) as

$$S = -m\int_1^2 ds\,\sqrt{\dot{x}_\mu\dot{x}^\mu} \equiv \int_1^2 ds\,L(s) \ , \tag{8}$$

[2] We work in natural units, $\hbar = c = 1$.

in order to introduce a Lagrangian $L(s)$ and the four velocity $\dot{x}^\mu \equiv dx^\mu/ds$. The latter obeys

$$\dot{x}^2 \equiv \dot{x}_\mu \dot{x}^\mu = 1 , \tag{9}$$

as the arc length provides a natural parametrization. Thus, $\dot{x}^\mu$ is a time-like vector, and we assume in addition that it points into the future, $\dot{x}^0 > 0$. In this way we guarantee relativistic causality ensuring that a real particle passing through a point P will always propagate into the future light cone based at P.

We proceed with the canonical formalism by calculating the canonical momenta as

$$p^\mu = -\frac{\partial L}{\partial \dot{x}_\mu} = m\dot{x}^\mu . \tag{10}$$

These are not independent, as can be seen by calculating the square using (9),

$$p^2 = m^2 \dot{x}^2 = m^2 , \tag{11}$$

which, of course, is the usual mass-shell constraint. This constraint indicates that the Lagrangian $L(s)$ defined in (8) is *singular*, so that its Hessian $W^{\mu\nu}$ with respect to the velocities,

$$W^{\mu\nu} \equiv \frac{\partial^2 L}{\partial \dot{x}_\mu \partial \dot{x}_\nu} = -\frac{m}{\sqrt{\dot{x}^2}} \left(g^{\mu\nu} - \frac{\dot{x}^\mu \dot{x}^\nu}{\dot{x}^2} \right) = -m \left(g^{\mu\nu} - \dot{x}^\mu \dot{x}^\nu \right) , \tag{12}$$

is degenerate. It has a zero mode given by the velocity itself,

$$W^{\mu\nu} \dot{x}_\nu = 0 . \tag{13}$$

The Lagrangian being singular implies that the velocities cannot be uniquely expressed in terms of the canonical momenta. This, however, is not obvious from (10), as we can easily solve for the velocities,

$$\dot{x}^\mu = p^\mu / m . \tag{14}$$

But if one now calculates the canonical Hamiltonian,

$$H_{\mathrm{c}} = -p_\mu \dot{x}^\mu - L = -m\dot{x}^2 + m\dot{x}^2 = 0 , \tag{15}$$

one finds that it is vanishing! It therefore seems that we do not have a generator for the time evolution of our dynamical system. In the following, we will analyze the reasons for this peculiar finding.

First of all we note that the Lagrangian is homogeneous of first degree in the velocity,

$$L(\alpha \dot{x}^\mu) = \alpha L(\dot{x}^\mu) . \tag{16}$$

Thus, under a reparametrization of the world-line,

$$s \mapsto s' , \quad x^\mu(s) \mapsto x^\mu\big(s'(s)\big) , \tag{17}$$

where the mapping $s \mapsto s'$ is one-to-one with $ds'/ds > 0$ (orientation conserving), the Lagrangian changes according to

$$L(dx^\mu/ds) = L\Big((dx^\mu/ds')(ds'/ds)\Big) = (ds'/ds)L(dx^\mu/ds') \,. \tag{18}$$

This is sufficient to guarantee that the action is invariant under (17), that is, *reparametrization invariant*,

$$S = \int_{s_1}^{s_2} ds\, L(dx^\mu/ds) = \int_{s_1'}^{s_2'} ds'\, \frac{ds}{ds'}\frac{ds'}{ds} L(dx^\mu/ds') \equiv S' \,, \tag{19}$$

if the endpoints remain unchanged, $s_{1,2} = s'_{1,2}$. On the other hand, L is homogeneous of the first degree if and only if Euler's formula holds, namely

$$L = \frac{\partial L}{\partial \dot{x}^\mu}\dot{x}^\mu = -p_\mu \dot{x}^\mu \,. \tag{20}$$

This is exactly the statement (15), the vanishing of the Hamiltonian. Furthermore, if we differentiate (20) with respect to $\dot{x}^\mu$, we recover (13) expressing the singular nature of the Lagrangian. Summarizing, we have found the general result [61,144] that if a Lagrangian is homogeneous of degree one in the velocities, the action is reparametrization invariant, and the Hamiltonian vanishes. In this case, the momenta are homogeneous of degree zero, which renders the Lagrangian singular.

The reparametrization invariance is generated by the first class constraint [44,144],

$$\theta \equiv p^2 - m^2 = 0 \,, \tag{21}$$

as can be seen as follows. From the canonical one form $-g_{\mu\nu}p^\mu dx^\nu$ we read off the Poisson bracket

$$\{x^\mu \,,\, p^\nu\} = -g^{\mu\nu} \,, \tag{22}$$

and calculate the change of the coordinate x^μ,

$$\begin{aligned} \delta x^\mu &= \{x^\mu \,,\, \theta\delta\epsilon\} = -2p^\mu\delta\epsilon = -2m\,\dot{x}^\mu\delta\epsilon \equiv \dot{x}^\mu\delta\tau \\ &= x^\mu(\tau+\delta\tau) - x^\mu(\tau) = x^\mu(\tau') - x^\mu(\tau) \,. \end{aligned} \tag{23}$$

Thus, the reparametrization (17) is indeed generated by the constraint (21).

Reparametrization invariance can be viewed as a gauge or *redundancy* symmetry. The redundancy consists in the fact that a single trajectory (worldline) can be described by an infinite number of different parametrizations. The physical objects, the trajectories, are therefore equivalence classes obtained by identifying ('dividing out') all reparametrizations. The method to do so is well known, namely gauge fixing. For the case at hand, this corresponds to a particular choice of parametrization, or, more physically, to the choice of a time parameter τ. This amounts to choosing a foliation of space-time into space *and* time. Minkowski space is thus decomposed into hypersurfaces of equal time, $\tau = const$, which in general are three-dimensional objects, and the time direction 'orthogonal' to them. The time development thus continuously evolves the hypersurface $\Sigma_0 : \tau = \tau_0$ into $\Sigma_1 : \tau = \tau_1 > \tau_0$. Put differently, initial conditions provided on Σ_0 together with the dynamical equations (being differential equations in τ) determine the state of the dynamical system on Σ_1.

Practically, the (3+1)-foliation is done as follows. We introduce some arbitrary coordinates, $\xi^\alpha = \xi^\alpha(x)$, which may be curvilinear. We imagine that three of these, say ξ^i, $i = 1, 2, 3$, parametrize the three-dimensional hypersurface Σ, so that the remaining one, ξ^0, represents the time variable, i.e. $\tau = \xi^0(x)$. This equation can equivalently be viewed as a gauge fixing condition.

The first question to be addressed is: what is a 'good' choice of time? There are two criteria to be met, namely existence and uniqueness. Existence means that the equal-time hypersurface Σ should intersect *any* possible world-line, while uniqueness requires that it does so once and only once. Mathematically, uniqueness can be analysed in terms of the Faddeev-Popov (FP) 'operator', which is given by the Poisson bracket of the gauge fixing condition with the constraint (evaluated on Σ),

$$\mathrm{FP} \equiv \{\xi^0(x)\,,\,\theta\} = \{\xi^0(x)\,,\,p^2\} = -2\,\frac{\partial \xi^0}{\partial x^\mu}\,p^\mu \equiv -2\,N\cdot p\,. \tag{24}$$

Here, we have introduced the normal N on Σ,

$$N^\mu(x) = \left.\frac{\partial \xi^0(x)}{\partial x^\mu}\right|_\Sigma\,, \tag{25}$$

which will be important later on. The statement now is that uniqueness is achieved (for a single degree of freedom) if the FP operator does not vanish, i.e. if $N \cdot p \neq 0$. Generically, this means that the particle trajectory must not be parallel to the hypersurface Σ of equal time.

As an aside we remark that this is completely analogous to the reasoning in standard gauge (field) theory. There, the constraint θ is given by Gauss's law which generates gauge transformations $A \to A + D\omega$, D denoting the covariant derivative. For a gauge fixing $\chi[A] = 0$, the equation corresponding to (24) becomes

$$\mathrm{FP} = \{\chi\,,\,\theta\} = \frac{\delta\chi}{\delta\omega} = \frac{\delta\chi}{\delta A}\frac{\delta A}{\delta\omega} = N\cdot D\,, \tag{26}$$

where all (functional) derivatives are to be evaluated on the gauge fixing hypersurface, $\chi = 0$.

Let us now perform an analysis of the canonical formalism for a general choice of hypersurface Σ. For this we need some notation. We write the line element as

$$ds^2 = g_{\mu\nu}dx^\mu dx^\nu = g_{\mu\nu}\frac{\partial x^\mu}{\partial \xi^\alpha}\frac{\partial x^\nu}{\partial \xi^\beta}d\xi^\alpha d\xi^\beta \equiv h_{\alpha\beta}(\xi)d\xi^\alpha d\xi^\beta\,. \tag{27}$$

Introducing a vierbein $e^{\mu}{}_{\alpha}(\xi)$, the metric $h_{\alpha\beta}(\xi)$ is alternatively given by

$$h_{\alpha\beta}(\xi) = g_{\mu\nu}\, e^{\mu}{}_{\alpha}(\xi)\, e^{\nu}{}_{\beta}(\xi) \,. \tag{28}$$

The transformation $x \to \xi$ is well known from general relativity, where it corresponds to the transformation from a local inertial frame described by the flat metric $g_{\mu\nu}$ to a noninertial frame with coordinate dependent metric $h_{\alpha\beta}(\xi)$. For our purposes we write this metric in a (3+1)-notation as follows,

$$h_{\alpha\beta} = \begin{pmatrix} h_{00} & h_{0i} \\ h_{i0} & h_{ij} \end{pmatrix} \equiv \begin{pmatrix} h_{00} & \boldsymbol{h}^T \\ \boldsymbol{h} & -H \end{pmatrix} . \tag{29}$$

Of particular interest is the component h_{00}, which explicitly reads

$$h_{00} = g_{\mu\nu} \frac{\partial x^{\mu}}{\partial \xi^0} \frac{\partial x^{\nu}}{\partial \xi^0} = g_{\mu\nu} e^{\mu}{}_0 e^{\nu}{}_0 \equiv n^2 \,, \tag{30}$$

where we have defined the unit vector in ξ^0-direction

$$n^{\mu} = \frac{\partial x^{\mu}}{\partial \xi^0} = e^{\mu}{}_0 \equiv \dot{x}^{\mu} \,, \tag{31}$$

which thus is the new four-velocity. It is related to the normal vector N^{μ} via

$$n \cdot N = e^{\mu}{}_0 e_{\mu}{}^0 = \frac{\partial \xi^0}{\partial x^{\mu}} \frac{\partial x^{\mu}}{\partial \xi^0} = 1 \,. \tag{32}$$

The normal vector N enters the inverse metric which we write as follows,

$$h^{\alpha\beta} = \begin{pmatrix} g_{00} & g_{0i} \\ g_{i0} & g_{ij} \end{pmatrix} = \begin{pmatrix} N^2 & \boldsymbol{g}^T \\ \boldsymbol{g} & -G \end{pmatrix} . \tag{33}$$

The h_{ij} are the metric components associated with the hypersurface. The invariant distance element (27) thus becomes (with $h_{0i} \equiv h_i$),

$$\begin{aligned} ds^2 &= h_{00} d\xi^0 d\xi^0 + 2h_{0i} d\xi^0 d\xi^i + h_{ij} d\xi^i d\xi^j \,, \\ &= \left(n^2 + 2h_i \frac{d\xi^i}{d\tau} + h_{ij} \frac{d\xi^i}{d\tau} \frac{d\xi^j}{d\tau} \right) d\tau^2 \equiv h(\tau) d\tau^2 \,, \end{aligned} \tag{34}$$

where, in the second step, we have used that $\xi^0 = \tau$. In the last identity we have defined a world-line metric or einbein

$$h(\tau) \equiv \dot{x}^2 = h_{\alpha\beta} \dot{\xi}^{\alpha} \dot{\xi}^{\beta} \,, \tag{35}$$

which expresses the arbitrariness in choosing a time by providing an (arbitrary) 'scale' for the velocity. Introducing the velocities expressed in the new coordinates, $w^i \equiv d\xi^i / d\tau$, the world-line metric can be written as

$$h(\tau) = n^2 + 2h_i w^i + h_{ij} w^i w^j \,. \tag{36}$$

Let us develop a canonical formalism for a general choice of the einbein $h(\tau)$ corresponding to the gauge fixing $\tau = \xi^0(x) = 0$. The Lagrangian becomes

$$L(\tau) = -m\sqrt{h(\tau)}\,, \tag{37}$$

leading to the canonical momenta

$$\pi_\alpha = -\frac{\partial L}{\partial \dot{\xi}^\alpha} = \frac{m}{\sqrt{h}}\, h_{\alpha\beta}\,\dot{\xi}^\alpha = e^\mu_\alpha\, p_\mu\,. \tag{38}$$

We see that the einbein h is appearing all over the place. The canonical Hamiltonian is expressed in terms of the inverse metric $h^{\alpha\beta}$,

$$H_c = -\pi_\alpha \dot{\xi}^\alpha - L = -\frac{\sqrt{h}}{m}(h^{\alpha\beta}\pi_\alpha\pi_\beta - m^2) = 0\,. \tag{39}$$

It vanishes (as it should) as it is proportional to the constraint,

$$\theta = h^{\alpha\beta}\pi_\alpha\pi_\beta - m^2 = p^2 - m^2 = 0\,. \tag{40}$$

The FP operator also depends on the entries of the inverse metric (33), in particular the normal vector N,

$$\mathrm{FP} = N^2\pi_0 + g^i\pi_i\,. \tag{41}$$

After gauge fixing, the generator of τ-evolution, $H_\tau \equiv \pi_0$, is obtained by solving the constraint (40) for π_0 which assumes the explicit form,

$$N^2\pi_0^2 + 2g^i\pi_i\pi_0 - G^{ij}\pi_i\pi_j - m^2 = 0\,. \tag{42}$$

Depending on the value of N^2, we thus have to consider two different cases. The generic one is that the normal N on Σ is *time-like*, $N^2 > 0$. In this case, the mass-shell constraint is of *second order* in π_0, so that there are two distinct solutions,

$$\pi_0 = \frac{1}{N^2}\left\{-(\boldsymbol{g},\boldsymbol{\pi}) \pm \sqrt{(\boldsymbol{g},\boldsymbol{\pi})^2 + N^2[(\boldsymbol{\pi}, G\boldsymbol{\pi}) + m^2]}\right\}\,. \tag{43}$$

Not unexpectedly, the 'problem' of two different signs in front of the square root arises [55]. Within quantum mechanics, this is somewhat difficult to interpret. Upon 'second quantization', i.e. in the context of quantum field theory, one has, of course, the natural explanation in terms of antiparticles. As we will not quantize the relativistic point particle, the sign 'problem' is of no concern to us. A possible arbitrariness will be removed ad hoc by demanding $\pi_0 > 0$. With this additional condition the FP operator becomes

$$\mathrm{FP} = -2\sqrt{(\boldsymbol{g},\boldsymbol{\pi})^2 + N^2[(\boldsymbol{\pi}, G\boldsymbol{\pi}) + m^2]}\,, \tag{44}$$

which is clearly nonvanishing for a massive particle, $m \neq 0$. A gauge fixing with $N^2 > 0$ is thus unique and leads to a well-defined description of the τ-evolution.

The second case to be considered is in a sense degenerate. It corresponds to a *light-like* normal, $N^2 = 0$. In this case, the constraint (42) is only of *first order* in π_0 leading to a single solution,

$$\pi_0 = \frac{(\boldsymbol{\pi}, G\boldsymbol{\pi}) + m^2}{(\boldsymbol{g}, \boldsymbol{\pi})} . \tag{45}$$

As a result, there is no 'sign problem' and no 'ugly' square root. Conservation of difficulties, however, is at work, because it is no longer obvious whether the FP operator,

$$\mathrm{FP} = -2(\boldsymbol{g}, \boldsymbol{\pi}) , \tag{46}$$

is different from zero. Clearly, this is absolutely necessary for (45) to represent a well-defined solution.

At this point, it should be mentioned that the results (43) and (45) are not yet the full story. The entries of the inverse metric, N^2, $\boldsymbol{g}$ and G should actually be expressed in terms of the quantities n^2, $\boldsymbol{h}$ and H defining the induced metric on Σ. So far, it is also not completely clear which choices of these parameters actually make sense physically. Of course, the normal N should not be space-like as this would imply that Σ contains time-like directions and thus possible particle trajectories. In the next subsection I will give some criteria for reasonable choices of time.

Before we come to that let us apply the general formalism to the standard choice of 'Galileian' time, $\tau = \xi^0(x) = x^0 = t$. In this case, the surface $\Sigma : t = 0$ is an entirely space-like hyperplane with constant normal vector $N = (1, \mathbf{0}) = n$. The other metric entries are $\boldsymbol{h} = \boldsymbol{g} = 0$ and $H = G = \mathbb{1}$. The world-line metric (35) thus becomes

$$h(t) = \dot{x}^2 = 1 - v^2 \equiv 1/\gamma^2 , \tag{47}$$

where γ is the usual Lorentz contraction factor. The Hamiltonian is obtained in line with the second-order case above,

$$H_t = N \cdot p = p^0 = \sqrt{\mathbf{p}^2 + m^2} \sim \mathrm{FP} . \tag{48}$$

It generates the dynamics via the basic Poisson bracket $\{x^i , p^j\} = \delta^{ij}$ leading to

$$\dot{x}^i = \{x^i , H_t\} = p^i/p^0 , \tag{49}$$

with p^0 given by (48). Note that a well-defined time evolution requires a nonvanishing FP operator (which is proportional to p^0).

As already announced, we will discuss alternatives to this standard choice of time in the next subsection.

2.2 Dirac's Forms of Relativistic Dynamics

To address this issue it is not sufficient to consider only the τ-development and the associated generator of time translations (i.e. the Hamiltonian) H_τ.

Instead, one has to refer to the full Poincaré group to be able to guarantee full relativistic invariance. The generators of the Poincaré group are the four momenta P^μ and the six operators $M^{\mu\nu}$ which combine angular momenta and boosts according to

$$L^i = \frac{1}{2}\epsilon^{ijk}M^{jk} , \tag{50}$$

$$K^i = M^{0i} , \tag{51}$$

with i, j, k = 1,2,3. These generators are elements of the Poincaré algebra which is defined by the Poisson bracket relations,

$$\begin{aligned} \{P^\mu , P^\nu\} &= 0 , \\ \{M^{\mu\nu} , P^\rho\} &= g^{\nu\rho}P^\mu - g^{\mu\rho}P^\nu , \\ \{M^{\mu\nu} , M^{\rho\sigma}\} &= g^{\mu\sigma}M^{\nu\rho} - g^{\mu\rho}M^{\nu\sigma} - g^{\nu\sigma}M^{\mu\rho} + g^{\nu\rho}M^{\mu\sigma} . \end{aligned} \tag{52}$$

It is well known that the momenta P^μ generate space-time translations and the $M^{\mu\nu}$ rotations and Lorentz boosts, cf. (50, 51). In the following we will only consider proper and orthochronous Lorentz transformations, i.e. we exclude space and time reflections.

Any Poincaré invariant dynamical theory describing e.g. the interaction of particles should provide a particular realization of the Poincaré algebra. For this purpose, the Poincaré generators are constructed out of the fundamental dynamical variables like positions, momenta, spins etc. An elementary realization of (52) is given as follows. Choose the space-time point x^μ and its conjugate momentum p^μ as canonical variables, i.e. adopt (22),

$$\{x^\mu , p^\nu\} = -g^{\mu\nu} . \tag{53}$$

The Poincaré generators are then found to be

$$P^\mu = p^\mu , \quad M^{\mu\nu} = x^\mu p^\nu - x^\nu p^\mu , \tag{54}$$

as is easily confirmed by checking (52) using (53). An infinitesimal Poincaré transformation is thus generated by

$$\delta G = -\tfrac{1}{2}\delta\omega_{\mu\nu}M^{\mu\nu} + \delta a_\mu P^\mu \tag{55}$$

in the following way,

$$\delta x^\mu = \{x^\mu , \delta G\} = \delta\omega^{\mu\nu}x_\nu + \delta a^\mu , \quad \delta\omega^{\mu\nu} = -\delta\omega^{\nu\mu} . \tag{56}$$

The action of the Poincaré group on some scalar function $F(x)$ is thus given by

$$\delta F = \{F , \delta G\} = \partial^\mu F\,\delta a_\mu - \tfrac{1}{2}(x^\mu\partial^\nu - x^\nu\partial^\mu)F\,\delta\omega_{\mu\nu} . \tag{57}$$

Though the realization (54) is covariant, it has several shortcomings. It does not describe any interaction; for several particles the generators are simply the sum of the single particle generators. This point, however, is of minor importance to us, and will only be touched upon at the end of this subsection. The solution of the problem, as already mentioned in the introduction, is the framework of local quantum field theory. More importantly, the representation (54) does not take into account the mass-shell constraint, $p^2 = m^2$, which we already know to guarantee relativistic causality as it generates the dynamics upon solving for H_τ.

To remedy the situation we proceed as before by choosing a time variable τ, i.e. a foliation of space-time into essentially space-like hypersurfaces Σ with time-like or light-like normals N. We have seen that Σ should be chosen in such a way that it intersects all possible world-lines once and only once (existence and uniqueness). Apart from this necessary consistency with causality the foliation appears quite arbitrary. However, given a particular foliation one can ask the question which of the Poincaré generators will leave the hypersurface Σ invariant. The set of all such generators defines a subgroup of the Poincaré group called the stability group G_Σ of Σ. The associated generators are called *kinematical*, the others *dynamical*. The latter map Σ onto another hypersurface Σ' and thus involve the development in τ. One thus expects that the dynamical generators will depend on the Hamiltonian (and, therefore, the interaction) which, by definition, is a dynamical quantity.

It is clear, however, that the stability group corresponding to a particular foliation will be empty if the associated hypersurface looks very irregular and thus does not have a high degree of symmetry. One therefore demands in addition that the stability group acts transitively on Σ: any two points on Σ can be connected by a transformation from G_Σ. With this additional requirement there are exactly five inequivalent classes of hypersurfaces [99] which are listed in Table 1.

Table 1. *All possible choices of hypersurfaces Σ: $\tau = const$ with transitive action of the stability group G_Σ. d denotes the dimension of G_Σ, that is, the number of kinematical Poincaré generators;* $\mathbf{x}_\perp \equiv (x^1, x^2)$.

name	Σ	τ	d
instant	$x^0 = 0$	t	6
light front	$x^0 + x^3 = 0$	$t + x^3/c$	7
hyperboloid	$x_0^2 - \mathbf{x}^2 = a^2 > 0,\ x^0 > 0$	$(t^2 - \mathbf{x}^2/c^2 - a^2/c^2)^{1/2}$	6
hyperboloid	$x_0^2 - \mathbf{x}_\perp^2 = a^2 > 0,\ x^0 > 0$	$(t^2 - \mathbf{x}_\perp^2/c^2 - a^2/c^2)^{1/2}$	4
hyperboloid	$x_0^2 - x_1^2 = a^2 > 0,\ x^0 > 0$	$(t^2 - x_1^2/c^2 - a^2/c^2)^{1/2}$	4

The first three choices have already been found by Dirac [43] in his seminal paper on 'forms of relativistic dynamics'. He called the associated forms the 'instant', 'front' and 'point' forms, respectively. These are the most important choices as the remaining two forms have a rather small stability group and thus are not very useful. We have only listed them for the sake of completeness.

It is important to note that for all forms one has $\lim_{c\to\infty} \tau = t$, which means that in the nonrelativistic case there is only one possible foliation leading to the absolute Galileian time t. This is consistent with the fact that there is no limiting velocity in this case implying that particle trajectories can have arbitrary slope (tangent vector). Therefore, the hypersurface $\Sigma_{\mathrm{nr}} : t = const$ is the only one intersecting all possible world-lines. For other choices, the criterion of existence introduced in the last subsection would be violated.

To decide which of the Poincaré generators are kinematical, we use the general formula (57) describing their action. Imagine that Σ is given in the form $\Sigma : \tau = \xi^0(x) \equiv F(x)$ as in Table 1. If P^μ or $M^{\mu\nu}$ are kinematical for some μ or ν, then, for these particular superscripts, the components of the gradient and rotor of F have to vanish on Σ,

$$\partial^\mu F = 0\,, \quad (x^\mu \partial^\nu - x^\nu \partial^\mu) F = 0\,. \tag{58}$$

In terms of the normal vector N these equations become

$$N^\mu = 0\,, \quad x^\mu N^\nu - x^\nu N^\mu = 0\,, \tag{59}$$

which again will hold for *some* of the superscripts μ and/or ν, if Σ has nontrivial stabilizer. The distinction between kinematical and dynamical is thus completely encoded in the normal vector N.

The choice of Galileian time $\tau = t$ is of course the most common one also in the relativistic case, and we have discussed it briefly at the end of the last subsection. To complete this discussion, we construct the associated representation of the Poincaré generators on $\Sigma : t = 0$. The idea is again to explicitly saturate the constraint $p^2 = m^2$ by solving for $H_t = p^0 = N \cdot p = (\mathbf{p}^2 + m^2)^{1/2}$, and setting $x^0 = 0$ in (54).

As a result, we obtain the following (3+1)-representation of the Poincaré generators,

$$\begin{aligned} P^i &= p^i\,, \ M^{ij} = x^i p^j - x^j p^i\,, \\ P^0 &= H_t\,, \ M^{i0} = x^i H_t\,. \end{aligned} \tag{60}$$

This outcome is as expected: Compared to (54), p^0 has been replaced by H_t, and x^0 has been set to zero. It should, however, be pointed out that for non-Cartesian coordinates the construction of the Poincaré generators is less straightforward.

Let us address the question of kinematical versus dynamical generators. In agreement with (58) and (59), one has

$$N^i = 0 = x^i N^j - x^j N^i\,, \quad i,j = 1,2,3\,, \tag{61}$$

so that Σ is both translationally and rotationally invariant confirming that the dimension of its stability group is six (cf. Table 1). On the other hand,

$$N^0 = 1 \qquad\qquad \neq 0 \, , \tag{62}$$
$$x^0 N^i - x^i N^0 = -x^i \neq 0 \, , \tag{63}$$

from which we read off that, apart from the Hamiltonian, also the boosts are dynamical, i.e., Σ is not boost invariant. The latter fact is, of course, well known because the boosts mix space and time. Under a boost along the **n**-direction with velocity **v**, t transforms as

$$t \to t' = t \cosh \omega + (\mathbf{n} \cdot \mathbf{x}) \sinh \omega \, , \tag{64}$$

where $\mathbf{n} = \mathbf{v}/v$ and ω is the rapidity, defined through $\tanh \omega = v$. From (64) it is evident that the hypersurface $\Sigma : t = 0$ is not boost invariant.

In obtaining the representation (60), we make the Poincaré algebra compatible with the instant-form gauge-fixing constraint, $x^0 = 0$. An elementary calculation, using $\{x^i \, , p^j\} = \delta^{ij}$, indeed shows that the generators (60) really obey the bracket relations (52). We have already seen in (49) that the Hamiltonian $P^0 = H_t$ generates the correct dynamics.

At this point it is getting time to really consider an alternative to the instant form in some detail.

2.3 The Front Form

For an arbitrary four-vector a we perform the following transformation to *light-cone* coordinates,

$$(a^0, a^1, a^2, a^3) \mapsto (a^+, a^1, a^2, a^-) \, , \tag{65}$$

where we have defined

$$a^+ = a^0 + a^3 \, , \quad a^- = a^0 - a^3 \, . \tag{66}$$

We also introduce the transverse vector part of a as

$$\mathbf{a}_\perp = (a^1, a^2) \, . \tag{67}$$

The metric tensor (29) becomes

$$h_{\alpha\beta} = \begin{pmatrix} n^2 & \boldsymbol{h}^T \\ \boldsymbol{h} & -H \end{pmatrix} = \begin{pmatrix} 0 & 0 & 0 & 1/2 \\ 0 & -1 & 0 & 0 \\ 0 & 0 & -1 & 0 \\ 1/2 & 0 & 0 & 0 \end{pmatrix} \, . \tag{68}$$

The entries 1/2 imply nonvanishing $\boldsymbol{h}$ and thus a slightly unusual scalar product,

$$a \cdot b = g_{\mu\nu} a^\mu b^\nu = \tfrac{1}{2} a^+ b^- + \tfrac{1}{2} a^- b^+ - a^i b^i \, , \quad i = 1, 2 \, . \tag{69}$$

According to Table 1, the front form is defined by choosing the hypersurface $\Sigma : x^+ = 0$, which is a plane tangent to the light-cone. It can equivalently be viewed as the wave front of a plane light wave traveling towards the positive z-direction. Therefore, Σ is also called a *light-front*. The normal vector is

$$N = (1, 0, 0, -1)\,, \quad N^2 = 0\,, \tag{70}$$

where N has been written in ordinary coordinates. We see that $N^+ = N^0 + N^3 = 0$ which implies that the normal N to the hypersurface lies *within* the hypersurface [114,129]. As N is a light-like or null vector, Σ is often referred to as a *null-plane* [114,35]. We have depicted the front-form hypersurface Σ together with the light-cone in Fig. 1.

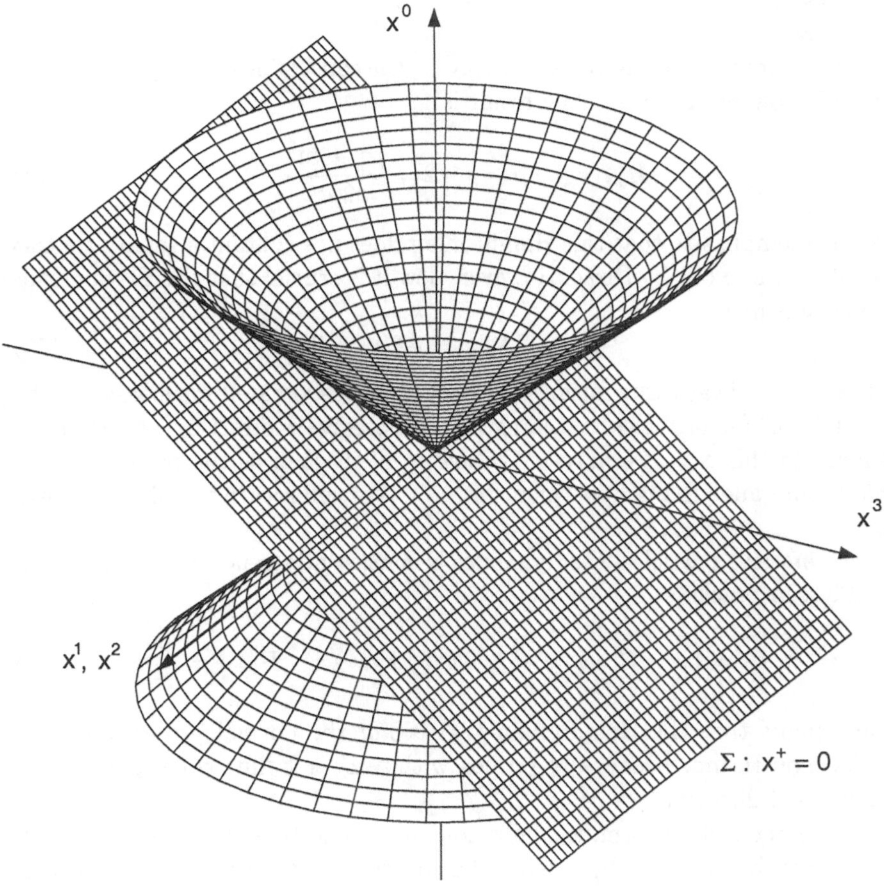

Fig. 1. *The hypersurface $\Sigma : x^+ = 0$ defining the front form. It is a null-plane tangential to the light-cone, $x^2 = 0$.*

As is already obvious from (68), the unit vector in x^+-direction is another null-vector,

$$n^\mu = \frac{\partial x^\mu}{\partial x^+} = \tfrac{1}{2}(1,0,0,1) \, , \tag{71}$$

so that $n \cdot N = 1$ as it should. Given the scalar product (69), we infer the invariant distance element

$$ds^2 = g_{\mu\nu} dx^\mu dx^\nu = dx^+ dx^- - dx^i dx^i = \left(\frac{dx^-}{dx^+} - \frac{dx^i}{dx^+}\frac{dx^i}{dx^+} \right) dx^+ dx^+ \tag{72}$$

from which the einbein h can be read off as

$$h(x^+) = \dot{x}^- - \dot{x}^i \dot{x}^i \equiv v^- - \mathbf{v}_\perp^2 \, . \tag{73}$$

Note that velocities are dimensionless, so that despite appearance the result is consistent (if you do not like it as it stands, just insert the appropriate factors of c).

The Hamiltonian is obtained by solving the constraint $p^+ p^- - p_\perp^2 - m^2 = 0$, which is now *linear* in p^-. The result is

$$H_{x^+} = n \cdot p = p^-/2 = \frac{p_\perp^2 + m^2}{2p^+} \, . \tag{74}$$

Let me reemphasize that this Hamiltonian does not contain a square root as already pointed out by Dirac. However, now it is crucial that the FP operator is nonvanishing,

$$\mathrm{FP} = -2N \cdot p = -2p^+ \neq 0 \, . \tag{75}$$

While this is always true for massive particles, it is violated for massless 'left-movers', i.e. for particles travelling in the negative z-direction at the speed of light. In this case, we have a 'Gribov problem' [59], as the particles move *within* our gauge-fixing hyperplane, $x^+ = 0$. We will return to this issue later on.

For massive particles, the dynamics is consistently generated by means of the Poisson brackets

$$v^- = \dot{x}^- = \{x^- \, , H_{x^+}\} = \frac{p^-}{p^+} \, , \quad v^i = \dot{x}^i = \{x^i \, , H_{x^+}\} = \frac{p^i}{p^+} \, . \tag{76}$$

Note finally, that the Hamiltonian (74) is *not* the normal projection $N \cdot p$ of the momentum, because $N \cdot p$ lies within Σ and thus corresponds to a kinematical direction.

As for the instant form, the light-cone representation of the Poincaré generators can be obtained by solving the constraint $p^2 = m^2$ for p^-, inserting the result into the elementary representation (54) of the generators and setting $x^+ = 0$. The kinematical generators are

$$\begin{aligned} P^i &= p^i \, , \quad P^+ = p^+ \, , \\ M^{+i} &= -x^i p^+ \, , \quad M^{12} = x^1 p^2 - x^2 p^1 \, , \quad M^{+-} = -x^- p^+ \, . \end{aligned} \tag{77}$$

They correspond to transverse and longitudinal translations within Σ (P^i and P^+, respectively), transverse boosts and rotations (M^{+i}), rotations around the z-axis (M^{12}) and boosts (!) in the z-direction (M^{+-}). The latter will be further analysed in a moment. We thus have found *seven* kinematical generators, so that the front form leads to the largest stability group among Dirac's forms of dynamics (cf. Table 1).

The dynamical generators are given by

$$P^- = \frac{p_\perp^2 + m^2}{p^+} , \quad M^{-i} = x^- p^i - x^i p^- . \tag{78}$$

As expected, the M^{-i} depend on the Hamiltonian, p^-. If we now consider rotations around the x- or y-axis, generated by

$$L^1 = M^{23} = \tfrac{1}{2}(M^{2+} - M^{2-}) , \tag{79}$$

$$L^2 = M^{31} = \tfrac{1}{2}(M^{+1} - M^{-1}) , \tag{80}$$

we note that they correspond to *dynamical* operations due to the appearance of M^{-i}. This leads to the notorious 'problem of angular momentum' within the front form, see e.g. [52]. Except for the free theory, it is very hard to write down states with good angular momentum as diagonalizing $\mathbf{L}^2$ is as difficult as solving the Schrödinger equation. A similar problem arises for parity. This exchanges light-cone space and time and thus also becomes dynamical [24]. For the kinematical component of the angular momentum, $L_z = M^{12}$, these difficulties do not arise.

Consider now the following boost in z-direction with rapidity ω written in instant-form coordinates,

$$t \to t\cosh\omega + z\sinh\omega , \tag{81}$$

$$z \to t\sinh\omega + z\cosh\omega . \tag{82}$$

As stated before, such a boost mixes space and time coordinates z and t. If we add and subtract these equations, we obtain the action of the boost for the front form,

$$x^+ \to e^{\omega} x^+ , \tag{83}$$

$$x^- \to e^{-\omega} x^- . \tag{84}$$

We thus find the important result that a boost in z-direction does not mix light-cone space and time but rather rescales the coordinates! Note that x^+ and x^- are rescaled inversely with respect to each other. The scaling factor can be written as

$$e^{\omega} = \sqrt{\frac{1-v}{1+v}} , \tag{85}$$

if the rapidity ω is defined in the usual manner in terms of the boost velocity v, $\tanh\omega = v$. One should note in particular, that one has the fixed point hypersurface $\Sigma : x^+ = 0$ which is mapped onto itself, so that the relevant generator, $M^{+-} = 2M^{30} = -2K^3$, is kinematical, in agreeement with (77). However, we see explicitly that this is no longer true for $x^+ \neq 0$, where we get a rescaling of x^+. Stated differently, the transformation to light-cone coordinates diagonalizes the boosts in z-direction. Therefore, the behavior under such boosts becomes especially simple. A pedagogical discussion and some elementary applications can be found in [116].

We are actually more interested in the transformation properties of the momenta, as these, being Poincaré generators, are more fundamental quantities than the coordinates, in particular in the quantum theory [99]. As P^μ transforms as a four-vector we just have to replace x^μ by P^μ in the boost transformations (83, 84) and obtain,

$$P^+ \to e^\omega P^+ \, , \tag{86}$$

$$P^- \to e^{-\omega} P^- \, . \tag{87}$$

We remark that $P^+ = 0$ is a fixed point under longitudinal boosts. In quantum field theory, it corresponds to the vacuum. For the transverse momentum, P^i, one finds a transformation law reminiscent of a Galilei boost,

$$P^i \to P^i + v^i P^+ \, . \tag{88}$$

In this identity, describing the action of M^{+i}, longitudinal and transverse momenta (which are both kinematical) get mixed.

We can now ask the question how to boost from (P^+, P^i) to momenta (Q^+, Q^i). This can be done by fixing the boost parameters ω and v^i as

$$\omega = -\log \frac{Q^+}{P^+} \, , \quad v^i = \frac{Q^i - P^i}{P^+} \, . \tag{89}$$

Obviously, this is only possible for $P^+ \neq 0$. We emphasize that in the construction above there is no dynamics involved. For the quantum theory, this means that we can build states of arbitrary light-cone momenta with very little effort. All we have to do is applying some kinematical boost operators. The simple behavior of light-cone momenta under boosts will be important for the discussion of bound states in Sect. 4.

The similarity between (88) and Galilei boosts is not accidental. This is exhibited by the following subalgebra of the light-cone Poincaré algebra. Consider the Poisson bracket relations of the seven generators P^μ, M^{12}, M^{+i},

$$\begin{aligned} \{M^{12}, M^{+i}\} &= \epsilon^{ij} M^{+j} \, , \\ \{M^{12}, P^i\} &= \epsilon^{ij} P^j \, , \\ \{M^{+i}, P^-\} &= -2P^i \, , \\ \{M^{+i}, P^j\} &= -\delta^{ij} P^+ \, . \end{aligned} \tag{90}$$

All other brackets of these generators vanish. Compare now with the two-dimensional Galilei group. Its generators (for a free particle of mass μ) are: two momenta k^i, one angular momentum $L = \epsilon^{ij}x^i k^j$, two Galilei boosts $G^i = \mu x^i$, the Hamiltonian $H = k^i k^i/2\mu$ and the mass μ, which is the Casimir generator. Upon using $\{x^i , k^j\} = \delta^{ij}$ and identifying $P^i \leftrightarrow k^i$, $M^{12} \leftrightarrow L$, $M^{+i} \leftrightarrow -2G^i$, $P^+ \leftrightarrow 2\mu$ and $P^- \leftrightarrow H$, one easily finds that (90) forms a subalgebra of the Poincaré algebra which is isomorphic to the Lie algebra of the two-dimensional Galilei group. (A second isomorphic subalgebra is obtained via identifying $M^{-i} \leftrightarrow 2G^i$ and exchanging P^+ with P^-.) The first two identities in (90), for instance, state that M^{+i} and P^i transform as ordinary two-dimensional vectors. P^+ can be interpreted as a variable Galilei mass which is also obvious from the nonrelativistic appearance of the light-cone Hamiltonian, $P^- = (P_\perp^2 + m^2)/P^+$ and the Galilei boost (88).

One thus expects that light-cone kinematics will partly show a nonrelativistic behavior which is associated with the transverse dimensions and governed by the two-dimensional Galilei group. This expectation is indeed realized and leads, for instance, to a separation of center-of-mass and relative dynamics in multi-particle systems. This will be discussed at length in the beginning of Sect. 4.

So far, our discussion of the Poincaré algebra was restricted to the free case. With the inclusion of interactions, one expects all dynamical Poincaré generators to differ from their free counterpart by some 'potential' term W. This has already been pointed out by Dirac [43], who also stated that finding potentials which are consistent with the commutation relations of the Poincaré algebra is the "real difficulty in the construction of a theory of a relativistic dynamical system" with a fixed number of particles.

It has turned out, however, that Poincaré invariance alone is not sufficient to guarantee a reasonable Hamiltonian formulation. There are no-go theorems both for the instant [96] and the front form [80], which state that the inclusion of any potential into the Poincaré generators, even if consistent with the commutation relations, spoils relativistic *covariance*. The latter is a stronger requirement as it enforces particular transformation laws for the particle coordinates. Thus, covariance imposes rather severe restrictions on the dynamical system [99].

The physical reason for these problems is that potentials imply an instantaneous interaction-at-a-distance which is in conflict with the existence of a limiting velocity and retardation effects. Relativistic causality is thus violated. This is equivalently obvious from the fact that a *fixed* number of particles is in conflict with the necessity of particle creation and annihilation and the appearance of antiparticles.

Nevertheless, with considerable effort, it is possible to construct dynamical quantum systems with a fixed number of constituents which are consistent with the requirements of Poincaré invariance and relativistic covariance [99,140,36].

At this point one might finally ask whether the different forms of relativistic dynamics are physically equivalent. From the point of view that different time choices correspond to different gauge fixings it is clear that equivalence must hold. After all, we are just dealing with different coordinate systems. People have tried to make this equivalence more explicit by working with coordinates which smoothly interpolate between the instant and the front form [124,125,91,75]. In the context of relativistic quantum mechanics, it has been shown that the Poincaré generators for different forms are unitarily equivalent [140].

We are, however, more interested in what might be called a 'top-down approach'. Our aim is to describe few-body systems not within quantum mechanics but quantum field theory to which we now turn.

3 Light-Cone Quantization of Fields

3.1 Construction of the Poincaré Generators

We want to derive the representation of the Poincaré generators within field theory and their dependence on the hypersurface Σ chosen to define the time evolution. To this end we follow [51] and describe the hypersurface mathematically through the equation

$$\Sigma : F(x) = \tau \; . \tag{91}$$

The surface element on Σ is implicitly defined via

$$\int_\Sigma d\sigma_\mu u(x) = \int d^4x \, N^\mu \delta(F(x) - \tau) u(x) \; , \tag{92}$$

where, as before, $N^\mu = \partial^\mu F(x)$ is the normal on Σ and u some integrable function. We will write this expression symbolically as

$$d\sigma_\mu = d^4x \, N^\mu \delta(F(x) - \tau) \; . \tag{93}$$

The central object of this subsection will be the energy-momentum tensor,

$$T^{\mu\nu} = \frac{\partial \mathcal{L}}{\partial(\partial_\mu \phi)} \partial^\nu \phi - g^{\mu\nu} \mathcal{L} \; , \tag{94}$$

with $\mathcal{L}$ being the Lagrangian depending on fields that are collectively denoted by ϕ. With the help of the energy-momentum tensor (94) we can define a generator

$$A[f] = \int_\Sigma d\sigma_\mu f_\nu(x) T^{\mu\nu}(x) \; , \tag{95}$$

where A and f can be tensorial quantities carrying dummy indices α, $\beta, \ldots$ which we have suppressed. $A[f]$ generates the infinitesimal transformations

$$\delta_f B(x) = f_\mu(x) \, \partial^\mu B(x) \; , \tag{96}$$

$$\delta_f x^\mu \;\; = f^\mu(x) \; , \tag{97}$$

where f is now understood as being infinitesimal. In the same way as for a finite number of degrees of freedom, the generator A is called kinematical, if it leaves Σ invariant, that is,

$$\delta_f F = f_\mu \partial^\mu F = f \cdot N = 0 \,. \tag{98}$$

Otherwise, A is dynamical. With the energy-momentum tensor $T^{\mu\nu}$ at hand, we can easily show that kinematical generators are interaction independent. We decompose $T^{\mu\nu}$,

$$T^{\mu\nu} = T_0^{\mu\nu} - g^{\mu\nu}\mathcal{L}_{\mathrm{int}} \,, \tag{99}$$

into a free part $T_0^{\mu\nu} = T^{\mu\nu}(g=0)$, g denoting the coupling, and an interacting part (we exclude the case of derivative coupling). If A is kinematical, we have from (93, 95, 98),

$$A_{\mathrm{int}}[f] = -\int d^4x\,\delta(F-\tau)\mathcal{L}_{\mathrm{int}} f^\mu \partial_\mu F = -\int d^4x\,\delta(F-\tau)\mathcal{L}_{\mathrm{int}}\delta_f F = 0 \,, \tag{100}$$

which indeed shows that A does not depend on the interaction. Dynamical operators, on the other hand, *will* contain interaction dependent pieces. Of course, we are particularly interested in the Poincaré generators, P^α and $M^{\alpha\beta}$. They correspond to the choices $f^\alpha_\mu = g^\alpha_\mu$ and $f^{\alpha\beta}_\mu = x^\alpha g^\beta_\mu - x^\beta g^\alpha_\mu$, respectively, so that, from (95), they are given in terms of $T^{\mu\nu}$ as

$$P^\alpha = \int_\Sigma d\sigma_\mu T^{\mu\alpha} \,, \tag{101}$$

$$M^{\alpha\beta} = \int_\Sigma d\sigma_\mu (x^\alpha T^{\mu\beta} - x^\beta T^{\mu\alpha}) \,. \tag{102}$$

From (98) it is easily seen that the Poincaré generators defined in (101, 102) act on $F(x) = \tau$ exactly as described in (57). The remarks of Sect. 2 on the kinematical or dynamical nature of the generators in the different forms are therefore equally valid in quantum field theory.

Let us first discuss the instant form. We recall the hypersurface of equal time, $\Sigma : F(x) \equiv N \cdot x \equiv x^0 = \tau$, which leads to a surface element

$$d\sigma^\mu = d^4x\, N^\mu \delta(x^0 - \tau) \,, \quad N^\mu = (1, 0) \,. \tag{103}$$

Using (101, 102), the Poincaré generators are obtained as

$$P^\mu = \int_\Sigma d^3x\, T^{0\mu} \,, \tag{104}$$

$$M^{\mu\nu} = \int_\Sigma d^3x\, \left(x^\mu T^{0\nu} - x^\nu T^{0\mu}\right) \,. \tag{105}$$

For the front form, quantization hypersurface and surface element are given by

$$\Sigma : F(x) \equiv N \cdot x \equiv x^+ = \tau \,, \quad d\sigma^\mu = d^4x\, N^\mu\, \delta(x^+ - \tau) \,, \tag{106}$$

where N is the light-like four-vector of (70). In terms of $T^{\mu\nu}$, the Poincaré generators are

$$P^\mu = \tfrac{1}{2}\int_\Sigma dx^- d^2x_\perp\, T^{+\mu}\,, \tag{107}$$

$$M^{\mu\nu} = \tfrac{1}{2}\int_\Sigma dx^- d^2x_\perp\, \left(x^\mu T^{+\nu} - x^\nu T^{+\mu}\right). \tag{108}$$

The somewhat peculiar factor 1/2 is the Jacobian which arises upon transforming to light-cone coordinates.

3.2 Schwinger's (Quantum) Action Principle

Our next task is to actually quantize the fields on the hypersurfaces $\Sigma : \tau = F(x)$ of equal time τ. There is more than one possibility to do so, and we will explain a few of these. We begin with a method that is essentially due to Schwinger [135–137]. We define a four-momentum density

$$\Pi^\mu = \frac{\partial \mathcal{L}}{\partial(\partial_\mu \phi)}\,, \tag{109}$$

so that the energy-momentum tensor $T^{\mu\nu}$ can be written as

$$T^{\mu\nu} = \Pi^\mu \partial^\nu \phi - g^{\mu\nu}\mathcal{L}\,. \tag{110}$$

In some sense, this can be viewed as a covariant generalization of the usual Legendre transformation between Hamiltonian and Lagrangian. Using the normal N^μ of the hypersurface Σ, we define the canonical momentum (density) as the projection of Π^μ,

$$\pi \equiv N \cdot \Pi\,. \tag{111}$$

Schwinger's action principle states that, upon variation, the action $S = \int d^dx\, \mathcal{L}$ changes at most by a surface term which (if Σ is not varied, i.e. $\delta x^\mu = 0$) is given by

$$\delta G(\tau) = \int_\Sigma d\sigma_\mu\, \Pi^\mu\, \delta\phi = \int d^dx\, \delta(\tau - F)\, \pi\, \delta\phi\,. \tag{112}$$

The quantity δG is interpreted as the generator of field transformations, so that we have

$$\delta\phi = \{\phi\,,\, \delta G\}\,, \tag{113}$$

in case that Σ is entirely space-like (with time-like normal) [135,136]. We note in passing that the generator δG is a field theoretic generalization of the canonical one-form $dG \equiv p_i dq^i$ used in analytical mechanics.

As in the preceding section we have to distinguish two cases depending on whether the normal vector N is time-like or space-like. For time-like N, the associated hypersurface is space-like. The basic example for this case is the instant form, to which we immediately specialize. The canonical momentum density is given by the velocity, $\pi = \dot{\phi}$, and the Lagrangian is *quadratic* in $\dot{\phi}$. The canonical Poisson bracket is derived from Schwinger's action principle using (113),

$$\begin{aligned} \delta\phi(x) &= \{\phi(x)\,,\,\delta G(\tau)\} \\ &= \int dy^0 \int d^3y\,\delta(x^0 - y^0)\,\{\phi(x)\,,\,\pi(y)\}\,\delta\phi(y) \\ &= \int d^3y\;\{\phi(x)\,,\,\dot{\phi}(y)\}\,\delta\phi(y)|_{x^0=y^0=\tau}\;. \end{aligned} \tag{114}$$

The canonical Poisson bracket, therefore, must be

$$\{\phi(x)\,,\,\dot{\phi}(y)\}_{x,y\in\Sigma} = \delta^3(\mathbf{x}-\mathbf{y})\;, \tag{115}$$

which, of course, is the standard result. The second case, N light-like, corresponds to the front form. With minor modifications, Schwinger's approach can also be used here, resulting in what is interchangably called light-cone, light-front or null-plane quantization. The canonical light-cone momentum is

$$\pi = N\cdot\Pi = N\cdot\partial\phi = \partial^+\phi \equiv 2\frac{\partial}{\partial x^-}\phi\;, \tag{116}$$

which is peculiar to the extent that it does not involve a (light-cone) time derivative. Therefore, π is a dependent quantity which does not provide additional information, being known on Σ when the field is known there. Thus, π is merely an abbreviation for $\partial^+\phi$ which is a *spatial* derivative. Again, the reason is that the normal N^μ of the null-plane Σ lies within Σ. This leads to the important consequence that the light-cone Lagrangian is *linear* in the velocity $\partial^-\phi$, or, put differently, that light-cone field theories are *first-order systems*. As a result, ϕ and $\partial^+\phi$ have to be treated on the same footing within Schwinger's approach which leads to an additional factor 1/2 compared to (113),

$$\tfrac{1}{2}\delta\phi = \{\phi\,,\,\delta G\}\;, \tag{117}$$

with a front-form generator

$$\delta G(x^+) = \tfrac{1}{2}\int_\Sigma dx^- d^2x_\perp \partial^+\phi\,\delta\phi\;. \tag{118}$$

The appearance of the peculiar factor 1/2 in (117) has been discussed at length by Schwinger [137] – see also [32]. Roughly speaking it stems from the fact that the independent field content within the front form is only one half of that in the instant form. The factor 1/2 cancels the light-cone Jacobian $J = 1/2$ in (118), so that we are left with the Poisson bracket,

$$\{\phi(x)\,,\,\pi(y)\}_{x,y\in\Sigma} = \delta(x^- - y^-)\delta^2(\mathbf{x}_\perp - \mathbf{y}_\perp)\;. \tag{119}$$

As usual, commutators are inferred from Poisson brackets by invoking Dirac's correspondence principle, that is, by replacing the bracket by i times the commutator. For arbitrary classical observables, A, B, this means explicitly,

$$[\hat{A}, \hat{B}] = i\widehat{\{A\,, B\}}\,. \tag{120}$$

We do not address the question of operator-ordering ambiguities at this point, as these will not be an issue in the applications to be discussed later on. One should, however, be aware of this problem, as it indeed *can* arise within the framework of light-cone quantization [70].

Using (120), the bracket (119) leads to the following commutator,

$$[\phi(x), \pi(y)]_{x^+=y^+=\tau} = i\delta(x^- - y^-)\delta^2(\mathbf{x}_\perp - \mathbf{y}_\perp)\,. \tag{121}$$

As the independent quantities are the fields themselves, we invert the derivative ∂^+ and obtain the more fundamental commutator

$$[\phi(x), \phi(y)]_{x^+=y^+=\tau} = -\tfrac{i}{4}\mathrm{sgn}(x^- - y^-)\delta^2(\mathbf{x}_\perp - \mathbf{y}_\perp)\,. \tag{122}$$

In deriving (122) we have chosen the anti-symmetric Green function $\mathrm{sgn}(x^-)$ satisfying

$$\frac{\partial}{\partial x^-}\mathrm{sgn}(x^-) = 2\delta(x^-)\,, \tag{123}$$

so that (121) is reobtained upon differentiating (122) with respect to y^-. We will see later that the *field* commutator (122) can be derived directly within Schwinger's method. Before that, however, let us study the relation between the choice of initializing hypersurfaces, the problem of field quantization and the solutions of the dynamical equations.

3.3 Quantization as an Initial- and/or Boundary-Value Problem

As a prototype field theory we consider a massive scalar field ϕ in 1+1 dimensions. Its dynamics is encoded in the action

$$S[\phi] = \int d^2x\,\mathcal{L} = \int d^2x\,\left(\tfrac{1}{2}\partial_\mu\phi\,\partial^\mu\phi - \tfrac{1}{2}m^2\phi^2 - \mathcal{V}[\phi]\right)\,, \tag{124}$$

where $\mathcal{V}$ is some interaction term like e.g. $\lambda\phi^4$ and $\mathcal{L} = \mathcal{L}_0 + \mathcal{V}$. By varying the free action in the standard way we obtain

$$\delta S = \int_{\partial M} d\sigma_\mu \Pi^\mu \delta\phi + \int_M \left[\frac{\partial\mathcal{L}_0}{\partial\phi} - \partial_\mu\frac{\partial\mathcal{L}_0}{\partial(\partial_\mu\phi)}\right]\delta\phi\,. \tag{125}$$

If we do not vary on the boundary of our integration region M, $\delta\phi|_{\partial M} = 0$, the surface term in δS (which is closely related to δG from (112)), vanishes and we end up with the (massive) Klein-Gordon equation in 1+1 dimensions,

$$(\Box + m^2)\phi = 0\,. \tag{126}$$

In this subsection, we will solve this equation by specifying initial and/or boundary conditions for the scalar field ϕ on different hypersurfaces Σ. In addition, we will clarify the relation between the associated initial value problems and the determination of 'equal-time' commutators.

It may look rather trivial to consider just the free theory, but this is not entirely true. Let us analyze what quantization of a field theory means in the light of the different forms of relativistic dynamics. One specifies canonical commutators like $[\phi(x), \phi(y)]_{x,y \in \Sigma}$, where the hypersurface Σ: $\tau = const$ defines the evolution parameter τ. As both x and y lie in Σ, the commutator is evaluated at 'equal time', which implies that it is a *kinematical* quantity. Therefore, it is the same for the free *and* the interacting theory.

Now, if ϕ *is* a *free* field, the commutator,

$$[\phi(x), \phi(0)] = i\Delta(x) , \tag{127}$$

is exactly known: it is the Pauli-Jordan or Schwinger function Δ [84,134] which is a special solution of the Klein-Gordon equation (126). It can be obtained directly from the action in a covariant manner as a Peierls bracket [119,40]. Alternatively, one can find it by evaluating the Fourier integral,

$$\begin{aligned} \Delta(x) &= -\frac{i}{2\pi} \int d^2p \, \delta(p^2 - m^2) \, \mathrm{sgn}(p^0) e^{-ip\cdot x} \\ &= -\tfrac{1}{2}\mathrm{sgn}(x^0) \, \theta(x^2) \, J_0(m\sqrt{x^2}) \\ &= -\tfrac{1}{4}\left[\mathrm{sgn}(x^+) + \mathrm{sgn}(x^-)\right] \, J_0(m\sqrt{x^+x^-}) , \end{aligned} \tag{128}$$

where I have given both the instant and front form representation [71]. We note that Δ is antisymmetric, $\Delta(x) = -\Delta(-x)$ and Lorentz invariant (under proper orthochronous transformations). Most important, it is causal, i.e. it vanishes outside the light-cone, $x^2 < 0$ (see Fig. 2).

If ϕ is an *interacting* field, causality, of course, must still hold. If x and y are space-like with respect to each other, the commutator thus still vanishes,

$$[\phi(x), \phi(y)]_{(x-y)^2<0} = 0 . \tag{129}$$

This expresses the fact that fields which are separated by a space-like distance cannot communicate with each other. For the front form, with the hypersurface $\Sigma : x^+ = 0$, (129) cannot be used to obtain the canonical commutators: In 1+1 dimensions, Σ is part of the light-cone and therefore entirely light-like. In higher dimensions, Σ still contains light-like directions namely where $x^- = \mathbf{x}_\perp = 0$. For this reason, the light-front commutator (122) of two free fields does *not* vanish identically.

Let me now discuss the explicit relation between the choice of equal-time commutators and the classical initial/boundary for the Klein-Gordon equation. Three examples are of interest.

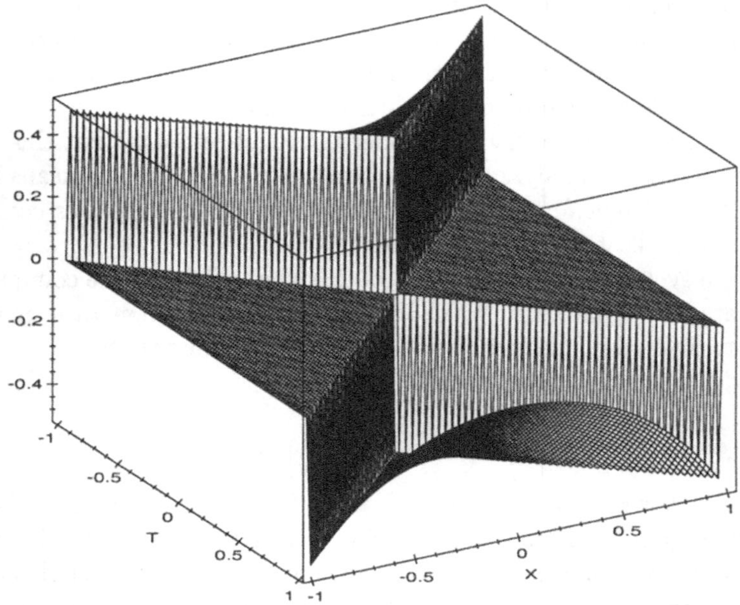

Fig. 2. *The Pauli-Jordan function as a function of* $T = mx^0/2$ *and* $X = mx^1/2$. *It vanishes outside the light-cone and oscillates inside.*

Cauchy Data: Instant Form. Conventional quantization on a space-like surface (based on the instant form) corresponds to a Cauchy problem: if one specifies the field ϕ and its time derivative $\dot{\phi}$ on $\Sigma : x^0 = t = 0$,

$$\phi(t = 0, x) = f(x) \,, \tag{130}$$

$$\dot{\phi}(t = 0, x) = g(x) \,, \tag{131}$$

where the functions f and g denote the initial data (depending on $x \equiv x^1$), the solution of the Klein-Gordon equation is uniquely determined. This can be checked by considering the Taylor expansion around $(t, x) \in \Sigma$, i.e. $t = 0$,

$$\phi(t, x) = \phi(0, x) + t\dot{\phi}(0, x) + \tfrac{1}{2}t^2\ddot{\phi}(0, x) + \ldots \,, \tag{132}$$

with the overdot denoting the time derivative. From this we see that one has to know *all* time derivatives of ϕ on Σ once the data f, g are given. If we calculate these,

$$
\begin{aligned}
\phi &= f \,, \\
\dot{\phi} &= g \,, \\
\ddot{\phi} &= \phi'' - m^2\phi = f'' - m^2 f \,, \\
\frac{\partial^3 \phi}{\partial t^3} &= \dot{\phi}'' - m^2\dot{\phi} = g'' - m^2 g \,, \\
&\vdots \quad .
\end{aligned} \tag{133}
$$

we find that indeed all time derivatives are given in terms of f and g and their known *spatial* derivatives, denoted by the prime. In the last two identities, we have made use of the equation of motion. As a result, we see that the Cauchy problem is well posed: the solution of the Klein-Gordon equation is uniquely determined by the data on Σ.

Upon quantization, this translates into the fact that the Fock operators can be expressed in terms of the data,

$$
a(p^1) = \int dx^1 e^{-ip^1x^1} \Big[\omega_p \, \phi(x^0 = 0, x^1) + i\dot{\phi}(x^0 = 0, x^1) \Big] \,, \tag{134}
$$

with $\omega_p = (p_1^2 + m^2)^{1/2}$. In addition, the canonical commutators can be viewed as the Cauchy data for the Pauli-Jordan function Δ,

$$
\Big[\phi(x), \phi(0) \Big]_{x^0=0} = i\Delta(x)|_{x^0=0} = 0 \,, \tag{135}
$$

$$
\Big[\dot{\phi}(x), \phi(0) \Big]_{x^0=0} = i\dot{\Delta}(x)|_{x^0=0} = -i\delta(x^1) \,. \tag{136}
$$

As stated above, the vanishing of the commutator (135) is due to causality.

The Characteristic Initial-Value Problem. In the following I will perform an analogous discussion for the hypersurfaces $\Sigma : x^\pm = const$, which, in $d = 1+1$, constitute the entire light-cone, $x^2 = 0$. In Dirac's classification, the light-cone corresponds to a degenerate point form with parameter $a = 0$ (see Table 1). One thus does not have transitivity as points on different 'legs' of the cone are not related by a kinematical operation. Still, it turns out that the associated initial-value problem is well posed [114,129]

The light-fronts $x^\pm = 0$ are characteristics of the Klein-Gordon equation [45]. Therefore, one is dealing with a characteristic initial-value problem [38], for which one has to provide the data

$$
\phi(x^+ = 0, x^-) = f(x^-) \,, \tag{137}
$$

$$
\phi(x^+, x^- = 0) = g(x^+) \,, \tag{138}
$$

$$
f(x^- = 0) = g(x^+ = 0) \,, \tag{139}
$$

where the last identity is a continuity condition. Consistency is again checked by Taylor expanding, this time around $(x^+, x^-) = 0$,

$$\begin{aligned}
\partial^+ \phi \quad &= \partial^+ f \equiv f' \, , \\
\partial^- \phi \quad &= \partial^- g \equiv \dot{g} \, , \\
\partial^+ \partial^- \phi \quad &= m^2 \phi = m^2 f = m^2 g \, , \\
\partial^+ \partial^+ \phi \quad &= f'' \, , \\
\partial^- \partial^- \phi \quad &= \ddot{g} \, , \\
\partial^+ \partial^+ \partial^- \phi &= m^2 f' \, , \\
\partial^- \partial^- \partial^+ \phi &= m^2 \dot{g} \, , \\
\vdots \qquad & \qquad .
\end{aligned} \tag{140}$$

Whereever a factor of m^2 appears we have made use of the Klein-Gordon equation. We thus note that the data (together with their known derivatives) determine *all* partial derivatives of ϕ at the vertex of the cone, $(x^+, x^-) = 0$. Intuitively, this corresponds to the fact that the information spreads from a source located at origin.

The characteristic initial-value problem amounts to quantization on *two* characteristics, $x^\pm = 0$, i.e., in $d = 1 + 1$, really *on* the light cone, $x^2 = 0$. The following two independent commutators,

$$[\phi(x), \phi(0)]_{x^\pm = 0} = i\Delta(x)|_{x^\pm = 0} = -\tfrac{i}{4} \mathrm{sgn}(x^\mp) \, , \tag{141}$$

are then characteristic data for the Pauli-Jordan function. It turns out that, in case the field ϕ is massless, the above quantization procedure is the only consistent one (in d=1+1), if one wants to use light-like hypersurfaces [17].

However, it is important to note that the characteristic initial-value problem does *not* correspond to light-cone quantization. One would need two Hamiltonians P^- *and* P^+, and, accordingly, two time parameters. This seems somewhat weird, to say the least, and will not be pursued any further.

Initial-Boundary-Data: Front-Form. In order to find the initial-value problem of the front form with a single time parameter x^+, let us naively try a straightforward analog of the Cauchy data and prescribe field and velocity on $\Sigma : x^+ = 0$,

$$\phi(x^+ = 0, x^-) \quad = f(x^-) \, , \tag{142}$$

$$\partial^- \phi(x^+ = 0, x^-) = g(x^-) \, . \tag{143}$$

It turns out, however, that this *overdetermines* the system. Namely, from the equation of motion

$$\partial^+ \partial^- \phi = m^2 \phi \, , \tag{144}$$

it is actually possible to obtain the velocity $\partial^- \phi$ by inversion of the *spatial* derivative ∂^+,

$$\partial^- \phi = m^2 (\partial^+)^{-1} \phi = m^2 (\partial^+)^{-1} f \, . \tag{145}$$

The last identity holds on Σ and implies that the data g are unnecessary (and will even lead to an inconsistency) as the velocity is already determined by f. This is confirmed by the Taylor expansion on Σ,

$$\begin{aligned} \phi &= f \, , \\ \partial^+ \phi &= \partial^+ f \, , \\ \partial^- \phi &= m^2 (\partial^+)^{-1} f \, , \\ \partial^- \partial^- \phi &= m^2 (\partial^+)^{-1} \partial^- \phi = m^4 (\partial^+)^{-2} f \, , \\ &\vdots \quad . \end{aligned} \tag{146}$$

It thus seems that the front form requires only half of the data as compared to the instant form. This appearance, however, is deceptive. Note that we have to invert the differential operator ∂^+. The inverse is nothing but the Green function G defined via

$$\partial^+ G(x^-) = \delta(x^-) \, . \tag{147}$$

Clearly, this Green function is determined only up to a homogeneous solution h satisfying

$$\partial^+ h = 0 \, , \tag{148}$$

i.e. up to a *zero mode* $h = h(x^+)$ of the operator ∂^+. Thus, in order to uniquely specify the Green function (147), we have to provide additional information in terms of *boundary conditions.* The standard choice is to demand antisymmetry in x^-, whence

$$G(x^-) = \frac{1}{4} \mathrm{sgn}(x^-) \, , \tag{149}$$

which we have already used in (122) and (123). Before we discuss the physical reason for demanding antisymmetry, let us briefly go to momentum space where we replace ∂^+ by ip^+. The equation (147) for the Green function becomes $ip^+ G(p^+) = 1$, which has the general solution

$$G(p^+) = -i/p^+ + h(p^-)\delta(p^+) \, . \tag{150}$$

In this identity, $1/p^+$ has to be viewed as a distribution corresponding to an arbitrary regularization of the singular function $1/p^+$ [53]. Any two regularizations differ by terms proportional to $\delta(p^+)$, i.e. a zero mode of p^+. Choosing an antisymmetric Green function uniquely yields a principal value prescription,

$$iG(p^+) = \mathcal{P}\frac{1}{p^+} = \frac{1}{2}\left(\frac{1}{p^+ + i\epsilon} + \frac{1}{p^+ - i\epsilon} \right) \, , \tag{151}$$

which is the canonical regularization of $1/p^+$.

Altogether we have seen that the front form corresponds to prescribing both initial and boundary conditions, so that one has a 'mixed' or *initial-boundary* value problem. What are the implications for quantization? We address this question by determining the Poisson brackets through the requirement that Euler-Lagrange and canonical equations should be equivalent. To this end we solve the Klein-Gordon equation (126) for the velocity $\dot{\phi} \equiv \partial\phi/\partial x^+$ as in (145). This gives

$$\dot{\phi}(x^+, x^-) = -\frac{m^2}{4} \int dy^- G(x^-, y^-) \phi(x^+, y^-) \, , \qquad (152)$$

using the Green function G from (149). The Hamiltonian equation of motion is given by the Poisson bracket with $H = \frac{1}{2} m^2 \int dx^- \phi^2$,

$$\dot{\phi}(x^+, x^-) = \frac{m^2}{2} \int dy^- \left\{ \phi(x^+, x^-) \, , \, \phi(x^+, y^-) \right\} \phi(x^+, y^-) \, , \qquad (153)$$

with the bracket of the fields ϕ to be determined. Clearly, Euler-Lagrange and Hamiltonian equation of motion, (152) and (153) become equivalent if one identifies

$$\left\{ \phi(x^+, x^-) \, , \, \phi(x^+, y^-) \right\} \equiv -\tfrac{1}{2} G(x^-, y^-) \, . \qquad (154)$$

We thus see that the fundamental Poisson bracket coincides with the Green function which, accordingly, justifies the requirement of antisymmetry. After quantization, (154) of course coincides with (122), the result from Schwinger's action principle, specialized to $d = 1 + 1$,

$$[\phi(x), \phi(0)]_{x^+=0} = i\Delta(x)|_{x^+=0} = -\frac{i}{4} \mathrm{sgn}(x^-) \, . \qquad (155)$$

From the momentum space perspective,

$$[\phi(p^+), \phi(0)] = \frac{i}{2} G(p^+) = \frac{1}{2} \mathcal{P} \frac{1}{p^+} \, , \qquad (156)$$

we conclude that, technically, light-cone quantization is the inversion of the longitudinal momentum p^+ and as such requires the specification of initial-*boundary* data. In some sense, this can also be viewed as an infrared regularization because one provides a prescription of dealing with a pole at vanishing longitudinal momentum, $p^+ = 0$. As is well known, a particularly nice way of regularizing in the infrared is to enclose the system under consideration in a finite spatial volume. This is the topic of the next subsection.

3.4 DLCQ – Basics

DLCQ is the acronym for 'discretized light-cone quantization', originally developed in [106,117,118]. The physical system under consideration is enclosed in a finite volume with discrete momenta and prescribed boundary conditions in x^-. Recently, there has been renewed interest in this method in the context of constructing a matrix model of M-theory [6,145].

Our starting point is the Fourier representation for a solution of the Klein-Gordon equation (still in infinite volume),

$$\begin{aligned}\phi(x) &= \int \frac{d^2p}{2\pi}\chi(p)\,\delta(p^+p^- - m^2)\,e^{-ip\cdot x} \\ &= \int \frac{dp^+dp^-}{4\pi}\chi(p^+,p^-)\frac{1}{|p^+|}\,\delta(p^- - m^2/p^+)\,e^{-ip\cdot x} \\ &= \int \frac{dp^+}{4\pi|p^+|}\chi(p^+,\hat{p}^-)\,e^{-i\hat{p}\cdot x} \\ &= \int_0^\infty \frac{dp^+}{4\pi p^+}\left[\chi(p^+,\hat{p}^-)e^{-i\hat{p}\cdot x} + \chi(-p^+,-\hat{p}^-)e^{i\hat{p}\cdot x}\right] \\ &\equiv \int \frac{dp^+}{4\pi p^+}\theta(p^+)\left[a(p^+)e^{-i\hat{p}\cdot x} + a^*(p^+)e^{i\hat{p}\cdot x}\right]\,. \end{aligned} \tag{157}$$

The following remarks are in order: we have defined the on-shell energy, $\hat{p}^- \equiv m^2/p^+$; contrary to the instant form, the integration over the positive and negative mass hyperboloid is achieved by a *single* delta function. Again, this is a consequence of the linearity of the mass-shell constraint in p^-. The two branches of the mass-shell correspond to positive and negative values of p^+ (and also $\hat{p}^-$), respectively. Associated with the two signs of the kinematical momentum p^+ are the positive and negative frequency modes a, a^*, defined in such a way that their argument p^+ is always positive (cf. the step function θ in the last line). This can equivalently be viewed as the reality condition

$$a^*(p^+) = a(-p^+)\,, \tag{158}$$

as is obvious from the last step in the derivation (157). Upon quantization this implies that annihilation operators with negative longitudinal momentum p^+ are actually creation operators for particles with positive p^+. The field commutator (155) is reproduced by demanding

$$[a(k^+), a^\dagger(p^+)] = 4\pi p^+\delta(k^+ - p^+)\,. \tag{159}$$

As already indicated, DLCQ amounts to compactifying the spatial light-cone coordinate, $-L \le x^- \le L$, and imposing periodic boundary conditions for the fields,

$$\phi(x^+, x^- = -L) = \phi(x^+, x^- = L)\,, \tag{160}$$

which are to hold for all light-cone times x^+. Space-time is thus endowed with the topology of a cylinder. This implies discrete longitudinal momenta, $k_n^+ = 2\pi n/L$, so that the Fock expansion (157) becomes

$$\phi(x^+ = 0, x^-) = a_0 + \sum_{n>0}\frac{1}{\sqrt{4\pi n}}\left(a_n e^{-in\pi x^-/L} + a_n^* e^{in\pi x^-/L}\right)\,, \tag{161}$$

Note that we have allowed for a zero momentum mode a_0. We will see in a moment that it actually vanishes in the free theory. Plugging (161) into the free Lagrangian

$$L_0[\phi] = \tfrac{1}{2}\int dx^- \left(\tfrac{1}{2}\partial^+\phi\,\partial^-\phi - \tfrac{1}{2}m^2\phi^2\right) , \quad (162)$$

we obtain (discarding a total time derivative)

$$L_0[a_n, a_0] = -i\sum_{n>0} a_n\dot{a}_n^* - m^2La_0^2 - \sum_{n>0}\frac{m^2L}{4\pi n}a_n^*a_n \equiv -i\sum_{n>0} a_n\dot{a}_n^* - H , \quad (163)$$

with H denoting the Hamiltonian and $\dot{a}_n^* = \partial a_n^*/\partial x^+$. From both representations (162) and (163) it is obvious that the light-cone Lagrangian is linear in the velocity ($\partial^-\phi$ and $\dot{a}_n^*$, respectively). A particularly suited method for quantization in this case is the one of Faddeev and Jackiw for first order systems [49,79]. It avoids many of the technicalities of the Dirac-Bergmann formalism and is in general more economic. It reduces phase-space right from the beginning as there are no 'primary constraints' introduced. The method is essentially equivalent to Schwinger's action principle, especially in the form presented in [137]. For the case at hand, the method basically boils down to demanding equivalence of the Euler-Lagrange and Hamiltonian equations of motion (cf. last subsection).

The former are given by

$$-i\dot{a}_n + \frac{m^2L}{4\pi n}a_n = 0 , \quad (164)$$

$$2m^2La_0 = 0 . \quad (165)$$

The first equation, (164), is just the free Klein-Gordon equation which can be easily seen upon multiplying by k_n^+. The second identity, (165), is nondynamical and thus a *constraint* which states the absence of a zero mode for free fields, $a_0 = 0$.

The canonical equations are

$$\dot{a}_n = \{a_n\, , H\} = \sum_{k>0}\frac{m^2L}{4\pi k}\{a_n\, , a_k^*\}\,a_k , \quad (166)$$

which obviously coincides with (164) if the canonical bracket is

$$\{a_k\, , a_n^*\} = -i\delta_{kn} . \quad (167)$$

The constraint (165) is obtained by differentiating the Hamiltonian,

$$\frac{\partial H}{\partial a_0} = 2m^2La_0 = 0 . \quad (168)$$

Let us briefly show that the approach presented above is equivalent to Schwinger's [137]. From (163) we read off a generator

$$\delta G = -i \sum_{n>0} a_n \delta a_n^* \tag{169}$$

effecting the transformation

$$\delta a_n^* = \{a_n^* \, , \delta G\} = -i \sum_{k>0} \{a_n^* \, , a_k\} \, \delta a_k^* \, , \tag{170}$$

which in turn implies the canonical bracket (167).

Quantization is performed as usual by employing the correspondence principle (120), so that, from (167), the elementary commutator is given by

$$[a_m, a_n^\dagger] = \delta_{mn} \, . \tag{171}$$

The Fock space expansion for the (free) scalar field ϕ thus becomes

$$\phi(x^+ = 0, x^-) = \sum_{n>0} \frac{1}{\sqrt{4\pi n}} \Big(a_n e^{-in\pi x^-/L} + a_n^\dagger e^{in\pi x^-/L} \Big) , \tag{172}$$

Like in the infinite-volume expression (157), the Fock 'measure' $1/\sqrt{4\pi n}$ does not involve any scale like the mass m or the volume L. This is at variance with the analogous expansion in the instant form which reads

$$\phi(x, t=0) = \frac{1}{\sqrt{2L}} \sum_n \frac{1}{\sqrt{2(k_n^2 + m^2)}} \left(a_n e^{ik_n x} + a_n^\dagger e^{-ik_n x} \right) , \tag{173}$$

where $-L \le x \le L$, $k_n = \pi n / L$, and $[a_n, a_m^\dagger] = \delta_{mn}$. Obviously, the 'measure' $(k_n^2 + m^2)^{1/2}$ *does* depend on m and L. We will discuss some consequences of this difference in Sect. 3.6.

We can use the results (171) and (172) to calculate the free field commutator at equal light-cone time x^+,

$$[\phi(x), \phi(0)]_{x^+=0} = \sum_{n \neq 0} \frac{1}{4\pi n} e^{-in\pi x^-/L} = -\frac{i}{2} \left[\frac{1}{2} \mathrm{sgn}(x^-) - \frac{x^-}{2L} \right] . \tag{174}$$

This coincides with (155) up to a finite size correction given by the additional term $x^-/2L$. The effect of this term is two-fold. First, it makes the sign function periodic (in the interval $-L \le x^- \le L$), and second, it guarantees the absence of a zero mode which must hold according to (165), (168), and

$$\int_{-L}^{L} dx^- [\phi(x), \phi(0)]_{x^+=0} = 0 \, . \tag{175}$$

One may equally think of this as the finite-volume analog of the principal value prescription.

The commutator (174) has originally been obtained in [106] using the Dirac-Bergmann algorithm for constrained systems. The Faddeev-Jackiw method, however, is much more economic and transparent. In particular, it makes clear that the basic canonical variables of a light-cone field theory are the Fock operators or their classical counterparts. The a_n with, say, $-N \leq n \leq N$ in (161) can be viewed as defining a $(2N+1)$-dimensional phase space. A phase space, however, should have even dimension. This is accomplished by choosing a polarization in terms of positions and momenta, here a_n and $a_n^\dagger$, with $n > 0$, and by the vanishing of the zero mode, $a_0 = 0$. It turns out that this vanishing is a peculiarity of the free theory as is discussed in [106,155,68].

At this point one should honestly state that the issue of zero modes is one of the unsolved problems of light-cone quantization. The constraint equations for the zero modes are in general very hard to solve unless one has some small parameter like in perturbation theory [70] or within a large-N expansion [19,20]. Using a path integral approach, it has recently been shown [72] that integrating out the zero modes constitutes a strong coupling problem. There are speculations that this problem might be less severe if one goes beyond quantum field theory, i.e. in string or M-theory [5].

In the last reference, the author also states that compactification in a light-like direction "is close to a space with periodic time" and thus "weird", in view of possible 'grandfather paradoxes'. Therefore, the natural question arises whether DLCQ is actually consistent with causality.

3.5 DLCQ – Causality

In this subsection I will address the question under which circumstances compactification of 'space' is compatible with the requirements of causality. The presented results are based on recent work with N. Scheu and H. Kröger [69].

In (128) and (129) we have seen that the (infinite-volume) commutator of two scalar fields vanishes whenever their space-time arguments are separated by a space-like distance (cf. Fig. 2). As already mentioned, this is a manifestation of the principle of microcausality, which is the general statement that the commutator of *any* two observables $\mathcal{O}_1(x)$ and $\mathcal{O}_2(y)$ must vanish whenever their separation $x - y$ is space-like. Physically, this implies that measurements of the observables $\mathcal{O}_1$ and $\mathcal{O}_2$ performed at x and y, do not interfere. Some consequences of this principle are the spin-statistics theorem, analyticity properties of Green functions leading to dispersion relations etc. [143].

Our starting point are the Fourier representations of the Pauli-Jordan function, both for the instant and front form (denoted IF and FF, respectively),

$$\text{IF:}\quad \Delta(x) = -\int \frac{dk^1}{2\pi\omega_k}\sin(k\cdot x) \equiv \int dk^1\, I(k^1)\,, \tag{176}$$

$$\text{FF:}\quad \Delta(x) = -\int_0^\infty \frac{dk^+}{2\pi k^+}\sin(k\cdot x) \equiv \int dk^+\, I(k^+)\,. \tag{177}$$

Both integrals yield the same result (128) for the Pauli-Jordan function. Note, however, that the integrand $I(k^+)$ is exploding and rapidly oscillating for $k^+ \to 0$ (see Fig. 3) so that the finite result for the integral is due to sizable cancellations that occur upon integration.

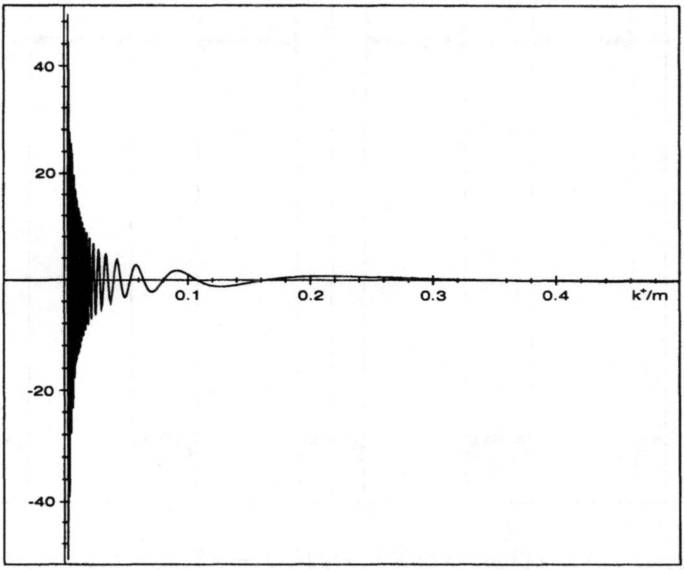

Fig. 3. The integrand $I(k^+)$.

To obtain the finite-volume representations for the commutator, one proceeds as in Sect. 3.4 by restricting the spatial coordinates, $-L \le x^1, x^- \le L$, and imposing periodic boundary conditions for the field ϕ. Momenta become discrete, $k_n^1 \equiv \pi n/L$, and $k_n^+ \equiv 2\pi n/L$. The finite-volume representations are *defined* by replacing the integrals (176) and (177) by the discrete sums,

$$\Delta_{\mathrm{IF}}(x) \equiv -\sum_{n=-N}^{N} \frac{1}{2\omega_n L}\sin(k_n\cdot x)\,, \tag{178}$$

$$\Delta_{\mathrm{FF}}(x) \equiv -\sum_{n=1}^{N} \frac{1}{2\pi n}\sin(k_n\cdot x)\,. \tag{179}$$

The on-shell energies for discrete momenta are defined as

$$\omega_n = (n^2\pi^2/L^2 + m^2)^{1/2} \quad \text{and} \quad \hat{k}_n^- = m^2 L/2\pi n \,. \tag{180}$$

For both functions, $\Delta_{\rm IF}$ and $\Delta_{\rm FF}$, the periodicity in x^1 and x^-, respectively, with periodicity length $2L$, is obvious. The limit $N \to \infty$ is understood unless we perform numerical calculations where N is kept finite.

The evaluation of the sums (178) and (179) is not straightforward. To gain some intuition, we evaluate them numerically beginning with the IF expression (178). The resulting $\Delta_{\rm IF}$ is plotted in Fig. 4.

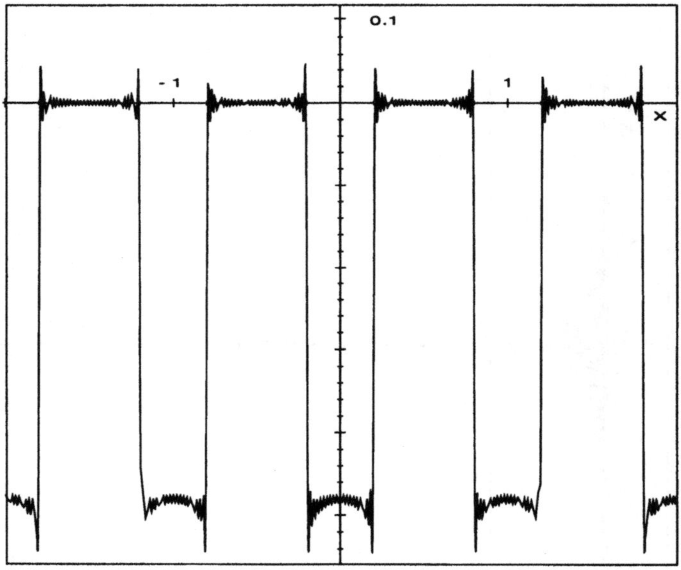

Fig. 4. $\Delta_{\rm IF}(X,T)$ as a function of $X = x^1/2L$. $T = X^0/2L = 0.2$, $mL = 1$, $N = 50$.

Upon inspection, one notes the following: Up to small oscillations stemming from the (unavoidable) Gibbs phenomenon, $\Delta_{\rm IF}$ vanishes outside the light-cone ($|x^0| < |x^1| < L$), and thus is causal even in finite volume. If we let the summation cutoff N go to infinity, $\Delta_{\rm IF}$ approaches the continuum Pauli-Jordan function Δ (for $-L < x^0, x^1 < L$). There is a clear physical picture behind these observations. One can imagine a periodic array of sources located at the quantization hypersurface $x^0 = 0$ at points $x^1 = 2Ln$. These sources 'emit' spherical 'waves' into their own future LCs which start to overlap after time $x^0 > L$. At this point the 'waves' emanating from the sources begin to interfere. Thus, the influence of the BC is felt only after a long time (as large as the spatial extension L of the system). This picture can be confirmed analytically. An application of the Poisson resummation formula

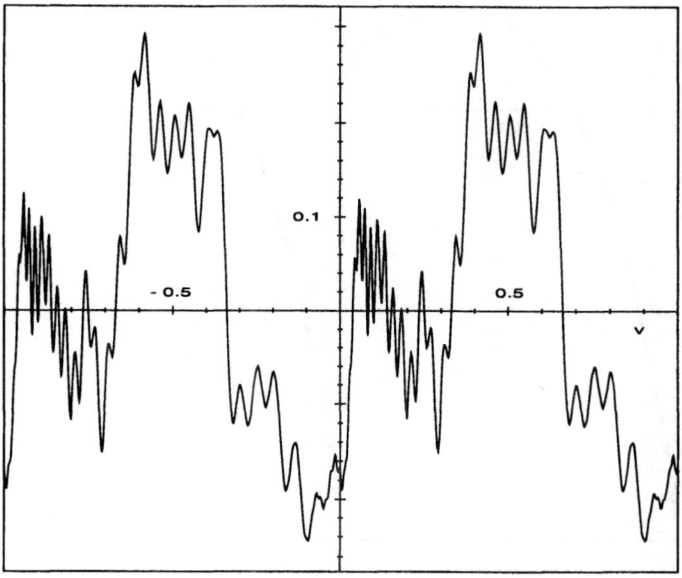

Fig. 5. $\Delta_{\mathrm{FF}}(v,w)$ as a function of $v = x^-/2L$. $w = 10000$, $N = 70$. It does not vanish outside the LC, $-1 < v < 0$.

yields

$$\Delta_{\mathrm{IF}}(x) = \sum_n \Delta(x^0, x^1 + 2Ln) , \tag{181}$$

i.e. a periodic array of Δ's which are nonoverlapping as long as $x^0 < L$.

For the front form, the situation turns out to be more complicated. Using Poisson resummation one can derive the finite-volume version of the canonical light-cone commutator, at $x^+ = 0$, which is a periodic sign function,

$$\Delta_{\mathrm{FF}}(x^+ = 0, x^-) = -\tfrac{1}{4} \sum_n \mathrm{sgn}(x^- + 2Ln) + x^-/4L . \tag{182}$$

This coincides with (174) if x^- is restricted to lie between $-L$ and L.

For $x^+ \neq 0$, I have evaluated Δ_{FF} numerically. The result is shown in Fig. 5 as a function of the dimensionless variables $v \equiv x^-/2L$, $w = m^2 L x^+/2$. For large values of w, Δ_{FF} attains a very irregular shape, though numerically the representation (179) converges to a periodic function. The most important observation, however, is that Δ_{FF} does not vanish outside the light-cone, i.e. for $x^- < 0$, if $x^+ > 0$ as in Fig. 5. This a clear *violation of microcausality* as has first been observed in [133].

As already stated, it is not straightforward to confirm these findings analytically. Poisson resummation does not work; first, because of the weak localization properties of Δ in x^- (asymptotically, is goes like $(x^-)^{-1/4}$); second, and even worse, because the zero mode $I(k^+ = 0)$ does not exist

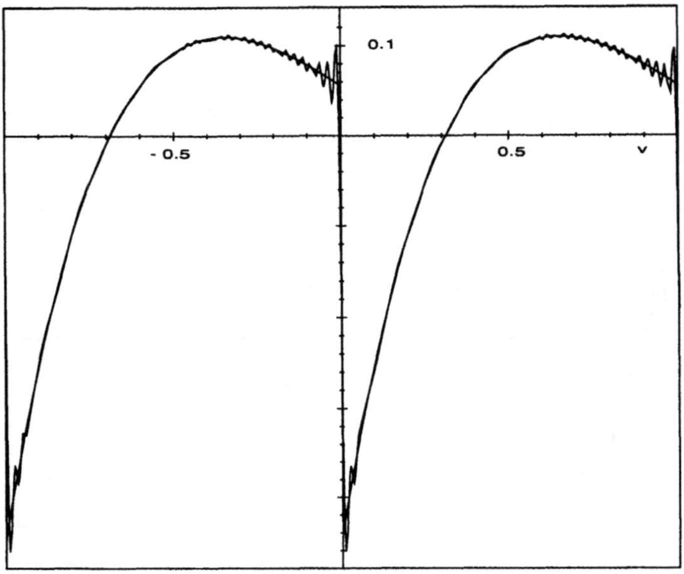

Fig. 6. Comparison of the Fourier representation (179) with the result of Bernoulli resummation (smooth, heavy line).

for $x^+ \neq 0$. Nevertheless, an independent confirmation of causality violation *can* be obtained from resumming (179) in terms of Bernoulli polynomials, thereby replacing the Fourier series by a (rapidly converging) power series in w. The result is shown in Fig. 6 for $w = 5$. There is nice agreement with the Fourier representation (179) (and no Gibbs phenomenon, as expected). Again, causality violation is obvious.

From a technical point of view, the violation of causality is not really astonishing. We are replacing the integral over the severely oscillating function $I(k^+)$ by a Riemann sum with equidistant grid points. In this way, we are sampling the integrand in such a way that the huge cancellations present in the *integral* do not take place. Instead, for small k^+, we replace $I(k^+)$ by a random 'staircase' function which in the end produces the 'noise' seen in Fig. 5.

At this point a natural questions arises: is there a remedy for the causality violation? The answer is positive. We have found two ways around the problem[3], both, however, with shortcomings of their own. The first way is to regularize the integral (177), replacing Δ by Δ_ϵ in such a way that the associated integrand satisfies $I_\epsilon(k^+ = 0) = 0$. One can chose e.g. a principal value regularization [76] or a more sophisticated prescription [110]. In this way, one suppresses the oscillations and the divergence at $k^+ = 0$ at the price of introducing a small causality violation of order ϵ. But now Δ_ϵ *can* be approximated by a discrete sum if the momentum grid is sufficiently

[3] For a third method, see [30].

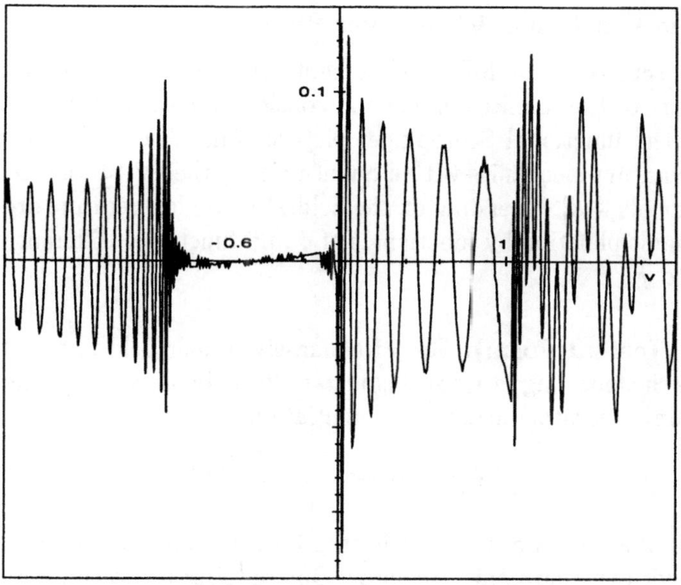

Fig. 7. The causal commutator Δ_c as a function of v. $x^+/2L = 0.2$, $mL = 50$, $N = 50$.

fine, $\triangle k^+ \ll \epsilon$. The order of limits, however, becomes important. First, one has to perform the infinite-volume limit, $\triangle k^+ \to 0$, $L \to \infty$, and only then the limit $\epsilon \to 0$. For this method, Poisson resummation should work [130]. However, it seems somewhat awkward and not very economic to perform *two* regularizations (finite L *and* ϵ).

An alternative way of resolving the problem is the following: instead of an equally spaced grid à la DLCQ (i.e. $\triangle k_n^+ = const$) one can chose an adapted momentum grid with spacing $\triangle k_n^+ \sim 1/n$ for small n. In this way, one is sampling the small-k^+-region of $I(k^+)$ in a more reasonable way. Practically, the method amounts to viewing $\Delta_{\rm IF}$ as the correct finite-volume expression and replacing x^0 and x^1 by $(x^+ \pm x^+)/2$, respectively. This is equivalent to introducing *new* discrete momenta, $k_n^\pm \equiv \omega_n \pm k_n^1$.

As a result, the point $k^+ = 0$ becomes an accumulation point of the momentum grid which leads to a causal finite-volume representation $\Delta_c(x^+, x^-)$ of Δ (see Fig. 7). This function, however, is no longer periodic in x^-. We thus find that, with a light-like direction being compactified, one cannot have both, periodicity *and* causality. On the other hand, the regularization method above seems to suggest that the causality violation is in some sense 'small' and thus may have a minor effect on the calculation of observables. Whether this statement is true has still to be worked out in detail.

3.6 The Functional Schrödinger Picture

Having discussed some difficulties of light-cone quantization in a finite 'box' we go back to the infinite volume and consider yet another *method of quantization*, the functional Schrödinger picture. The idea of this method is to mimic quantum mechanics within quantum field theory. States are described as functionals $\Psi[\phi]$ depending on the field(s) ϕ, while operators are combinations of multiplication by functionals of ϕ and functional differentiation with respect to ϕ.

Bosons (Instant Form). For a free massive scalar field in two dimensions one finds the following results using instant form dynamics [78]. The canonical momentum (operator) acts via differentiation,

$$\hat{\pi}(x)\Psi[\phi] = -i\frac{\delta}{\delta\phi(x)}\Psi[\phi] \, . \tag{183}$$

The ground state Ψ_0 of the system, the Fock vacuum, is obtained by direct analogy with the harmonic oscillator. One rewrites the vacuum annihilation condition $a(k)|0\rangle = 0$ as a functional differential equation,

$$\left[\int dy \Omega(x-y)\phi(y) + \frac{\delta}{\delta\phi(x)}\right]\Psi_0[\phi] = 0 \, , \tag{184}$$

which is solved by a Gaussian,

$$\Psi_0[\phi] \sim \exp\left[-\tfrac{1}{2}(\phi, \Omega\phi)\right] \, . \tag{185}$$

Here we have introduced the 'quadratic form',

$$(\phi, \Omega\phi) \equiv \int dx\, dy\, \phi(x)\Omega(x-y)\phi(y) \, , \tag{186}$$

using the kernel (or *covariance*) $\Omega(x-y)$. The latter is defined by its Fourier transform,

$$\Omega(k) \equiv \sqrt{k^2+m^2} \equiv \omega_k \, , \tag{187}$$

which is nothing but the on-shell energy of a free massive scalar. For later use, it is important to note that the instant form covariance Ω is explicitly mass dependent. As a consequence, if we have two free scalars of masses m_1 and m_2, respectively, their Fock vacua are related by a Bogolubov transformation,

$$|\Omega_2\rangle = U_{21}|\Omega_1\rangle \, , \tag{188}$$

with the unitary operator U_{21} explicitly given by

$$U_{21} = \exp\int\frac{dk}{4\pi}\,\theta_k[a_1(k)a_1(-k) - a_1^\dagger(k)a_1^\dagger(-k)] \, . \tag{189}$$

θ_k is the Bogolubov angle. We mention in passing that in order to properly define U_{21} as an operator one should work in finite volume to avoid infrared singularities [67]. Otherwise the two vacuum states have vanishing overlap. Within the functional Schrödinger picture this has been analysed in [78].

Fermions (Instant Form). For fermions, the situation is slightly more involved. Note that even within the instant form, the Dirac Lagrangian represents a first-order system, so that one expects some similarities with light-cone quantization in this case. This expectation will indeed turn out to be true. The instant form fermionic field operators are given by [86]

$$\hat{\psi}_\alpha = \frac{1}{\sqrt{2}}\left(u_\alpha + \frac{\delta}{\delta u_\alpha^*}\right) , \quad \hat{\psi}_\alpha^\dagger = \frac{1}{\sqrt{2}}\left(u_\alpha^* + \frac{\delta}{\delta u_\alpha}\right) , \tag{190}$$

and thus are linear combinations of multiplication by and differentiation with respect to the complex-valued Grassmann functions $u_\alpha(x)$ and $u_\alpha^*(x)$. These functions characterize the states, for example the ground state (Fock vacuum) which is again Gaussian,

$$\Psi_0[u, u^*] \sim \exp(u^*, \Omega u) . \tag{191}$$

For $2d$ massive fermions, the covariance is found to be

$$\Omega(k) = \frac{1}{\sqrt{k^2 + m^2}}(k\sigma^1 - m\sigma^3) , \tag{192}$$

with σ^1 and σ^3 the standard Pauli matrices. Again, Ω is explicitly mass dependent. In the massless case, $m = 0$, it becomes particularly simple,

$$\Omega(k) = \text{sgn}(k)\sigma^1 , \tag{193}$$

or, after Fourier transformation,

$$\Omega(x - y) = \frac{i}{\pi}\mathcal{P}\frac{1}{x - y}\sigma^1 . \tag{194}$$

Here we have once more made use of the fact that the principle value $\mathcal{P}(1/x)$ is the Fourier transform of the sign function.

Bosons (Front Form). As stated above, the latter case is somewhat similar to the generic situation in light-cone quantization. Let us again consider a massive free scalar field ϕ in $2d$ with Fock expansion (157). We decompose it into positive and negative frequency part,

$$\phi = \phi^+[a] + \phi^-[a^*] \equiv u + u^* , \tag{195}$$

where $\phi^- = (\phi^+)^*$ as ϕ is real. Quantization is performed by defining field *operators* such that the canonical light-cone commutator (122) is reproduced. The solution turns out to be somewhat more complicated than for instant form fields, namely

$$\hat{\phi}^+(x^-) = u(x^-) + \frac{1}{2}\int dy^- i\Delta_+(x^- - y^-)\frac{\delta}{\delta u^*(y^-)} , \tag{196}$$

$$\hat{\phi}^-(x^-) = u^*(x^-) + \frac{1}{2}\int dy^- i\Delta_-(x^- - y^-)\frac{\delta}{\delta u(y^-)} . \tag{197}$$

Δ_+ and Δ_- are distributions that sum up to Δ, $\Delta_+ + \Delta_- = \Delta$, and are explicitly given by

$$i\Delta_\pm(x^-) = \mp \ln(\pm x^- - i\epsilon) = \mp \ln|x^-| \pm \frac{i}{4}\theta(\mp x^-) \, . \tag{198}$$

The ground state is annihilated by $\hat{\phi}^+$, $\hat{\phi}^+\Psi_0[u,u^*] = 0$, which yields a functional differential equation, again with Gaussian solution,

$$\Psi_0[u,u^*] = \exp[-(u^*, \Omega u)] \, . \tag{199}$$

The covariance is given by

$$\Omega(x^-) = 2i\partial^+\delta(x^-) \, , \quad \Omega(k^+) = 2k^+ \, , \tag{200}$$

and thus is a *local* expression which is a very peculiar finding. In momentum space, the ubiquitous longitudinal momentum k^+ appears. One thing that is particularly obvious from (200) is the fact that the light-cone Fock vacuum is mass independent, $\Psi_0(m_1) = \Psi_0(m_2)$ which means that the analog of the Bogolubov transformation (189) is trivial, i.e. $U_{21} = \mathbb{1}$. This has been checked explicitly for several examples, including the Nambu-Jona-Lasinio model [42] and bosons and fermions coupled to external sources [67].

I believe that the locality of the covariance has far reaching consequences which, however, are still to be worked out in the present framework. Within ordinary Fock space language, some properties of the light-cone vacuum will be discussed in what follows.

3.7 The Light-Cone Vacuum

One of the basic axioms of quantum field theory states that the spectrum of the four-momentum operator is contained within the closure of the forward light-cone [143,18]. The four-momentum P^μ of any physical, that is, observable particle thus obeys

$$P^2 \geq 0 \, , \quad P^0 \geq 0 \, , \tag{201}$$

which is, of course, consistent with the mass-shell constraint, $p^2 = m^2$. The tip of the cone, the point $P^2 = P^0 = 0$, corresponds to the vacuum. From the spectrum condition (201) we infer that

$$P_0^2 - P_3^2 \geq P_\perp^2 \geq 0 \quad \text{or} \quad P^0 \geq |P^3| \, . \tag{202}$$

This implies for the longitudinal light-cone momentum,

$$P^+ = P^0 + P^3 \geq |P^3| + P^3 \geq 0 \, . \tag{203}$$

We thus have the important kinematical constraint that physical states must have nonnegative longitudinal momentum,

$$\langle \text{phys}|P^+|\text{phys}\rangle \geq 0 \, . \tag{204}$$

The spectrum of P^+ is thus bounded from below. Due to Lorentz invariance, the vacuum $|0\rangle$ must have vanishing four-momentum, and in particular

$$P^+|0\rangle = 0 \, . \tag{205}$$

Therefore, the vacuum is an eigenstate of P^+ with the lowest possible eigenvalue, namely zero. We will be interested in the phenomenon of spontaneous symmetry breaking, i.e. in the question whether – roughly speaking – the vacuum is degenerate. Let us thus analyse whether there is another state, $|p^+ = 0\rangle$, having the same eigenvalue, $p^+ = 0$, as the vacuum. If so, it must be possible to create this state from the vacuum with some operator U,

$$|p^+ = 0\rangle = U|0\rangle \, , \tag{206}$$

where U must not produce any longitudinal momentum. Note that within ordinary quantization such a construction is straightforward and quite common, for example in BCS theory. A state with vanishing three-momentum can be obtained via

$$|\mathbf{p} = 0\rangle = \int d^3k f(\mathbf{k}) a^\dagger(\mathbf{k}) a^\dagger(-\mathbf{k})|0\rangle \, , \tag{207}$$

where f is an arbitrary wave function. Evidently, the contributions from modes with positive and negative momenta cancel each other. It is obvious as well, that within light-cone quantization things must be different as there cannot be an analogous cancellation for the longitudinal momenta which are always nonnegative. Instead, one could imagine something like

$$|p^+ = \mathbf{p}_\perp = 0\rangle = \int_0^\infty dk^+ \int d^2k_\perp \, f(\mathbf{k}_\perp)\delta(k^+) a^\dagger(k^+, \mathbf{k}_\perp) a^\dagger(k^+, -\mathbf{k}_\perp)|0\rangle \, . \tag{208}$$

The problem thus boils down to the question whether there are Fock operators carrying light-cone momentum $k^+ = 0$. As we have seen in Sect. 3.4, there are no such operators, and a construction like (208) is impossible.

The only remaining possibility is that, if U contains a creation operator $a^\dagger(k^+ > 0)$ carrying longitudinal momentum $k^+ \neq 0$, there must be annihilators that annihilate exactly the same amount k^+ of momentum. Thus, after Wick ordering, U must have the general form

$$\begin{aligned} U = \langle 0|U|0\rangle &+ \int\limits_{k^+>0} dk^+ \, f_2(k^+) a^\dagger(k^+) a(k^+) \\ &+ \int\limits_{p^+>0} dp^+ \int\limits_{k^+>0} dk^+ \, f_3(k^+, p^+) a^\dagger(p^+ + k^+) a(p^+) a(k^+) \\ &+ \int\limits_{p^+>0} dp^+ \int\limits_{k^+>0} dk^+ \, \tilde{f}_3(k^+, p^+) a^\dagger(p^+) a^\dagger(k^+) a(k^+ + p^+) \\ &+ \ldots \, . \end{aligned} \tag{209}$$

It follows that the light-cone vacuum $|0\rangle$ is an eigenstate of U,

$$U|0\rangle = \langle 0|U|0\rangle|0\rangle \ . \tag{210}$$

As we only deal with rays in Hilbert space, the action of U on the vacuum does not create a state distinct from the vacuum. One says that the vacuum is *trivial.* Put differently, 'there is no vacuum but the Fock vacuum'. Note that this is actually consistent with our findings in the last subsection: the light-cone vacuum is the same irrespective of the masses of the particles; the Bogolubov transformation present in the instant form becomes trivial.

Let us analyse the dynamical implications of the general result (210). Any quantity that is obtained by integrating some functional of the fields over x^-, i.e.,

$$F[\phi] = \int dx^- \mathcal{F}[\phi] \ , \tag{211}$$

is of the form (209), because the integration can be viewed as a projection onto the longitudinal momentum sector $k^+ = 0$. The most important examples for such quantities are the Poincaré generators, as is obvious from the representations (107, 108). This implies in particular that the trivial light-cone vacuum is an eigenstate of the fully interacting light-cone Hamiltonian P^-,

$$P^-|0\rangle = \langle 0|P^-|0\rangle|0\rangle \ . \tag{212}$$

This can be seen alternatively by considering

$$P^+P^-|0\rangle = P^-P^+|0\rangle = 0 \ , \tag{213}$$

which says that $P^-|0\rangle$ is a state with $k^+ = 0$, so that P^- must have a Fock representation like U in (209).

The actual value of $\langle 0|P^-|0\rangle$ is not important at this point as it only defines the zero of light-cone energy. Note that, within the instant form, the Fock or trivial vacuum is *not* an eigenstate of the full Hamiltonian as the latter usually contains terms with only creation operators where positive and negative three-momenta compensate to zero as in (207). The instant-form vacuum thus is unstable under time evolution. Such a vacuum, a typical example of which is provided by (207), is called 'nontrivial'.

This concludes the general discussion of light-cone quantum field theory. Having understood the foundations of the approach we are now heading for the applications. We begin with a survey of the light-cone Schrödinger equation.

4 Light-Cone Wave Functions

In this section we collect some basic facts about the eigenvalue problem of the light-cone Hamiltonian, or, in other words, about the light-cone Schrödinger equation and its solutions, the light-cone wave functions. Throughout this section, I will use the conventions of [22].

4.1 Kinematics

To set the stage for the definition of light-cone wave functions let me first introduce some relevant kinematical variables. Consider a system of many particles which, for the time being, will be assumed as noninteracting. Let the ith particle have mass m_i and light-cone four-momentum

$$p_i = (p_i^+, \mathbf{p}_{\perp i}, p_i^-) \, . \tag{214}$$

As the particles are free, the total four-momentum is conserved and thus given by the sum of the individual momenta,

$$P = \sum_i p_i \, . \tag{215}$$

The individual particles are on-shell, so their four-momentum squared is

$$p_i^2 = p_i^+ p_i^- - p_{\perp i}^2 = m_i^2 \, . \tag{216}$$

The square of the total four-momentum, on the other hand, defines the *free invariant mass squared*,

$$P^2 = P^+ P^- - P_\perp^2 \equiv M_0^2 \, , \tag{217}$$

a quantity that will become important later on. We introduce *relative* momentum coordinates x_i and $\mathbf{k}_{\perp i}$ via

$$p_i^+ \equiv x_i P^+ \, , \tag{218}$$

$$\mathbf{p}_{\perp i} \equiv x_i \mathbf{P}_\perp + \mathbf{k}_{\perp i} \, . \tag{219}$$

Thus, x_i and $\mathbf{k}_{\perp i}$ denote the longitudinal momentum fraction and the relative transverse momentum of the ith particle, respectively. Comparing with (215) we note that these variables have to obey the constraints

$$\sum_i x_i = 1 \, , \quad \sum_i \mathbf{k}_{\perp i} = 0 \, . \tag{220}$$

A particularly important property of the relative momenta is their boost invariance. To show this we calculate, using (86),

$$x_i' = e^\omega p_i^+ / e^\omega P^+ = x_i \, . \tag{221}$$

From this and (88) we find in addition

$$\mathbf{k}_{\perp i}' = \mathbf{p}_{\perp i}' - x_i \mathbf{P}_\perp' = \mathbf{p}_{\perp i} + \mathbf{v}_\perp p_i^+ - x_i(\mathbf{P}_\perp + \mathbf{v}_\perp P^+) = \mathbf{k}_{\perp i} \, , \tag{222}$$

which indeed proves the frame independence of x_i and $\mathbf{k}_{\perp i}$.

Let us calculate the total light-cone energy of the system in terms of the relative coordinates. Making use of the constraints (220), we obtain

$$\begin{aligned} P^- &= \sum_i p_i^- = \sum_i \frac{p_{\perp i}^2 + m_i^2}{p_i^+} = \sum_i \frac{(x_i \mathbf{P}_\perp + \mathbf{k}_{\perp i})^2 + m_i^2}{x_i P^+} \\ &= \frac{1}{P^+}\left(P_\perp^2 + \sum_i \frac{k_{\perp i}^2 + m_i^2}{x_i}\right) \equiv P_{\mathrm{CM}}^- + P_{\mathrm{r}}^- \,. \end{aligned} \tag{223}$$

This is another important result: the light-cone Hamiltonian P^- separates into a center-of-mass term,

$$P_{\mathrm{CM}}^- = P_\perp^2 / P^+ \,, \tag{224}$$

and a term containing only the relative coordinates,

$$P_{\mathrm{r}}^- = \frac{1}{P^+}\left(\sum_i \frac{k_{\perp i}^2 + m_i^2}{x_i}\right) = \frac{M_0^2}{P^+} \,. \tag{225}$$

The second identity, which states that P_{r}^- is essentially the free invariant mass squared, follows upon multiplying (223) by P^+,

$$P^+ P_{\mathrm{r}}^- = P^+ P^- - P_\perp^2 = M_0^2 = \sum_i \frac{k_{\perp i}^2 + m_i^2}{x_i} \,. \tag{226}$$

To simplify things even more, one often goes to the 'transverse rest frame' where $\mathbf{P}_\perp$ and therefore the center-of-mass Hamiltonian P_{CM}^- from (224) vanish.

In the interacting case, the dynamical Poincaré generators acquire 'potential' terms as I have shown in Sect. 3.1. The light-cone Hamiltonian, e.g. becomes $P^- = P_0^- + V$, leading to a four-momentum squared

$$P^2 = P^+(P_0^- + V) - P_\perp^2 \equiv M^2 \,. \tag{227}$$

Subtracting (217) we obtain the useful relation,

$$M^2 - M_0^2 = P^+ V \equiv W \,. \tag{228}$$

In the quantum theory, this operator identity, when applied to physical states, is nothing but the light-cone Schrödinger equation.

Summarizing we note that the special behavior under boosts together with the transverse Galilei invariance leads to frame independent relative coordinates and a separation of the center-of-mass motion, reminiscent of ordinary nonrelativistic physics. This is at variance with the instant form, where the appearance of the notorious square root in the energy, $P^0 = (\mathbf{P}^2 + M_0^2)^{1/2}$, prohibits a similar separation of variables.

4.2 Definition of Light-Cone Wave Functions

Let us first stick to a discrete notation and, for the time being, stay in 1+1 dimensions. We thus have a Fock basis of states

$$\begin{aligned} |n\rangle &= a_n^\dagger |0\rangle \,, \\ |m,n\rangle &= a_m^\dagger a_n^\dagger |0\rangle \,, \\ &\vdots \\ |n_1,\ldots,n_N\rangle &= a_{n_1}^\dagger \ldots a_{n_N}^\dagger |0\rangle \,. \end{aligned} \tag{229}$$

This leads to a completeness relation defining the unit operator in Fock space,

$$\begin{aligned} \mathbb{1} &= |0\rangle\langle 0| + \sum_{n>0} |n\rangle\langle n| + \tfrac{1}{2} \sum_{m,n>0} |m,n\rangle\langle m,n| + \ldots \\ &= |0\rangle\langle 0| + \sum_{N=1}^{\infty} \frac{1}{N!} \sum_{n_1,\ldots,n_N>0} |n_1,\ldots,n_N\rangle\langle n_1,\ldots,n_N| \,. \end{aligned} \tag{230}$$

An arbitrary state $|\psi\rangle$ can thus be expanded as

$$|\psi\rangle = \sum_{n>0} \langle n|\psi\rangle |n\rangle + \tfrac{1}{2} \sum_{m,n>0} \langle m,n|\psi\rangle |m,n\rangle + \ldots \,. \tag{231}$$

The sums are such that the longitudinal momenta in each Fock sector add up to the total longitudinal momentum of $|\psi\rangle$. Note that the vacuum state does not contribute as it is orthogonal to any particle state, $\langle 0|\psi\rangle = 0$. The normalization of this state is obtained as

$$\begin{aligned} \langle\psi|\psi\rangle &= \sum_{n>0} |\langle n|\psi\rangle|^2 + \tfrac{1}{2} \sum_{m,n>0} |\langle m,n|\psi\rangle|^2 + \ldots \\ &= \sum_{N=1}^{\infty} \frac{1}{N!} \sum_{n_1,\ldots,n_N>0} |\langle n_1,\ldots,n_N|\psi\rangle|^2 \,. \end{aligned} \tag{232}$$

Let us assume that the state $|\psi\rangle$ corresponds to a bound state obeying the light-cone Schrödinger equation derived from (228),

$$(M^2 - \hat{M}_0^2)|\psi\rangle = \hat{W}|\psi\rangle \,. \tag{233}$$

We want to project this equation onto the different Fock sectors. For this we need the eigenvalues of the free invariant mass squared when applied to an N-particle state

$$|N\rangle \equiv |n_1,\ldots,n_N\rangle \,. \tag{234}$$

We find that $|N\rangle$ is an eigenstate of $\hat{M}_0^2$,

$$\hat{M}_0^2|N\rangle = \sum_{i=1}^{N} \frac{m_i^2}{x_i} |N\rangle \equiv M_N^2 |N\rangle \,. \tag{235}$$

The light-cone Schrödinger equation thus becomes a system of coupled eigenvalue equations,

$$\begin{bmatrix} (M^2 - M_1^2)\, \langle l|\psi\rangle \\ (M^2 - M_2^2)\, \langle kl|\psi\rangle \\ \vdots \end{bmatrix} = \begin{bmatrix} \langle l|W|m\rangle & \langle l|W|mn\rangle & \ldots \\ \langle kl|W|m\rangle & \langle kl|W|mn\rangle & \ldots \\ \vdots & \vdots & \ddots \end{bmatrix} \begin{bmatrix} \langle l|\psi\rangle \\ \langle kl|\psi\rangle \\ \vdots \end{bmatrix} . \tag{236}$$

Clearly, this represents an infinite number of equations which in general will prove impossible to solve unless the matrix is very sparse and/or the matrix elements are small. The former condition is usually fulfilled as the interaction W in the light-cone Hamiltonian typically changes particle number at most by two[4]. Assuming the matrix elements to be small amounts to dealing with a perturbative situation. This will be true for nonrelativistic bound states of heavy constituents, but not for light hadrons which we are mainly interested in. There are, however, situations where the magnitude of the amplitudes decreases enormously with the particle number N, so that it is a good approximation to restrict to the lowest Fock sectors. In instant form field theory this has long been known as the Tamm-Dancoff method [148,39].

Let us turn to the more realistic case of 3+1 dimensions in a continuum formulation. The invariant normalization of a momentum eigenstate $|P^+, \mathbf{P}_\perp\rangle \equiv |\boldsymbol{P}\rangle$ is given by

$$\langle \boldsymbol{P}|\boldsymbol{K}\rangle = 16\pi^3 P^+ \delta^3(\boldsymbol{P} - \boldsymbol{K}) . \tag{237}$$

We already know that the bare Fock vacuum is an eigenstate of the interacting Hamiltonian. It thus serves as an appropriate ground state on top of which we can build a reasonable Fock expansion. If we specialize immediately to the case of QCD, we are left with the Fock basis states

$$\begin{aligned} &|0\rangle , \\ &|q\bar{q} : \boldsymbol{k}_i, \alpha_i\rangle = b^\dagger(\boldsymbol{k}_1, \alpha_1) d^\dagger(\boldsymbol{k}_2, \alpha_2)|0\rangle , \qquad (238) \\ &|q\bar{q}g : \boldsymbol{k}_i, \alpha_i\rangle = b^\dagger(\boldsymbol{k}_1, \alpha_1) d^\dagger(\boldsymbol{k}_2, \alpha_2) a^\dagger(\boldsymbol{k}_3, \alpha_3)|0\rangle , \\ &\vdots \qquad (239) \end{aligned}$$

In these expressions, $b^\dagger$, $d^\dagger$ and $a^\dagger$ create quarks q, antiquarks $\bar{q}$ and gluons g with momenta $\boldsymbol{k}_i$ from the trivial vacuum $|0\rangle$. The α_i denote all other relevant quantum numbers, like helicity, polarization, flavor and color.

[4] Note that terms with only creation operators are forbidden by k^+-conservation. Still, in 1+1 dimensions, things can become messy as interactions with polynomials of arbitrary powers in ϕ are allowed without spoiling renormalizability [159]

In a more condensed notation we can thus describe, say, a pion with momentum $\boldsymbol{P} = (P^+, \mathbf{P}_\perp)$, as

$$|\pi(\boldsymbol{P})\rangle = \sum_{n,\lambda_i} \int \overline{\prod_i} dx_i \frac{d^2 k_{\perp i}}{16\pi^3} \psi_{n/\pi}(x_i, \mathbf{k}_{\perp i}, \lambda_i) \Big| n : x_i P^+, x_i \mathbf{P}_\perp + \mathbf{k}_{\perp i}, \lambda_i \Big\rangle \,, \tag{240}$$

where we have suppressed all discrete quantum numbers apart from the helicities λ_i. The integration measure takes care of the constraints (220) which the relative momenta in each Fock state (labeled by n) have to obey,

$$\overline{\prod_i} dx_i \equiv \prod_i dx_i \, \delta\Big(1 - \sum_j x_j\Big) \,, \tag{241}$$

$$\overline{\prod_i} d^2 k_{\perp i} \equiv 16\pi^3 \prod_i d^2 k_{\perp i} \, \delta^2\Big(\sum_j \mathbf{k}_{\perp j}\Big) \,. \tag{242}$$

As a mnemonic rule, we note that any measure factor $d^2 k_{\perp i}$ in (240) is always accompanied by $1/16\pi^3$.

The most important quantities in (240) are the *light-cone wave functions*

$$\psi_{n/\pi}(x_i, \mathbf{k}_{\perp i}, \lambda_i) \equiv \langle n : x_i P^+, x_i \mathbf{P}_\perp + \mathbf{k}_{\perp i}, \lambda_i | \pi(\boldsymbol{P}) \rangle \,, \tag{243}$$

which are the amplitudes to find n constituents with relative momenta $p_i^+ = x_i P^+$, $\mathbf{p}_{\perp i} = x_i \mathbf{P}_\perp + \mathbf{k}_{\perp i}$ and helicities λ_i in the pion. Due to the separation properties of the light-cone Hamiltonian the wave functions do not depend on the total momentum $\boldsymbol{P}$ of the pion. Applying (237) to the pion state (240), we obtain the normalization condition

$$\sum_{n,\lambda_i} \int \overline{\prod_i} dx_i \frac{d^2 k_{\perp i}}{16\pi^3} |\psi_{n/\pi}(x_i, \mathbf{k}_{\perp i}, \lambda_i)|^2 = 1 \,. \tag{244}$$

The light-cone bound-state equation for the pion is a straightforward generalization of (236),

$$\begin{bmatrix} (M^2 - M^2_{q\bar{q}}) \, \langle q\bar{q}|\pi\rangle \\ (M^2 - M^2_{q\bar{q}g}) \, \langle q\bar{q}g|\pi\rangle \\ \vdots \end{bmatrix} = \begin{bmatrix} \langle q\bar{q}|W|q\bar{q}\rangle & \langle q\bar{q}|W|q\bar{q}g\rangle & \dots \\ \langle q\bar{q}g|W|q\bar{q}\rangle & \langle q\bar{q}g|W|q\bar{q}g\rangle & \dots \\ \vdots & \vdots & \ddots \end{bmatrix} \begin{bmatrix} \langle q\bar{q}|\pi\rangle \\ \langle q\bar{q}g|\pi\rangle \\ \vdots \end{bmatrix} . \tag{245}$$

If a constituent picture for the pion were true, the valence state would dominate,

$$|\psi_{2/\pi}|^2 \gg |\psi_{n/\pi}|^2 \,, \quad n > 2 \,, \tag{246}$$

and, in the extreme case, the pion wave function would be entirely given by the projection $\langle q\bar{q}|\pi\rangle$ onto the valence state. All the higher Fock contributions would vanish and the unitarity sum (244) would simply reduce to

$$\sum_{\lambda\lambda'} \int_0^1 dx \int \frac{d^2 k_\perp}{16\pi^3} |\psi_{q\bar{q}/\pi}(x, \mathbf{k}_\perp, \lambda, \lambda')|^2 = 1 \,. \tag{247}$$

We will later discuss a model where this is indeed a good approximation to reality.

4.3 Properties of Light-Cone Wave Functions

Let us rewrite the light-cone bound-state equation (236) by collecting all light-cone wave functions $\psi_n = \langle n|\psi\rangle$ into a vector Ψ,

$$\Psi = \frac{W\Psi}{M^2 - M_0^2} \,. \tag{248}$$

From this expression it is obvious that all light-cone wave functions tend to vanish whenever the denominator

$$\epsilon \equiv M^2 - M_0^2 = M^2 - \Big(\sum_i p_i\Big)^2 = M^2 - \sum_i \frac{k_{\perp i}^2 + m_i^2}{x_i} \tag{249}$$

becomes very large. This quantity measures how far off energy shell the total system, i.e. the bound state is,

$$P^- - \sum_i p_i^- = \epsilon / P^+ \,. \tag{250}$$

For this reason, ϵ is sometimes called the 'off-shellness' [112,89]. We thus learn from (248) that there is only a small overlap of the bound state with Fock states that are far off shell. This implies the limiting behavior

$$\psi(x_i, \mathbf{k}_{\perp i}, \lambda_i) \to 0 \quad \text{for } x_i \to 0 \,,\; k_{\perp i}^2 \to \infty \,. \tag{251}$$

These boundary conditions are related to the self-adjointness of the light-cone Hamiltonian and to the finiteness of its matrix elements. Analogous criteria have been used recently to relate wave functions of different Fock states n [3] and to analyse the divergence structure of light-cone perturbation theory [25].

Omitting spin, flavor and color degrees of freedom, a light-cone wave function will be a scalar function $\phi(x_i, \mathbf{k}_{\perp i})$ of the parameter ϵ. This is used for building models, the most common one being to assume a Gaussian behavior, originally suggested in [149],

$$\phi(x_i, \mathbf{k}_{\perp i}) = N \exp(-|\epsilon|/\beta^2) \,, \tag{252}$$

where β measures the size of the wave function in momentum space. Note, however, that a Gaussian ansatz is in conflict with perturbation theory which is the appropriate tool to study the high-$\mathbf{k}_\perp$ behavior and indicates a power decay of the *renormalized* wave function – up to possible logarithms [22]. For the unrenormalized wave functions the boundary conditions (251) are violated unless one uses a cutoff as a regulator (see next section).

As the off-shellness ϵ is the most important quantity characterizing a light-cone wave function let us have a closer look by specializing to the simplest possible system, namely two bound particles of equal mass. One can think of this, for instance, as the valence wave function of the pion or positronium. The off-shellness becomes

$$\epsilon = M^2 - \frac{k_\perp^2 + m^2}{x(1-x)} = -\frac{1}{x(1-x)}\Big[M^2\big(x-\tfrac{1}{2}\big)^2 + \underbrace{\frac{4m^2 - M^2}{4}}_{\geq 0} + k_\perp^2\Big] . \quad (253)$$

The second term in square brackets is positive because, for a bound state, the binding energy,

$$E = M - 2m , \quad (254)$$

is negative so that $2m > M$. As a result, the off-shellness is always negative. Only for free particles it is zero, because *all* momentum components (including the energy) sum up to the total momentum. In this case, each individual term in (253) vanishes,

$$M = 2m , \quad x = \tfrac{1}{2} , \quad \mathbf{k}_\perp = 0 . \quad (255)$$

It follows that the light-cone wave function of a two-particle system (composed of equal-mass constituents) with the binding energy switched off 'adiabatically', is of the form

$$\phi(x, \mathbf{k}_\perp) \sim \delta(x - 1/2)\, \delta^2(\mathbf{k}_\perp) . \quad (256)$$

4.4 Examples of Light-Cone Wave Functions

From the discussion above, one expects that for weak binding, in particular for nonrelativistic systems, the wave functions will be highly peaked around $x = 1/2$ (in the equal mass case) and $\mathbf{k}_\perp = 0$. Let us check this explicitly for hydrogen-like systems which constitute our first example [93].

Example 1: Hydrogen-Like Systems. Let me recall the ordinary Schrödinger equation of the Coulomb problem written in momentum space,

$$\left(E - \frac{\mathbf{p}^2}{2m}\right)\psi(\mathbf{p}) = \int \frac{d^3k}{(2\pi)^3} V(\mathbf{p}-\mathbf{k})\psi(\mathbf{k}) = \int \frac{d^3k}{(2\pi)^3} \frac{4\pi\alpha}{(\mathbf{p}-\mathbf{k})^2}\psi(\mathbf{k}) . \quad (257)$$

This integral equation looks very similar to a light-cone Schrödinger equation within a two-particle truncation. The Coulomb kernel is due to the exchange of an instantaneous photon having a propagator proportional to $\delta(x^0)$. One can actually solve the Coulomb problem directly in momentum space [50,14] but for our purposes it is simpler just to Fourier transform the ground state wave function $\psi_0(r) = N \exp(-m\alpha r)$, yielding

$$\psi_0(\mathbf{p}) = 8\pi N \frac{m\alpha}{(\mathbf{p}^2 + m^2\alpha^2)^2} , \quad (258)$$

with $\alpha = e^2/4\pi = 1/137$ being the fine structure constant, m the reduced mass and $\mathbf{p}$ the relative momentum.

How does this translate into the light-cone language? To answer this question, we go to the particle rest frame with $\mathbf{P} = 0$ or $P^+ = P^- = M$ and $\mathbf{P}_\perp = 0$, implying $\mathbf{p}_{\perp i} = \mathbf{k}_{\perp i}$. In this frame, the nonrelativistic limit is defined by the following inequalities for the constituent masses and momenta (in ordinary instant-form coordinates),

$$p_i^0 - m_i \simeq \frac{\mathbf{p}_i^2}{2m_i} \ll |\mathbf{p}_i| \ll m_i \,. \tag{259}$$

The prototype systems in this class are of hydrogen type where we have for binding energy and r.m.s. momentum,

$$|E| = \frac{\langle \mathbf{p}^2 \rangle}{2m} = \frac{m\alpha^2}{2} \,, \tag{260}$$

$$\langle p \rangle = m\alpha \,. \tag{261}$$

In this case, the hierarchy (259) becomes

$$\frac{\alpha^2}{2} \ll \alpha \ll 1 \,, \tag{262}$$

which is fulfilled to a very good extent in view of the smallness of α.

Consider now the longitudinal momentum of the ith constituent,

$$p_i^+ = p_i^0 + p_i^3 \simeq m_i + \frac{\mathbf{p}_i^2}{2m_i} + p_i^3 = x_i P^+ = x_i M \,. \tag{263}$$

We thus find that we should replace p_i^3 in instant-form nonrelativistic wave functions by

$$p_i^3 = x_i M - m_i \,, \tag{264}$$

where we neglect terms of order $\mathbf{p}_i^2/m_i$. Let us analyze the consequences for the off-shellness. The latter is in ordinary coordinates

$$\epsilon = M^2 - M_0^2 = (M + \sum_i p_i^0)(M - \sum_i p_i^0) \,. \tag{265}$$

We thus need

$$\sum_i p_i^0 \simeq \sum_i m_i + \sum_i \frac{\mathbf{p}_i^2}{2m_i} = M - E + \sum_i \frac{\mathbf{p}_i^2}{2m_i} \,, \tag{266}$$

with $E = M - \sum m_i$ denoting the mass-defect, which is a small quantity, $E \ll M$. The off-shellness (265), therefore, becomes

$$\epsilon \simeq 2M\Big(E - \sum_i \mathbf{p}_i^2/2m_i\Big) \simeq -2M \sum_i \frac{k_{\perp i}^2 + (Mx_i - m_i)^2}{2m_i} \,, \tag{267}$$

where we have performed the replacement (264) in the second identity. The light-cone wave functions will be peaked where the off-shellness is small, that is, for

$$x_i = m_i/M\,, \quad \text{and} \quad \mathbf{k}_{\perp i} = 0\,, \tag{268}$$

as expected from the noninteracting case.

To be explicit, we consider the ground state wave function of positronium, given by (258) with the reduced mass m being half the electron mass m_e. Using the replacement prescription (264) once more, we obtain

$$\psi(x, \mathbf{k}_\perp) = 8\pi N \frac{m\alpha}{\left[k_\perp^2 + (xM - m_e)^2 + m^2\alpha^2\right]^2}\,, \tag{269}$$

where $M \simeq 2m_e$ is the bound state mass. This result is valid for small momenta, i.e. when $k_\perp^2, (xM - m_e)^2 \ll m_e^2$. It is obvious from (269) that the positronium wave function is sharply peaked around $x = m_e/M \simeq 1/2$ and $k_\perp^2 = 0$.

Example 2: 't Hooft Model. The 't Hooft model [146,147] is QCD in two space-time dimensions with the number N_C of colors being infinite. The Lagrangian is

$$\mathcal{L} = \bar{\psi}(i\partial\!\!\!/ - m)\psi - \frac{1}{4}F_{\mu\nu}F^{\mu\nu}\,. \tag{270}$$

The limit of large N_C is taken in such a way that the expression $g^2 N_C$, g denoting the Yang-Mills coupling, stays finite. In two dimensions, g has mass dimension one, which renders the theory superrenormalizable and provides a basic unit of mass, namely,

$$\mu_0 \equiv \sqrt{g^2 N_C/2\pi}\,. \tag{271}$$

The model is interesting because it contains physics analogous or similar to what one finds in 'real' QCD[5]. First of all, the model is (trivially) confining due to the linear rise of the Coulomb potential in $2d$. This is most easily exhibited by working in light-cone gauge, $A^+ = 0$, and eliminating A^- via Gauss's law. In this way it becomes manifest that there are no dynamical gluons in $2d$. Within covariant perturbation theory, the Coulomb potential can be understood as resulting from the exchange of an instantaneous gluon.

In the next section, we will discuss the spontaneous breakdown of chiral symmetry in QCD. It turns out that in the 't Hooft model a similar phenomenon occurs: chiral symmetry is 'almost' spontaneously broken [154,158]. As a consequence, there arises a *massless* bound state in the chiral limit of vanishing quark mass [146,147] which we will call the 'pion' for brevity[6]. Furthermore, there is a nonvanishing quark condensate in the model which has

[5] For a recent review on the 't Hooft model, see [1].

[6] As explained in [154,158], this is not in contradiction with Coleman's theorem [37] as the 'pion' is not a Goldstone boson.

first been calculated by Zhitnitsky [158],

$$\langle 0|\bar{\psi}\psi|0\rangle / N_{\mathrm{C}} = -0.28868 \,. \tag{272}$$

Note that the condensate is proportional to N_{C} as it involves a color trace.

It turns out that the 't Hooft model has one big advantage for the application of light-cone techniques which is due to the large-N_{C} limit. The matrix elements entering the light-cone Schrödinger equation in the two-particle sector have the following N_{C}-dependence,

$$\langle 2|W|2n\rangle \sim \left(\frac{g^2 N_{\mathrm{C}}}{N_{\mathrm{C}}}\right)^n \sim N_{\mathrm{C}}^{-n} \,. \tag{273}$$

As a result, those diagrams which correspond to a change in particle number (like $2 \to 4$, $2 \to 6, \ldots$) are suppressed by powers of $1/N_{\mathrm{C}}$. The truncation to the two-particle sector therefore becomes exact: the 'pion' is a pure quark-antiquark state; there are no admixtures of higher Fock states. A constituent picture is thus realized, and we are left with the light-cone Schrödinger equation,

$$\left[M^2 - \frac{m^2}{x(1-x)}\right]\phi(x) = \mathcal{P}\int_0^1 dy\, \frac{\phi(x)-\phi(y)}{(x-y)^2} \,. \tag{274}$$

This expression defines the 'Coulomb problem' of the 't Hooft model. It corresponds to the first line of (245) where, as a result of the truncation, only the matrix element $W_2 \equiv \langle q\bar{q}|W|q\bar{q}\rangle$ has been retained. We will refer to (274) as the 't Hooft equation in what follows. $\phi(x)$ denotes the valence part of the 'pion' wave function, x and y are the momentum fractions of the two quarks (with equal mass m) in the meson. The symbol $\mathcal{P}$ indicates that the integral is defined as a principal value [146,53]. It regularizes[7] the Coulomb singularity $1/(x-y)^2$ in the matrix element W_2. All masses are expressed as multiples of the basic scale μ_0 defined in (271). The eigenvalue M denotes the mass of the lowest lying bound state (the 'pion'). Our objective is to calculate M and ϕ.

In his original work on the subject, [146,147], 't Hooft used the following ansatz for the wave function

$$\phi(x) = x^\beta (1-x)^\beta \,. \tag{275}$$

This ansatz is symmetric in $x \leftrightarrow 1-x$ (charge conjugation odd), and β is supposed to lie between zero and one so that the endpoint behavior is nonanalytic. As a nontrivial boundary condition, one has the exact solution of the massless case, $m = 0$ (the 'chiral limit'),

$$M^2 = 0 \,, \quad \text{and} \quad \phi(x) = 1 \,, \quad i.e. \quad \beta = 0 \,. \tag{276}$$

[7] [156] has suggested a theoretical alternative to the principal value which nowadays is called 'Leibbrandt-Mandelstam prescription [102,90]. It leads to completely different physics. This apparent contradiction has only recently been clarified [9].

In this limit, the 'pion' wave function is constant, i.e. the pion has no structure and is point-like. The main effect of a nonzero quark mass is the vanishing of the wave functions at the endpoints implying a nonzero β. This suggests the following series expansion for β,

$$\beta(m) = \beta_1 m + \beta_2 m^2 + \beta_3 m^3 + O(m^4) , \tag{277}$$

and for the 'pion' mass squared,

$$M^2 = M_1 m + M_2 m^2 + M_3 m^3 + O(m^4) . \tag{278}$$

As is obvious from the last two expressions, we are working in the limit of small quark mass m. The expansion (277) shows that also β is small in this case so that the wave function will be rather flat (for intermediate values of x). On the other hand we know from Example 1 that light-cone wave functions are highly peaked near $x = 1/2$ in case the binding is weak. This suggests that, for small quark mass m, the 'pion' is rather strongly bound.

The exponent β in (275) can actually be determined exactly by studying the small-x behavior of the bound state equation (274). To this end we evaluate the principal value integral for $x \to 0$ and plug it into (274). This yields the transcendental equation [146],

$$m^2 - 1 + \pi\beta \cot \pi\beta = 0 . \tag{279}$$

Using this expression we can determine β either numerically for arbitrary m or analytically for small m, which yields the coefficients of (277),

$$\beta = \frac{\sqrt{3}}{\pi} m + O(m^3) . \tag{280}$$

The 'pion' mass is determined by calculating the expectation value of the light-cone Hamiltonian (274) in the state given by 't Hooft's ansatz (275)[8]. This yields M^2 as a function of β and m,

$$M^2 = \frac{2}{\beta_1} m + O(m^2) . \tag{281}$$

Upon comparing with (277) and (280) the lowest order coefficient M_1, which is the slope of $M^2(m)$ at $m = 0$, is found to be

$$M_1 = 2\pi/\sqrt{3} . \tag{282}$$

Note that M^2 indeed vanishes in the chiral limit.

Having obtained an approximate solution for the mass and wave function of the 'pion' we are in the position to calculate 'observables'. It turns out that,

[8] The relevant integrals can be found in [7] and [63].

for small m, all of them can be expressed in terms of the lowest order coefficient, M_1. We begin with the 'pion decay constant' [26,158], which is given by the 'wave function at the origin', i.e. the integral over the (momentum space) wave function

$$f_\pi \equiv \langle 0|\bar{\psi} i\gamma_5 \psi|\pi\rangle = \sqrt{\frac{N_C}{\pi}\frac{M^2}{2m}} \int dx\, \phi(x) = \sqrt{\frac{N_C}{4\pi}} M_1 \, . \tag{283}$$

The quark condensate is obtained via a sum rule using the chiral Ward identity [158,66],

$$\langle 0|\bar{\psi}\psi|0\rangle = -m\frac{f_\pi^2}{M^2} = -\frac{N_C}{4\pi} M_1 \, . \tag{284}$$

Inserting the value (282) for M_1 this coincides with (272). The last identity can actually be viewed as the 'Gell-Mann-Oakes-Renner relation' [54] of the 't Hooft model,

$$M^2 = -4\pi\, m\, \langle 0|\bar{\psi}\psi|0\rangle / N_C + O(m^2) \, , \tag{285}$$

which provides a relation between the particle spectrum (the 'pion' mass) and a *ground state* property (the condensate). This is conceptually important because it implies that we can circumvent the explicit construction of a nontrivial vacuum state by calculating the spectrum of excited states, i.e. by solving the light-cone Schrödinger equation. The eigenvalues and wave functions actually contain information about the structure of the vacuum! This point of view has been adopted long ago in the context of chiral symmetry breaking within the (light-cone) parton model [28]: "In this framework the spontaneous symmetry breakdown must be attributed to the properties of the hadron's wave function and not to the vacuum" [29]. Related ideas have been put forward more recently in [91].

In the above, we have been using the value for β given in (280). One can equally well use β as a variational parameter and minimize the expectation value of the mass operator with respect to it. This yields the same result for M_1, namely (282). The variational method, however, is better suited if one wants to go beyond the leading order in expansion (278). This has been done in [62]. The results are shown in Table 2 where we list the expansion coefficients M_i of the 'pion' mass squared, M^2, as they change with increasing number of variational parameters, (a, b, c, d).

Interestingly, the value of M_1 does not change at all by enlarging the space of trial functions. M_2 and M_3, on the other hand, do change and show rather good convergence. For M_2 we finally have a relative accuracy of $8 \cdot 10^{-7}$, and for M_3 of $4 \cdot 10^{-5}$. Furthermore, the coefficients are getting *smaller* if one adds more basis functions, in accordance with the variational principle. The associated light-cone wave functions are shown in Fig. 8. There are only minor changes upon including more variational parameters. In Fig. 9 we display the eight lowest excited states. They have been obtained using the position of the nodes as additional variational parameters [142].

Table 2. *Expansion coefficients of M^2 for the 't Hooft model obtained by successively enlarging the space of variational parameters. M_1 is the standard 't Hooft result (282). Note the good convergence towards the bottom of the table.*

Ansatz	$M_1 = 2\pi/\sqrt{3}$	M_2	M_3
't Hooft	3.62759873	3.61542218	0.043597197
a	3.62759873	3.58136872	0.061736701
b	3.62759873	3.58107780	0.061805257
c	3.62759873	3.58105821	0.061795547
d	3.62759873	3.58105532	0.061793082

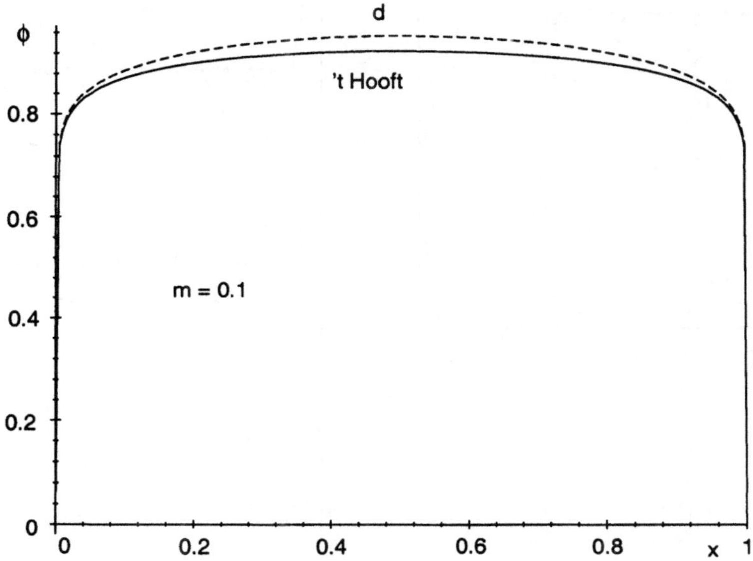

Fig. 8. *The light-cone wave function of the 't Hooft model 'pion' for $m = 0.1$. The solid curve represents the result from 't Hooft's original ansatz, the dashed curve our best result (with maximum number of variational parameters). At the given resolution, however, the curves of all extensions of 't Hooft's ansatz (a, b, c, d) lie on top of each other.*

Example 3: Gaussian Model. A very simple and intuitive example of a light-cone wave function is provided by the Gaussian model in the form presented in [93],

$$\phi(x, \mathbf{k}_\perp) = N \exp\left[-\frac{1}{\Lambda^2} \frac{k_\perp^2 + m^2}{x(1-x)} \right] . \tag{286}$$

This is a scalar two-particle wave function that drops exponentially with the free invariant mass squared of two constituents with equal mass m. The

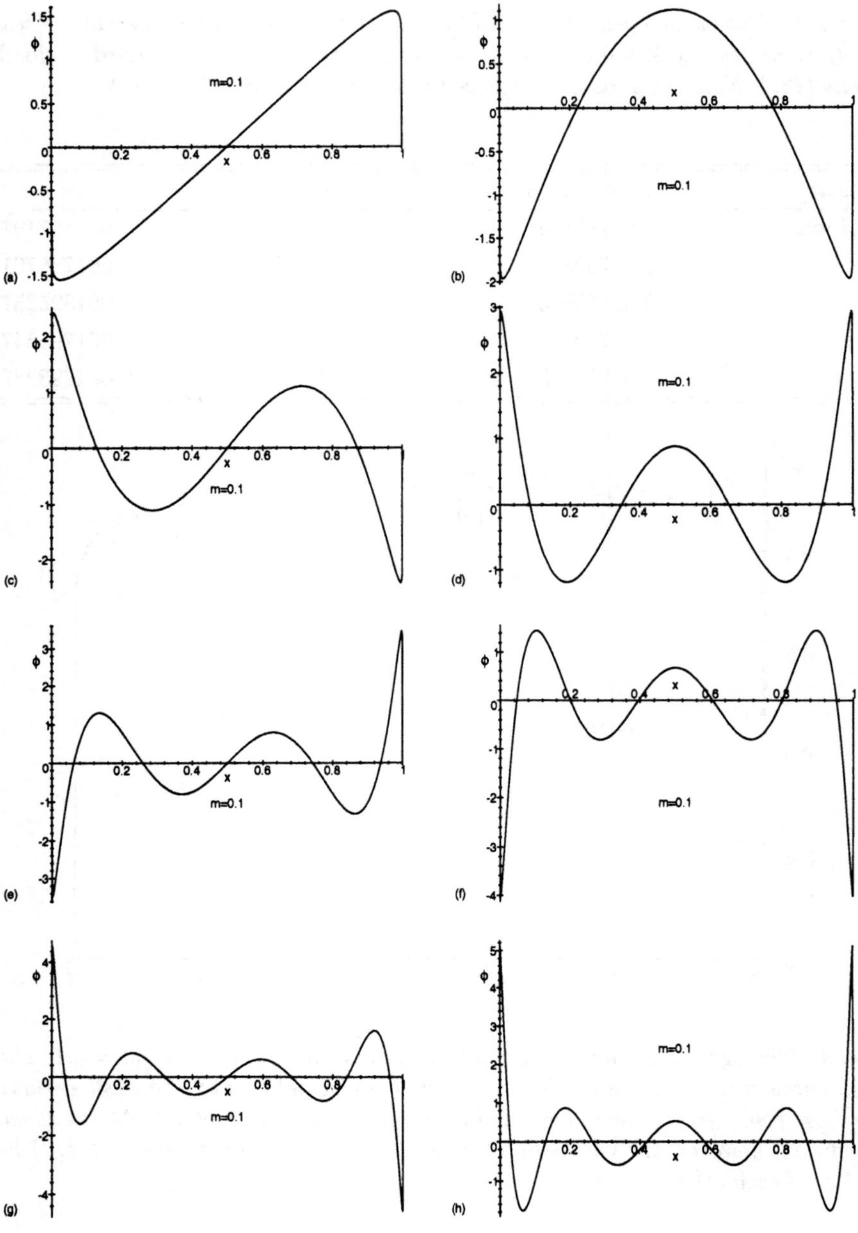

Fig. 9. Wave functions of the first eight excited states in the 't Hooft model as obtained via variational methods [142]. Note that all wave functions vanish at the end points, $x = 0$ and $x = 1$.

shortcoming of the model is that the wave function is not derived dynamically, i.e. as a solution of a light-cone Schrödinger equation. However, it satisfies all the requirements derived in Sect. 4.3. The benefit of the model is its simplicity. The latter is even enhanced if one neglects the constituent mass m by assuming $m \ll \Lambda$. Then one is left with only two parameters, the normalization N and the transverse size Λ.

One enforces a constituent picture by normalizing to unity,

$$||\phi(x, \mathbf{k}_\perp)||^2 \equiv \int_0^1 dx \int \frac{d^2 k_\perp}{16\pi^3} \phi^2(x, \mathbf{k}_\perp) = \frac{N^2 \Lambda^2}{192\pi^2} \stackrel{!}{=} 1 \, . \tag{287}$$

This obviously relates N and Λ. If we now calculate the r.m.s. transverse momentum in the bound state described by (286), we find using (287),

$$\langle k_\perp^2 \rangle \equiv \int_0^1 dx \int \frac{d^2 k_\perp}{16\pi^3} \, k_\perp^2 \, \phi^2(x, \mathbf{k}_\perp) = \frac{N^2 \Lambda^4}{1920\pi^2} = \Lambda^2/10 \, . \tag{288}$$

If we view the Gaussian wave function as a crude model for, say, the pion we can actually estimate the width parameter Λ. The pion is highly relativistic, so we expect its r.m.s. transverse momentum to be of the order of the constituent quark mass, $\langle k_\perp^2 \rangle^{1/2} \simeq m_Q \simeq 330$ MeV. This leads to a typical width of $\Lambda \simeq 1$ GeV.

Having thus determined the two parameters of the model we could go on and calculate observables [93]. This will actually be done in the next section using a more realistic model for the (determination of) the pion wave function.

5 The Pion Wave Function in the NJL Model

The ultimate goal of light-cone field quantization is to derive and solve light-cone Schrödinger equations, in particular the one of QCD. This would yield hadron masses and light-cone wave functions and thus detailed information on the internal structure of mesons and baryons. In order to successfully pursue this program, a number of problems has to be overcome.

In dealing with gauge theories in a Hamiltonian framework one has to solve Gauss's law together with the light-cone specific constraints. The only known solution so far is in the light-cone gauge $A^+ = 0$ [150,27,93] which, however, is beset by infrared problems of its own – see e.g. [8][9]. A Hamiltonian formulation analogous to the Weyl gauge, where one fixes the residual gauge freedom and solves Gauss's law *after* quantization, seems particularly difficult [65].

Even after successful light-cone quantization of QCD one encounters a severe problem: the theory has to be renormalized. Otherwise, the light-cone

[9] Recently, some progress has also been made in the covariant gauge [141].

wave function will not be normalizable and physically meaningful. Renormalization in a Hamiltonian framework presents enormous difficulties as there is no explicit covariance. In addition, rotational invariance is not manifest within the front form (see Sect. 2). Due to the lack of these important symmetries, there is an abundance of possible counterterms which even can be nonlocal, e.g. behave like $\sim 1/k^+$. As a result, to the best of my knowledge, the renormalization program has not been extended beyond one loop – with one notable exception in QED [23].

The issue of renormalization is, of course, delicately intertwined with solving the light-cone bound state equation. The latter attempt will in general only be feasible if a Tamm-Dancoff truncation in particle number is performed. This again violates important symmetries (even gauge symmetry). One hopes, however, that a Wilsonian renormalization group explicitly taylored for this case will restore these symmetries [153]. The present status of this program is nicely reviewed in [120].

A conceptual problem has already been mentioned. The light-cone vacuum is trivial as argued in Sect. 3. On the other hand, it is well known that the instant form vacuum is populated by all sorts of nonperturbative quantum fluctuations leading to nonvanishing vacuum expectation values or condensates. The whole business of QCD sum rules [138] is based upon this picture. How do we reconcile this with the triviality of the light-cone vacuum? A possible resolution to this 'triviality problem' has been given in the last section. The spectrum of excited states actually carries implicit information on the structure of the vacuum. The task then is to make this information *explicit*.

The list of problems just given is yet another manifestation of the 'principle of conservation of difficulties'. In a first attempt to tackle these problems I will simply side-step most of them by considering an (instructive) model instead of QCD. This model, however, is designed to capture some important physical features of 'real' QCD. The idea is originally due to Nambu and Jona-Lasinio (NJL), who, back in 1961, invented a "Dynamical Model of Elementary Particles Based on an Analogy with Superconductivity" [111]. It was meant to provide a microscopic mechanism for the generation of *nucleon* masses, with the mass gap being the analog of the BCS energy gap in a superconductor. Nowadays, with the nucleons replaced by quarks, the model serves as a low-energy effective theory of QCD explaining the spontaneous breakdown of chiral symmetry. Let me thus give a brief introduction to the latter phenomenon before I come to the detailed explanation of the model.

5.1 A Primer on Spontaneous Chiral Symmetry Breaking

If we have a look at Table 3, which provides a list of all quark flavors, we realize that there are large differences in the quark masses as they appear in the QCD Hamiltonian. In particular, there is a hierarchy,

$$\underbrace{m_u, m_d \ll m_s}_{\text{light quarks}} \ll \underbrace{m_c, m_b, m_t}_{\text{heavy quarks}} \ . \tag{289}$$

As the masses of heavy and light quarks are separated by the very same scale ($\simeq 1$ GeV) as the perturbative and nonperturbative regime, one expects different physics associated with those two kinds of quarks. This expectation turns out to be true. The physics of heavy quarks is governed by a symmetry called 'heavy quark symmetry' leading to a very successful 'heavy quark effective theory' [113]. The physics of light quarks, on the other hand, is governed by chiral symmetry which we are now going to explain.

Table 3. *The presently observed quark flavors. Q/e is the electric charge in units of the electron charge. The (scale dependent!) quark masses are given for a scale of 1 GeV.*

flavor		Q/e	mass	
down	d	$-1/3$	10	MeV
up	u	$+2/3$	5	MeV
strange	s	$-1/3$	150	MeV
charm	c	$+2/3$	1.5	GeV
bottom	b	$-1/3$	5.1	GeV
top	t	$+2/3$	180	GeV

Let us write the QCD Hamiltonian in the following way,

$$H_{\text{QCD}} = H_\chi + \bar{\psi}\mathcal{M}\psi \ , \tag{290}$$

$\mathcal{M} = diag(m_u, m_d, m_s)$ being the mass matrix for the light flavors. To a good approximation, one can set $\mathcal{M} = 0$. In this case, the QCD Hamiltonian H_χ is invariant under the symmetry group $SU(3)_R \otimes SU(3)_L$, the chiral flavor group. Under the action of this group, the left and right handed quarks independently undergo a chiral rotation. Due to Noether's theorem, there are sixteen conserved quantities, eight vector charges and, more important for us, eight pseudo-scalars, the chiral charges Q_5^a satisfying

$$[Q_5^a, H_\chi] = 0 \ . \tag{291}$$

This states both that the chiral charges are conserved, and that H_χ is chirally invariant. Under parity, $Q_5^a \to -Q_5^a$. Now, if $|A\rangle$ is an eigenstate of H_χ, so is $Q_5^a|A\rangle$ with the same eigenvalue. Thus, one expects (nearly) degenerate parity doublets in nature, which, however, do not exist empirically. The only explanation for this phenomenon is that chiral symmetry is spontaneously broken. In contradistinction to the Hamiltonian, the QCD ground state (the vacuum) is *not* chirally invariant,

$$Q_5^a|0\rangle \neq 0 \, . \tag{292}$$

For this reason, there must exist a nonvanishing vacuum expectation value, the quark condensate,

$$\langle \bar{\psi}\psi \rangle = \langle \bar{\psi}_R \psi_L + \bar{\psi}_L \psi_R \rangle \, . \tag{293}$$

This condensate is not invariant (it mixes left and right) and therefore serves as an order parameter of the symmetry breaking. Note that in QCD the quark condensate is a renormalization scale dependent quantity. A recent estimate can be found in [46], with a numerical value,

$$\langle 0|\bar{\psi}\psi|0\rangle(1 \text{ GeV}) \simeq (-229 \text{ MeV})^3 \, . \tag{294}$$

The spontaneous breakdown of chiral symmetry thus implies that the (QCD) vacuum is nontrivial: it must contain quark-antiquark pairs with spins and momenta aligned in a way consistent with vacuum quantum numbers. A possible analog is the BCS ground state given in (207).

In terms of the full quark propagator,

$$S(p) = \frac{\not{p} + M(p)}{p^2 - M^2(p)} \, , \tag{295}$$

where we have allowed for a momentum dependent (or 'running') mass $M(p)$, the quark condensate is given by

$$\langle 0|\bar{\psi}\psi|0\rangle = -i \int \frac{d^4p}{(2\pi)^4} \mathrm{tr} S(p) = -4iN_C \int \frac{d^4p}{(2\pi)^4} \frac{M(p)}{p^2 - M^2(p)} \, . \tag{296}$$

We thus see that the involved Dirac trace yields a nonvanishing condensate only if the effective quark mass $M(p)$ is nonzero. This links the existence of a quark condensate to the mechanism of dynamical mass generation. In this way we have found another argument that the bare quarks appearing in the QCD Hamiltonian (290) indeed acquire constituent masses. Of course we are still lacking a microscopic mechanism for that.

Goldstone's theorem [57] now states that for any symmetry generator which does not leave the vacuum invariant, there must exist a massless boson with the quantum numbers of this generator. This results in the prediction that in massless QCD one should have an octet of massless pseudoscalar mesons. In reality one finds what is listed in Table 4.

The nonvanishing masses of these mesons are interpreted as stemming from the nonvanishing quark masses in the QCD Hamiltonian which break chiral symmetry explicitly. Being small, they can be treated as perturbations.

Table 4. *Masses of the pseudoscalar octet mesons (in MeV).*

meson	π^0	$\pi^\pm$	$K^0, \bar{K}^0$	$K^\pm$	η
mass	135	140	500	494	549

It should be stressed that chiral symmetry has nothing to say about the mechanism of confinement which presumably is a totally different story. This is also reflected within a QCD based derivation of chiral symmetry breaking in terms of the instanton model, as presented e.g. in [41]. This model explains many facts of low-energy hadronic physics but is known not to yield confinement. It is therefore possible that confinement is not particularly relevant for the understanding of hadron structure [41].

The instanton vacuum actually leads to an effective theory very close to the NJL model, the basics of which are our next topic.

5.2 NJL Folklore

Before I consider the light-cone formulation of the model, let me briefly recall its main physical features[10]. In its standard form, the NJL model has a chirally invariant four-fermion interaction, which can be imagined as the result of 'integrating out' the gluons in the QCD Lagrangian. For simplicity I concentrate on the case of one flavor. Extension to several flavors is straightforward. In the chiral limit (quark mass $m_0 = 0$), the Lagrangian is

$$\mathcal{L} = \bar{\psi} i \partial\!\!\!/ \psi - G(\bar{\psi}\gamma_\mu\psi)^2 \equiv \mathcal{L}_0 + \mathcal{L}_{\text{int}} \, . \tag{297}$$

Its four-fermion interaction is chirally symmetric under $U(1)_L \times U(1)_R$. We shall see in a moment that this symmetry is spontaneously broken.

It is important to observe that the coupling G of the model has negative mass dimension, $[G] = -2$, hence, it is not renormalizable. Accordingly, it requires a cutoff which is viewed as a parameter of the model that is to be fixed by phenomenology. We thus follow the general spirit of effective field theory [73,85,103]. From the point of view of the light-cone formulation to be developed later, the nonrenormalizability is an advantage: it enables us to 'circumvent' the difficulties of the light-cone renormalization program. There simply is no 'need' to renormalize.

Following the standard approach [111,151] we treat the model in mean-field approximation (which actually coincides with the large-N limit). We begin with a Fierz transformation by schematically rewriting the interaction Lagrangian in Fierz symmetric form

$$\mathcal{L}_{\text{int}} = G \sum_i c_i (\bar{\psi}\Gamma_i\psi)^2 = G(\bar{\psi}\psi)^2 + \dots \, , \tag{298}$$

[10] For recent reviews, see [151,87,64].

where $i = S, P, V, A$ enumerates the different Dirac bilinears (scalar, pseudoscalar, vector and axial vector, respectively). In (298), I have only displayed the scalar part of the interaction because only the scalar density $S \equiv \bar{\psi}\psi$ can have a vacuum expectation value, the quark condensate,

$$\langle S \rangle = \langle 0|\bar{\psi}\psi|0\rangle \ . \tag{299}$$

Let us determine this quantity in mean-field approximation. To define the latter we calculate,

$$S^2 = (S - \langle S \rangle + \langle S \rangle)^2 = (S - \langle S \rangle)^2 + 2S\langle S \rangle - \langle S \rangle^2 \simeq 2S\langle S \rangle + const \ . \tag{300}$$

Thus, by neglecting quadratic fluctuations of S around its expectation value, we linearize the interaction and obtain the mean-field Lagrangian,

$$\mathcal{L}_{\mathrm{MFA}} = \bar{\psi}(i\partial\!\!\!/ + 2G\langle S \rangle)\psi \ . \tag{301}$$

The mean-field solution has a very intuitive explanation. One essentially argues that the main effect of the interaction is to generate the mass of the quarks which become quasi-particles that interact only weakly. Neglecting this interaction entirely, one can view the process of mass generation as the transition of quarks with mass $m_0 = 0$ to mass m resulting in a mass term $m\bar{\psi}\psi$ in the Lagrangian (301). We thus find that the dynamically generated mass is determined by the *gap equation*

$$m = -2G\langle 0|\bar{\psi}\psi|0\rangle \ . \tag{302}$$

How do we actually calculate the condensate $\langle 0|\bar{\psi}\psi|0\rangle$? To this end we go back to (296) and express the condensate in terms of the full propagator 'at the origin', i.e. at space-time point $x = 0$,

$$\langle 0|\bar{\psi}\psi|0\rangle_m = -i\,\mathrm{tr} S_F(x=0) = -i\,\mathrm{tr} \int \frac{d^4p}{(2\pi)^4} \frac{1}{p\!\!\!/ - m + i\epsilon} \ . \tag{303}$$

The following remarks are in order. First we note that within mean-field approximation the dynamically generated mass m is constant, i.e. independent of the momentum p. Furthermore, the full propagator S_F is defined in terms of the constituent mass m, so that the gap equation (302) becomes an implicit self-consistency condition where the mass m to be determined appears on both sides. This equation will be solved in a moment. The integral appearing on the r.h.s. of (303) is quadratically divergent[11] and has to be regulated. If we use a cutoff Λ we have, for dimensional reasons,

$$\langle 0|\bar{\psi}\psi|0\rangle \sim \Lambda^2 m \ . \tag{304}$$

[11] As the condensate involves the product of field operators at coinciding space-time points, this clearly is a short-distance singularity.

A particularly intuitive way to calculate the condensate (which will also be used in the light-cone case) is based on the Hellmann-Feynman theorem. This states that 'the derivative of an expectation value is the expectation value of the derivative', if the expectation is taken between normalizable states. If applied to the vacuum expectation value of our mean-field Hamiltonian,

$$\mathcal{H}(m) = \mathcal{H}_0 + m\bar{\psi}\psi \,, \tag{305}$$

which is the vacuum energy density $\mathcal{E}(m) \equiv \langle 0|\mathcal{H}(m)|0\rangle$, the theorem yields,

$$\langle 0|\bar{\psi}\psi|0\rangle_m = \frac{\partial}{\partial m}\mathcal{E}(m) = -\frac{\partial}{\partial m}\int \frac{d^3p}{(2\pi)^3} 2\sqrt{\mathbf{p}^2 + m^2} \,. \tag{306}$$

The factor of two in the integrand is due to the fermion spin degeneracy. If we choose a noncovariant three-vector cutoff, $|\mathbf{p}| < \Lambda_3$, the result for the condensate is of the form (304),

$$\langle 0|\bar{\psi}\psi|0\rangle_m = -\frac{m\Lambda_3^2}{2\pi^2}\left[1 + \frac{\delta^2}{2}\ln\delta^2 + O(\delta^2)\right] , \tag{307}$$

with $\delta \equiv m/\Lambda_3$. The dynamical mass is found by inserting this result into the gap equation (302). A nontrivial solution $m = m(G) \neq 0$ arises above a critical coupling G_c which is determined by the identity $m(G_c) = 0$. It is known from the theory of critical phenomena that the mean-field approximation leads to a square root behaviour of the mass around $G_c = 4\pi^2/\Lambda^2$,

$$m(G) \sim (G - G_c)^{1/2} \,, \quad G \geq G_c \,. \tag{308}$$

We thus have seen that the NJL model describes the transformation of bare quarks of mass $m_0 (= 0)$ into dressed (or constituent) quarks (Q) of mass $m \neq 0$. By considering the bound state equation in the pseudoscalar channel one can also verify Goldstone's mechanism: there is a massless $Q\bar{Q}$ bound state, which is identified with the pion, exactly if the gap equation holds [111,151]. This concludes the presentation of spontaneous chiral symmetry breaking within the model. One should keep in mind that it follows the same pattern as in QCD.

Let me conclude the NJL 'crash course' by reemphasizing that the model does not confine. This means in particular, that there is a nonvanishing probability for mesons to decay into constituent quarks. We thus cannot expect to obtain reliable estimates for strong decay widths of mesons or other quantities that are not dominated by their chiral properties.

It is getting time to discuss the light-cone formulation of the model. For a 'light-cone physicist' the model is interesting for several reasons. I have already pointed out that the lack of renormalizability is welcome because we only have to worry about a proper regularization. Furthermore, the model addresses the conceptually important questions of spontaneous symmetry breaking and condensates. Are these in conflict with a trivial vacuum? Finally, a constituent picture seems to be realized which should make a truncation of the light-cone Schrödinger equation feasible. Now, we again encounter the standard rule of physics that there is nothing like free lunch. Here, this is mainly due to the appearance of complicated constraints for part of the fermionic degrees of freedom. Let me thus make a small aside on the special features of light-cone fermions.

Light-Cone Fermions. The solution of the Dirac equation (for free fermions of mass m_0) has the following light-cone Fock expansion at $x^+ = 0$,

$$\psi(\boldsymbol{x}, 0) = \sum_\lambda \int_0^\infty \frac{dk^+}{k^+} \int \frac{d^2k_\perp}{16\pi^3} \left[b(\boldsymbol{k}, \lambda) u(\boldsymbol{k}, \lambda) e^{-i\boldsymbol{k}\cdot\boldsymbol{x}} + d^\dagger(\boldsymbol{k}, \lambda) v(\boldsymbol{k}, \lambda) e^{i\boldsymbol{k}\cdot\boldsymbol{x}} \right], \tag{309}$$

where we recall the notations

$$\boldsymbol{k} \equiv (k^+, \mathbf{k}_\perp) \, , \quad \boldsymbol{x} \equiv (x^-, \mathbf{x}_\perp) \, , \quad \boldsymbol{k}\cdot\boldsymbol{x} \equiv \tfrac{1}{2} k^+ x^- - \mathbf{k}_\perp \cdot \mathbf{x}_\perp \, . \tag{310}$$

Like for scalars, the Fock measure is independent of the mass m_0. The Fock operators satisfy the canonical anti-commutation relations,

$$\left\{ b(\boldsymbol{k}, \lambda) \, , \, b^\dagger(\boldsymbol{p}, \lambda') \right\} = \left\{ d(\boldsymbol{k}, \lambda) \, , \, d^\dagger(\boldsymbol{p}, \lambda') \right\} = 16\pi^3 k^+ \delta^3(\boldsymbol{k} - \boldsymbol{p}) \, , \tag{311}$$

The basis spinors u and v obey the Dirac equations,

$$(\not{k} - m_0)\, u(\boldsymbol{k}, \lambda) = 0 \, , \tag{312}$$

$$(\not{k} + m_0)\, v(\boldsymbol{k}, \lambda) = 0 \, . \tag{313}$$

and are explicitly given by

$$u(\boldsymbol{k}, \lambda) = \frac{1}{\sqrt{k^+}} (k^+ + \beta m + \alpha_i k_i) X_\lambda \, , \tag{314}$$

$$v(\boldsymbol{k}, \lambda) = \frac{1}{\sqrt{k^+}} (k^+ - \beta m + \alpha_i k_i) X_{-\lambda} \, . \tag{315}$$

The four-spinor X will be defined in a moment; α_i and β are the standard hermitean Dirac matrices. The crucial point is the decomposition of the fermion field, $\psi = \psi_+ + \psi_-$ into 'good' (+) and 'bad' (−) components, $\psi_\pm \equiv \Lambda_\pm \psi$, by means of the projection matrices,

$$\Lambda_\pm \equiv \frac{1}{4} \gamma^\mp \gamma^\pm = \frac{1}{2} \begin{pmatrix} \mathbb{1} & \pm\sigma_3 \\ \pm\sigma_3 & \mathbb{1} \end{pmatrix} \, . \tag{316}$$

The spinor X appearing in (314) and (315) is an eigenspinor of Λ_+, $\Lambda_+ X_\lambda = X_\lambda$. The Dirac equation decomposes accordingly into *two* equations. The one for ψ_- reads

$$2i\partial^+ \psi_- = (-i\gamma^i \partial^i + m_0)\gamma^+ \psi_+ \, , \tag{317}$$

and does not contain a light-cone time derivative. Therefore, the bad component ψ_- is constrained; it can be expressed in terms of the good component, i.e. $\psi_- = \psi_-[\psi_+]$. Again, this requires the inversion of the notorious spatial derivative ∂^+. As a result, the free Dirac Hamiltonian (density) only depends on ψ_+,

$$\mathcal{H} = \psi_+^\dagger \, \frac{-\partial_\perp^2 + m_0^2}{i\partial^+} \, \psi_+ \equiv \mathcal{H}[\psi_+] \, , \tag{318}$$

where we easily recognize the light-cone energy, $(k_\perp^2 + m_0^2)/k^+$.

It turns out that in case of a four-fermion interaction like in the NJL model the constraint becomes rather awkward to solve. In particular, its solution has to be consistent with the mean field approximation employed [42]. This can be achieved most elegantly by using the large-N expansion [12,77]. Nevertheless, the light-cone Hamiltonian of the model is a rather complicated expression. We will therefore follow an alternative road which is the topic of the next subsection.

5.3 Schwinger–Dyson Approach

The first derivation, analysis and solution of a light-cone bound-state equation appeared in 't Hooft's original paper on what is now called the 't Hooft model [146]. We have discussed this model in the last section where we also rederived 't Hooft's solution. Interestingly, 't Hooft did not use the light-cone formalism in the manner we presented it and which nowadays might be called standard. This amounts to deriving the canonical light-cone Hamiltonian and setting up the associated system of bound-state equations by projecting on the different sectors of Fock space (cf. Sect. 4). Instead, he started from covariant equations, namely the Schwinger–Dyson equations for the quark propagator (or self-energy), and the Bethe-Salpeter equation for the bound-state amplitude, which needs the quark self-energy as an input. The light-cone Schrödinger equation was then obtained by projecting the Bethe-Salpeter equation onto hypersurfaces of equal light-cone time. In this way, one avoids to explicitly derive the light-cone Hamiltonian, which, as explained above, can be a tedious enterprise in view of complicated constraints one has to solve. Let us therefore have a closer look at this way of proceeding.

The Schwinger–Dyson Equation for the Propagator. The first step in the program[12] is to solve the Schwinger–Dyson equation for the propagator, or, equivalently, for the quark self-energy. As this cannot be done exactly, one resorts to mean-field (or large-N) approximation. This is essentially what has been done in the last subsection. Let us rewrite this in terms of Schwinger–Dyson equations. The one for the full propagator S reads

$$S = S_0 + S_0 \Sigma S \, , \tag{319}$$

[12] For recent literature on the Schwinger–Dyson approach, see e.g. [127,128,105].

and is formally solved by

$$S(p) = \frac{1}{S_0^{-1}(p) - \Sigma(p)} , \tag{320}$$

where S_0 is the free propagator,

$$S_0(p) = \frac{1}{\not{p} - m_0} , \tag{321}$$

and Σ the quark self-energy. In mean-field approximation, it is momentum independent and defines the constituent mass through the gap equation, $\Sigma = const = m$, see (302) and (303).

To solve the latter in the light-cone framework, we basically just have to calculate the condensate. As in the standard approach, this can be obtained via the Feynman-Hellman theorem by differentiating the energy density of the quasi-particle Dirac sea,

$$\begin{aligned} \langle 0|\bar{\psi}\psi|0\rangle_m &= \frac{\partial}{\partial m}\mathcal{E}(m) = \frac{\partial}{\partial m}\int\limits_{-\infty}^{0} dk^+ \int \frac{d^2k_\perp}{16\pi^3} 2\frac{m^2 + k_\perp^2}{k^+} \\ &= -\frac{m}{4\pi^3}\int\limits_0^\infty \frac{dk^+}{k^+}\int d^2k_\perp . \end{aligned} \tag{322}$$

Again, as it stands, the integral is divergent and requires regularization. In the most straightforward manner one might choose $m^2/\Lambda \le k^+ \le \Lambda$ and $|\mathbf{k}_\perp| \le \Lambda$, so that the condensate becomes

$$\langle 0|\bar{\psi}\psi|0\rangle_m = -\frac{m}{4\pi^2}\int\limits_{m^2/\Lambda}^{\Lambda} \frac{dk^+}{k^+}\int\limits_0^{\Lambda^2} d(k_\perp^2) = -\frac{m}{4\pi^2}\Lambda^2 \ln\frac{\Lambda^2}{m^2} . \tag{323}$$

Plugging this result into the gap equation (302) one finds for the dynamical mass squared,

$$m^2(G) = \Lambda^2 \exp\left(-\frac{2\pi^2}{G\Lambda^2}\right) . \tag{324}$$

The critical coupling is determined by the vanishing of this mass, $m(G_c) = 0$, and from (324) we find the surprising result

$$G_c = 0 . \tag{325}$$

This result, however, is wrong since one knows from the conventional treatment of the model that the critical coupling is finite of the order π^2/Λ^2, both for covariant and noncovariant cutoff [111]. In addition, it is quite generally clear that in the *free* theory ($G = 0$) chiral symmetry is not broken (as

$m_0 = 0$) and, therefore, this should not happen for arbitrarily small coupling, either, cf. (308). The remedy is to use the *invariant-mass cutoff* [92],

$$M_0^2 \equiv \frac{k_\perp^2 + m^2}{x(1-x)} \le \Lambda^2 \,, \tag{326}$$

where we have defined the longitudinal momentum fraction, $x \equiv k^+/\Lambda$. This provides a cutoff both in x (or k^+) *and* $k_\perp$,

$$0 \quad \le k_\perp^2 \le \Lambda^2 x(1-x) - m^2 \,, \tag{327}$$

$$x_0 \quad \le x \le x_1 \,, \tag{328}$$

$$x_{0,1} \equiv \tfrac{1}{2}(1 \mp \sqrt{1 - 4\epsilon^2}) \,, \tag{329}$$

with $\epsilon^2 \equiv m^2/\Lambda^2$. Note that the transverse cutoff becomes a polynomial in x. The $k_\perp$-integration thus has to be performed before the x-integration.

For the condensate (322) the invariant-mass cutoff (326) yields an analytic structure different from (323),

$$\langle 0|\bar{\psi}\psi|0\rangle_m = -\frac{m\Lambda^2}{8\pi^2}\left(1 + 2\epsilon^2 \ln \epsilon^2 + O(\epsilon^2)\right) \,, \tag{330}$$

where we have neglected sub-leading terms in ϵ^2. The result (330) coincides with the standard one, (307), if one identifies the noncovariant cutoffs according to

$$\Lambda^2 \equiv 4(\Lambda_3^2 + m^2) \,. \tag{331}$$

This has independently been observed in [12]. From (330), one infers the correct cutoff dependence of the critical coupling,

$$G_c = \frac{4\pi^2}{\Lambda^2} \,. \tag{332}$$

The moral of this calculation is that even in a nonrenormalizable theory like the NJL model, the light-cone regularization prescription is a subtle issue.

In the NJL model with its second-order phase transition of mean-field type, the usual analogy with magnetic systems can be made. Chiral symmetry corresponds to rotational symmetry, the vacuum energy density to the Gibbs free energy, and the mass m to an external magnetic field. The order parameter measuring the rotational symmetry breaking is the magnetization. It is obtained by differentiating the free energy with respect to the external field. This is the analogue of expression (322) derived from the Feynman-Hellmann theorem.

The Bound-State Equation. Once the physical fermion mass m is known by solving the gap equation, it can be plugged into the Bethe-Salpeter equation for quark-antiquark bound states (mesons), given by

$$\chi_{BS} = S_1 S_2 K \chi_{BS} \,. \tag{333}$$

S_1, S_2 denote the full propagators of quark and anti-quark, K the Bethe-Salpeter kernel, and χ_{BS} the Bethe-Salpeter amplitude. From the latter, one obtains the light-cone wave function via integration over the energy variable k^- [108,21,31,100],

$$\phi_{\mathrm{LC}}(\boldsymbol{k}) = \int \frac{dk^-}{2\pi} \chi_{\mathrm{BS}}(k) \, , \quad \boldsymbol{k} = (k^+, \mathbf{k}_\perp) \, . \tag{334}$$

In ladder approximation (again equivalent to the large-N limit), (333) and (334) become

$$\phi_{\mathrm{LC}}(\boldsymbol{k}) = \int \frac{dk^-}{2\pi} S(k) S(k-P) \int \frac{d^3p}{(2\pi)^3} \int \frac{dp^-}{2\pi} K(k,p) \chi_{\mathrm{BS}}(p) \, , \tag{335}$$

with P denoting the bound-state four-momentum. On the left-hand-side, the projection onto $x^+ = 0$ (i.e. the k^--integration) has already been carried out. On the right-hand-side, the two integrations over k^- and p^- still have to be performed. Whether this can easily be done depends of course crucially on the kernel K, which in principle is a function of both energy variables. For the NJL model, however, K assumes the very simple form,

$$K(k,p) = 2\gamma_5 \otimes \gamma_5 - \gamma_\mu \gamma_5 \otimes \gamma^\mu \gamma_5 \, , \tag{336}$$

i.e. it is momentum independent due to the four-point contact interaction,

$$W \sim \int d^4x \int d^4y \, (\bar\psi \Gamma \psi)(x) \, \delta^4(x-y) \, (\bar\psi \Gamma \psi)(y) \, . \tag{337}$$

Thus, the p^--integration immediately yields ϕ_{LC}, and the k^--integration can be performed via residue techniques and is completely determined by the poles of the propagators, $S(k)$ and $S(k-P)$. As a result, one finds a nonvanishing result only if $0 \le k^+ \le P^+$, and *one* of the two particles is put on-shell, e.g. $k^2 = m^2$, as already observed by Gross [60].

The upshot of all this is nothing but the light-cone bound-state equation, which explicitly reads

$$\begin{aligned}
\phi_{\mathrm{LC}}(x, \mathbf{k}_\perp) = & - \frac{2G}{x(1-x)} \frac{(\hat{\not k} + m)\gamma_5(\hat{\not k} - \not P + m)}{M^2 - M_0^2} \\
& \times \int_0^1 dy \int \frac{d^2p_\perp}{8\pi^3} \mathrm{tr}\left[\gamma_5 \phi_{\mathrm{LC}}(y, \mathbf{p}_\perp)\right] \theta_\Lambda(y, \mathbf{p}_\perp) \\
& + \frac{G}{x(1-x)} \frac{(\hat{\not k} + m)\gamma_\mu\gamma_5(\hat{\not k} - \not P + m)}{M^2 - M_0^2} \\
& \times \int_0^1 dy \int \frac{d^2p_\perp}{8\pi^3} \mathrm{tr}\left[\gamma^\mu \gamma_5 \phi_{\mathrm{LC}}(y, \mathbf{p}_\perp)\right] \theta_\Lambda(y, \mathbf{p}_\perp) \, .
\end{aligned} \tag{338}$$

Here we have defined the longitudinal momentum fractions $x = k^+/P^+$, $y = p^+/P^+$, the on-shell momentum $\hat{k} = (\hat{k}^-, \mathbf{k}_\perp, k^+)$ with $\hat{k}^- = (k_\perp^2 + m^2)/k^+$, and the bound state mass squared, $M^2 = P^2$, which is the eigenvalue to be solved for. $\theta_\Lambda(x, \mathbf{p}_\perp)$ denotes the invariant mass cutoff (326).

Now, while (338) may appear somewhat complicated it is actually very simple; indeed, it is basically already the solution of the problem. The crucial observation is to note that the two integral expressions are mere normalization constants,

$$C_\Lambda \equiv \int_0^1 dy \int \frac{d^2p_\perp}{8\pi^3} \mathrm{tr}\,[\gamma_5 \phi_{\mathrm{LC}}(y, \mathbf{p}_\perp)]\, \theta_\Lambda(y, \mathbf{p}_\perp) \,, \tag{339}$$

$$D_\Lambda P^\mu \equiv \int_0^1 dy \int \frac{d^2p_\perp}{8\pi^3} \mathrm{tr}\,[\gamma^\mu \gamma_5 \phi_{\mathrm{LC}}(y, \mathbf{p}_\perp)]\, \theta_\Lambda(y, \mathbf{p}_\perp) \,. \tag{340}$$

Thus, the solution of the light-cone bound-state equation (338) is

$$\phi_{\mathrm{LC}}(x, \mathbf{k}_\perp) = -\frac{2GC_\Lambda}{x(1-x)} \frac{(\hat{\not k} + m)\gamma_5(\hat{\not k} - \not P + m)}{M^2 - M_0^2} + \frac{GD_\Lambda}{x(1-x)} \frac{(\hat{\not k} + m)\not P \gamma_5(\hat{\not k} - \not P + m)}{M^2 - M_0^2} \,, \tag{341}$$

with yet undetermined normalization constants C_Λ and D_Λ. As a first check of our bound-state wave function (341) we look for a massless pion in the chiral limit. To this end we decompose the light-cone wave function into Dirac components according to Lucha et al. [101],

$$\phi_{\mathrm{LC}} = \phi_{\mathrm{S}} + \phi_{\mathrm{P}}\gamma_5 + \phi_{\mathrm{A}}^\mu \gamma_\mu \gamma_5 + \phi_{\mathrm{V}}^\mu \gamma_\mu + \phi_{\mathrm{T}}^{\mu\nu} \sigma_{\mu\nu} \,. \tag{342}$$

Multiplying (341) with γ_5, taking the trace and integrating over $\boldsymbol{k}$ we find

$$C_\Lambda = -\frac{GC_\Lambda}{2\pi^3} \int_0^1 dx \int d^2k_\perp \frac{M^2 x + M_0^2(1-x)}{x(1-x)(M^2 - M_0^2)} \theta_\Lambda(x, \mathbf{k}_\perp) + \frac{GD_\Lambda}{2\pi^3} M^2 \int_0^1 dx \int d^2k_\perp \theta_\Lambda(x, \mathbf{k}_\perp) \,. \tag{343}$$

In the chiral limit one expects a solution for $M = 0$, the Goldstone pion. In this case one obtains

$$1 = \frac{G}{2\pi^3} \int_0^1 dx \int d^2k_\perp \theta_\Lambda(x, \mathbf{k}_\perp) \,. \tag{344}$$

This is exactly the gap equation (302) using the definition (322) of the condensate (with the invariant-mass cutoff (326) understood in both identities). Note once more the light-cone peculiarity that the (Fock) measure in (344) is entirely mass independent. All the mass dependence, therefore, has to come from the (invariant-mass) cutoff. Otherwise one will get a wrong behavior of the dynamical mass m as a function of the coupling G, as was the case in (324).

With this in mind, we see that the Goldstone pion is a solution of the light-cone bound-state equation exactly if the gap equation holds. This provides additional evidence for the self-consistency of the procedure. The deeper reason for the fact that the quark self-energy and the bound-state amplitude satisfy essentially the same equation, is the chiral Ward identity relating the quark propagator and the pseudoscalar vertex [132].

Our next task is to actually evaluate the solution (341) of the bound-state equation. ϕ_{LC} is a Dirac matrix and therefore is not yet a light-cone wave function as defined in Sect. 4. The relation between the two quantities has been given in [100],

$$2P^+\psi(x,\mathbf{k}_\perp,\lambda,\lambda') = \bar{u}(xP^+,\mathbf{k}_\perp,\lambda)\gamma^+\phi_{\mathrm{LC}}(\boldsymbol{k})\gamma^+ v(\bar{x}P^+,-\mathbf{k}_\perp,\lambda') \,, \qquad (345)$$

where we have denoted $\bar{x} \equiv 1-x$ to save space. A somewhat lengthy calculation yields the result [67],

$$\psi(x,\mathbf{k}_\perp,\lambda,\lambda') = \frac{2GP^+/\sqrt{x\bar{x}}}{M^2-M_0^2}\left(\frac{2C_\Lambda}{M}\bar{u}_\lambda M\gamma_5 v_{\lambda'} - D_\Lambda \bar{u}_\lambda \not{P}\gamma_5 v_{\lambda'}\right) \,, \qquad (346)$$

with the arguments of the spinors $\bar{u}$ and v suppressed. At this point we have to invoke another symmetry principle. Ji et al. [82] have pointed out that the spin structure $(\bar{u}\Gamma v)$ should be consistent with the one obtained form the instant form spinors via a subsequent application of a Melosh transformation [107] and a boost. Using this recipe, one obtains the following relation between the constants C_Λ and D_Λ,

$$2C_\Lambda/M = -D_\Lambda \equiv N/2G \,. \qquad (347)$$

As a result, the spin structure in (346) coincides with the standard one used e.g. in [48,34,81,83,82]. The NJL wave function of the pion thus becomes

$$\psi(x,\mathbf{k}_\perp,\lambda,\lambda') = \frac{NP^+/\sqrt{x\bar{x}}}{M^2-M_0^2}\,\bar{u}_\lambda(M+\not{P})\,\gamma_5\, v_{\lambda'}\,\theta(\Lambda^2-M_0^2) \,. \qquad (348)$$

Not surprisingly, the off-shellness $M^2 - M_0^2$ appears in the denominator. N is the normalization parameter defined in (347), and the spin (or helicity) structure is given by

$$\begin{aligned}&\bar{u}(xP^+,\mathbf{k}_\perp,\lambda)\,(M+\not{P})\,\gamma_5\, v(\bar{x}P^+,-\mathbf{k}_\perp,\lambda') = \\ &= \frac{1}{\sqrt{x\bar{x}P^+}}\left[\lambda\left(mM+m^2-\mathbf{k}_\perp^2+M^2x\bar{x}\right)\delta_{\lambda,-\lambda'} - k_{-\lambda}(M+2m)\delta_{\lambda\lambda'}\right] \end{aligned} \qquad (349)$$

where we have used (314), (315) and denoted $k_\lambda \equiv k^1 + i\lambda k^2$. The first term with spins anti-parallel corresponds to $L_z = 0$, the second one (with spins parallel) to $L_z = \pm 1$. It has already been pointed out by Leutwyler that both spin alignments should contribute to the pion wave function [97,98]. Note that the latter is a cutoff dependent quantity. This is necessary in order to render the wave function normalizable. A single power of the off-shellness in the denominator is not sufficient for that. Only the cutoff guarantees the boundary conditions (251) so that the wave function drops off sufficiently fast in x and $k_\perp$.

As we are interested in analyzing the quality of a constituent picture, we approximate the pion by its valence state, denoting $\psi \equiv \psi_2$,

$$|\pi : \boldsymbol{P}\rangle = \sum_{\lambda,\lambda'} \int_0^1 dx \int \frac{d^2 k_\perp}{16\pi^3} \, \psi_2(x, \mathbf{k}_\perp, \lambda, \lambda') \, |q\bar{q} : x, \mathbf{k}_\perp, \lambda, \lambda'\rangle \tag{350}$$

which should be compared with the general expression (240). The normalization of this state is given by (237),

$$\langle \pi : \boldsymbol{P}' | \pi : \boldsymbol{P} \rangle = 16\pi^3 P^+ \delta^3(\boldsymbol{P} - \boldsymbol{P}') \, . \tag{351}$$

As usual we work in a frame in which the total transverse momentum vanishes, i.e. $\boldsymbol{P} = (P^+, \mathbf{P}_\perp = 0)$. Expression (351) yields the normalization (247) of the wave function,

$$\sum_{\lambda\lambda'} \int_0^1 dx \int \frac{d^2 k_\perp}{16\pi^3} |\psi_2(x, \mathbf{k}_\perp, \lambda, \lambda')|^2 \equiv \|\psi_2\|^2 = 1 \, . \tag{352}$$

It is of course a critical assumption that the probability to find the pion in its valence state is one. In this way we enforce a constituent picture *by fiat*, and it is clear that such an assumption has to be checked explicitly by comparing with phenomenology.

5.4 Observables

With the light-cone wave function at hand, we are in the position to calculate observables. To proceed we will employ the following two simplifications. First of all, we will always work in the chiral limit of vanishing quark mass, $m_0 = 0$, which, as we have seen, leads to a massless Goldstone pion, $M = 0$. We write the pion wave function as a matrix in helicity space,

$$\begin{aligned} \psi_2(x, \mathbf{k}_\perp) &= \begin{pmatrix} \psi_{2\uparrow\uparrow} & \psi_{2\uparrow\downarrow} \\ \psi_{2\downarrow\uparrow} & \psi_{2\downarrow\downarrow} \end{pmatrix} \\ &= -\frac{N}{k_\perp^2 + m^2} \begin{pmatrix} -2m(k^1 - ik^2) & m^2 - k_\perp^2 \\ k_\perp^2 - m^2 & -2m(k^1 + ik^2) \end{pmatrix} \theta(\Lambda^2 - M_0^2) \, . \end{aligned} \tag{353}$$

Note that, in the chiral limit, the wave function becomes independent of x (apart from cutoff effects). This actually agrees with our findings in the two-dimensional 't Hooft model, cf. (276).

The diagonal terms in (353) correspond to parallel spins, the off-diagonal ones to anti-parallel spins. The different components are related by the symmetry properties,

$$\psi_{2\downarrow\downarrow} = \psi^*_{2\uparrow\uparrow} , \quad \psi_{2\downarrow\uparrow} = -\psi_{2\uparrow\downarrow} . \tag{354}$$

Second, we will go to the large-cutoff limit, that is, we will keep only the leading order in $\epsilon^2 = m^2/\Lambda^2$. We thus assume that the cutoff is large compared to the constituent mass. From the standard values, $\Lambda \simeq 1$ GeV, $m \simeq 300$ MeV, we expect that this assumption should induce an error of the order of 10%. The technical advantage of the large-cutoff limit is a simple analytic evaluation of all the integrals we will encounter. Furthermore, the leading order will be independent of the actual value of the constituent mass. It should be mentioned that the same procedure has been used in calculations based on the instanton model of the QCD vacuum [122]. There, the ratio ϵ^2 can be related to parameters of the instanton vacuum, namely

$$\epsilon^2 = (m\rho)^2 \sim (\rho/R)^4 , \tag{355}$$

where $\rho \simeq 1/3$ fm is the instanton size and R the mean distance between instantons. Thus, ϵ^2 can be identified with the 'diluteness parameter' or 'packing fraction' of the instanton vacuum and hence is parametrically small, $\epsilon^2 \simeq 1/4$.

The upshot of all this is that we work with the extremely simple model wave function [126]

$$\psi_{2\uparrow\downarrow} \simeq N\,\theta(\Lambda^2 - M_0^2) , \quad \psi_{2\uparrow\uparrow} = 0 , \tag{356}$$

which is entirely determined by two parameters, the normalization constant N and the cutoff Λ. We thus need two constraints on the wave function to fix our two parameters.

Normalization. As announced, we enforce a constituent picture by demanding (352) which decomposes into

$$1 = ||\psi_2||^2 = ||\psi_{2\uparrow\downarrow}||^2 + ||\psi_{2\uparrow\uparrow}||^2 + ||\psi_{2\downarrow\uparrow}||^2 + ||\psi_{2\downarrow\downarrow}||^2 . \tag{357}$$

Explicitly, one finds

$$\begin{aligned} ||\psi_{2\uparrow\downarrow}||^2 = ||\psi_{2\downarrow\uparrow}||^2 &= N^2 \int_0^1 dx \int \frac{d^2k_\perp}{16\pi^3}\,\theta(\Lambda^2 - M_0^2) \\ &= \frac{N^2}{16\pi^2} \int_0^1 dx \int_0^{\Lambda^2 x(1-x)} dk_\perp^2 = \frac{N^2\Lambda^2}{96\pi^2} \stackrel{!}{=} 1/2 , \end{aligned} \tag{358}$$

while the components with parallel spins have vanishing norm in the large-cutoff limit, $||\psi_{2\uparrow\uparrow}||^2 = ||\psi_{2\downarrow\downarrow}||^2 = 0$, cf. (356).

Pion Decay Constant. A second constraint on the wave function is provided by the pion decay constant f_π, which appears in the semi-leptonic process $\pi \to \mu\nu$. The relevant matrix element is

$$\langle 0|\bar{\psi}_{\bar{d}}(0)\gamma^+\gamma_5\psi_u(0)|\pi^+(P^+)\rangle = i\sqrt{2}P^+ f_\pi \, , \tag{359}$$

$\bar{\psi}_{\bar{d}}$ and ψ_u denote the field operators of the $\bar{d}$ and u quark in the pion. If we insert all quantum numbers, the pion state to the right of the matrix element is given by

$$|\pi^+\rangle = \psi_{\bar{d}u} \otimes \frac{1}{\sqrt{6}}\left(|\bar{d}_{c\uparrow}u_{c\downarrow}\rangle - |\bar{d}_{c\downarrow}u_{c\uparrow}\rangle\right) . \tag{360}$$

The spatial (or internal) structure of the state is encoded in the light-cone wave function $\psi_2 \equiv \psi_{\bar{d}u}$. If we insert the Fock expansions (309) for $\bar{\psi}_{\bar{d}}$ and ψ_u as well as the pion state (350), we obtain the following constraint on the pion wave function,

$$\int_0^1 dx \int \frac{d^2k_\perp}{16\pi^3}\, \psi_{2\uparrow\downarrow}(x, \mathbf{k}_\perp) = \frac{f_\pi}{2\sqrt{3}} \, . \tag{361}$$

The left-hand-side is basically the (position space) 'wave function at the origin'. Quark (u) and antiquark ($\bar{d}$) thus have to sit on top of each other in order to have sizable probability for decay. Note that only the $L_z = 0$ component contributes. Concerning the effect of higher Fock states, it can be shown [93,22] that indeed only the valence wave function contributes to (361). This constraint is therefore *exact* and holds beyond a constituent picture. Empirically, the pion decay constant is $f_\pi = 92.4$ MeV [73].

As already stated, this is our second source of phenomenological information to fix cutoff and normalization. Using the explicit form (356) of the wave function, the constraint (361) becomes

$$\int_0^1 dx \int \frac{d^2k_\perp}{16\pi^3}\, \psi_{2\uparrow\downarrow}(x, \mathbf{k}_\perp) = \frac{N\Lambda^2}{96\pi^2} \stackrel{!}{=} \frac{f_\pi}{2\sqrt{3}} \, . \tag{362}$$

With (362) and (358) we now have two equations for our two parameters which accordingly are determined as

$$N = \sqrt{3}/f_\pi \, , \tag{363}$$

$$\Lambda = 4\pi f_\pi \simeq 1.16 \text{ GeV} \, . \tag{364}$$

The value (364) for the cutoff Λ is the standard scale below which chiral effective Lagrangians are believed to make sense [104]. It is reassuring that within our approximations we get exactly this value. This means that we are not doing something entirely stupid. A more severe test of consistency is provided by the next constraint to be satisfied.

Constraint from $\pi^0 \to 2\gamma$. This constraint has also been derived by Brodsky and Lepage [93] within an analysis of the $\pi\gamma$ transition form factor. It assumes the very simple form,

$$\int_0^1 dx\, \psi_{2\uparrow\downarrow}(x, \mathbf{0}_\perp) = \frac{\sqrt{3}}{f_\pi} . \tag{365}$$

Inserting the light-cone wave function (356), the right-hand-side simply becomes the normalization N which is indeed consistent with our findings (363) and (364). We mention in passing that the constraint (365) usually is the simplest way to fix the normalization N. Its derivation, however, is more complicated than that of (361).

Pion Form Factor. We proceed by calculating the pion electromagnetic formfactor. It is defined by the matrix element of the electromagnetic current J^μ_{em} between pion states,

$$\langle \pi : \boldsymbol{P} | J^\mu_{\text{em}} | \pi : \boldsymbol{P}' \rangle = 2(P + P')^\mu F(Q^2) , \quad Q^2 \equiv -(P - P')^2 . \tag{366}$$

Considering $\mu = +$ in a frame where $\boldsymbol{P} = (P^+, \mathbf{0})$ and $\boldsymbol{P}' = (P^+, \mathbf{q}_\perp)$ one is led to the the Drell-Yan formula [47,22],

$$F(\mathbf{q}_\perp^2) = \sum_{\lambda\lambda'} \int_0^1 dx \int \frac{d^2 k_\perp}{16\pi^3}\, \psi^*(x, \mathbf{k}'_\perp, \lambda, \lambda')\, \psi(x, \mathbf{k}_\perp, \lambda, \lambda') . \tag{367}$$

The transverse momentum of the struck quark is $\mathbf{k}'_\perp = \mathbf{k}_\perp + (1-x)\mathbf{q}_\perp$. The formula (367) with its overlap of two wave functions on the right-hand-side is rather similar to the nonrelativistic result as will be shown in what follows.

The form factor of a nonrelativistic system is given by the Fourier transform of the charge distribution (normalized to one), that is,

$$F(\mathbf{p}) = \int d^3 r\, \psi^*(\mathbf{r})\, \psi(\mathbf{r})\, e^{i\mathbf{p}\cdot\mathbf{x}} = \int \frac{d^3 k}{(2\pi)^3}\, \psi^*(\mathbf{k} + \mathbf{p})\, \psi(\mathbf{k}) . \tag{368}$$

It is important to note that $\mathbf{k}$ and $\mathbf{k}' \equiv \mathbf{k} + \mathbf{p}$ are *relative* momenta,

$$\mathbf{k} \equiv \frac{1}{M}(m_2 \mathbf{k}_1 - m_1 \mathbf{k}_2) , \tag{369}$$

(and analogously for $\mathbf{k}'$), so that the $\mathbf{k}_{1,2}$ are the actual particle momenta. Accordingly, $\mathbf{p}$ is the *relative* momentum transfer,

$$\mathbf{p} = \mathbf{k}' - \mathbf{k} = \frac{m_2}{M}(\mathbf{k}'_1 - \mathbf{k}_1) \equiv \frac{m_2}{M}\mathbf{q} = x_2 \mathbf{q} , \tag{370}$$

where, in the last step, we have used (268). Plugging this into the form factor (368) we obtain the formula,

$$F(\mathbf{q}) = \int \frac{d^3 k}{(2\pi)^3}\, \psi^*(\mathbf{k} + x_2 \mathbf{q})\, \psi(\mathbf{k}) , \tag{371}$$

which, as promised, is quite similar to (367).

If one sets the momentum transfer $\mathbf{q}_\perp$ in (367) equal to zero, one is left with the normalization integral (352), so that the form factor is automatically normalized to one (the same is true in the nonrelativistic case).

We will use the Drell-Yan formula for the form factor to determine the pion charge radius r_π, which is given by the slope of the form factor at vanishing momentum transfer,

$$F(q_\perp^2) \equiv 1 - \frac{r_\pi^2}{6} q_\perp^2 + O(q_\perp^4) \, . \tag{372}$$

Using (367) this results in the nice explicit formula,

$$r_\pi^2 = -\frac{3}{2} \int_0^1 dx \int \frac{d^2 k_\perp}{16\pi^3} \frac{\partial^2}{\partial q_i \partial q_i} \psi_2^*(\mathbf{k}_\perp + \bar{x}\mathbf{q}_\perp) \Big|_{q_\perp = 0} \psi(\mathbf{k}_\perp) \, . \tag{373}$$

Upon inserting the wave function (356), however, one encounters a problem. The sharp cutoff (corresponding to a step function) is too singular to lead to a reasonable result. The derivatives in (373) are concentrated at the boundary of the support of the wave function which in the end leads to artificial infinities. Thus, for the time being, we resort to a smooth cutoff,

$$\theta_\Lambda^s(x, \mathbf{k}_\perp) \equiv \exp\left[-\frac{k_\perp^2}{\Lambda^2 x(1-x)} \right] , \tag{374}$$

which basically transforms the sharp-cutoff wave function (356) to the Gaussian (286) (with m set to zero). Plugging this into (373) yields the pion charge radius

$$r_\pi^2 = \frac{12}{\Lambda^2} = \frac{3}{4\pi^2 f_\pi^2} = (0.60 \text{ fm})^2 \, . \tag{375}$$

This is the standard result for the NJL model [16] and has also been obtained within the instanton model [41]. It is slightly smaller than the experimental value, $r_\pi = 0.66$ [2], a discrepancy which is usually attributed to the use of the large-cutoff limit. A pole fit using our value of the pion charge radius is displayed in Fig. 10.

Transverse Size. As in Sect. 4, Example 3, we can use the light-cone wave function (356) to calculate the r.m.s. transverse momentum which leads to

$$\langle k_\perp^2 \rangle \equiv \int_0^1 dx \int \frac{d^2 k_\perp}{16\pi^3} k_\perp^2 \, ||\psi_{2\uparrow\downarrow}(x, \mathbf{k}_\perp)||^2 = \frac{\Lambda^2}{10} \simeq (370 \text{ MeV})^2 \, . \tag{376}$$

This actually coincides with the result (288) for the smooth-cutoff Gaussian wave function (286) or (374). Therefore, unlike the charge radius r_π, the r.m.s. transverse momentum is insensitive to the details of the cutoff procedure. We thus have $\langle k_\perp^2 \rangle^{1/2} \simeq m > M_\pi$, which confirms that the pion is highly relativistic.

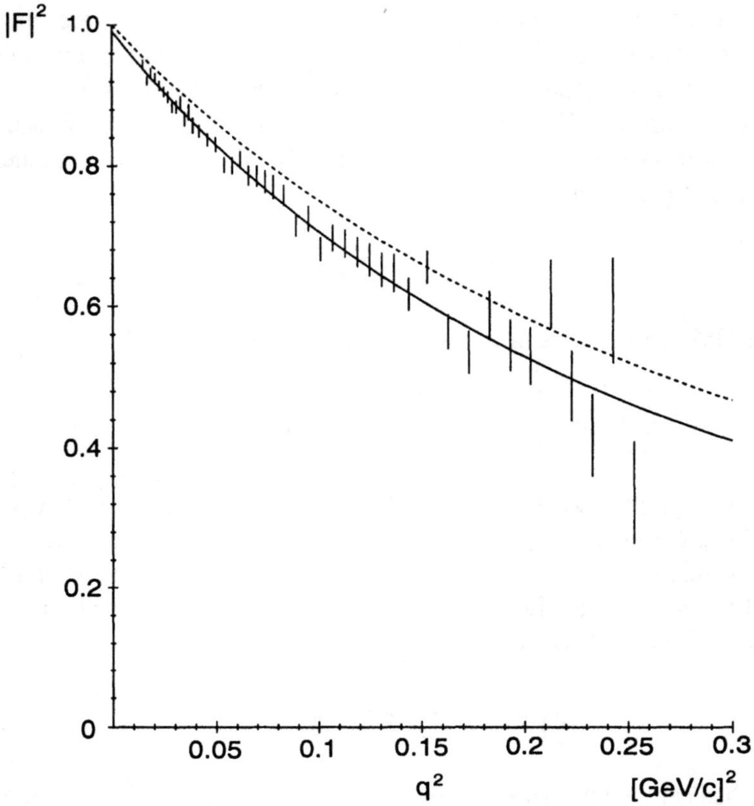

Fig. 10. *The pion form factor squared vs. momentum transfer $q^2 \equiv q_\perp^2$. The full line is the monopole fit of [2], $|F|^2 = n/(1 + q_\perp^2 r_\pi^2/6)^2$ with $n = 0.991$, $r_\pi^2 = 0.431$ fm^2; the dashed line is the same fit with our values, $n = 1$, $r_\pi^2 = 0.36$ fm^2. The agreement is consistent with the expected accuracy of 10%.*

The r.m.s. transverse momentum can easily be translated into a transverse size scale $R_\perp$,

$$R_\perp^2 \equiv 1/\langle k_\perp^2 \rangle \simeq (0.54 \text{ fm})^2 . \tag{377}$$

This is slightly smaller than the charge radius which we attribute to the fact that the charge distribution measured by the charge radius does not coincide with the distribution of baryon density. The 'core radius' $R_\perp$ is sometimes related to the decay constant f_π via the dimensionless quantity $C = f_\pi R_\perp$ [152]. In constituent quark models one typically gets $C \simeq 0.4$. This implies the fairly large value $R_\perp \simeq 0.8$ fm. Using standard many-body techniques, Bernard et al. have calculated this quantity in a model treating the pion as a collective excitation of the QCD vacuum, and find $C \simeq 0.2$ [13]. This result is close to what we get from (377),

$$C = f_\pi R_\perp \simeq 0.25 . \tag{378}$$

Thus, though we work within a constituent picture, we do get a reasonable value. We believe that this is due to the intrinsic consistency of the light-cone framework with the requirements of relativity, a feature that is lacking in the constituent quark model.

(Valence) Structure Function. The pion structure function arises in the description of deep inelastic scattering off charged pions. In terms of light-cone wave functions it is defined as the momentum fraction x times the sum of 'quark distributions' f_q weighted by the quark charges e_q. The quark distributions are given by the squares of light-cone wave functions integrated over $k_\perp$. For the valence structure function of the pion we thus have the formula [93,22],

$$\begin{aligned} F_2^v(x) &= x\left[e_u^2 f_u^v(x) + e_{\bar d}^2 f_{\bar d}^v(x)\right] = \frac{5}{9}\, x f^v(x) \\ &= \frac{5}{9}\, x \sum_{\lambda\lambda'} \int \frac{d^2k_\perp}{16\pi^3} \left|\psi_2(x, \mathbf{k}_\perp, \lambda, \lambda')\right|^2 , \end{aligned} \tag{379}$$

where the f^v denote the valence quark distributions. For the model wave function (356) the structure function becomes

$$F_2^v(x) = \frac{10}{3} x^2 (1-x) , \tag{380}$$

which in turn leads to the (valence) quark distribution

$$f^v(x) \equiv f_u^v(x) = f_{\bar d}^v(x) = 6\, x(1-x) . \tag{381}$$

The following consistency checks can be made. The probability to find a valence quark in the pion,

$$\int_0^1 dx\, f^v(x) = 1 , \tag{382}$$

is unity, as it should. For the mean value of the momentum fraction x carried by one of the quarks one finds

$$\langle x \rangle \equiv \int_0^1 dx\, x f^v(x) = 1/2 . \tag{383}$$

Thus, on average, the quarks share an equal amount of longitudinal momentum, which again, of course, is the correct result.

If one compares with other NJL calculations of the structure function [139,12] and with the empirical parton distributions in the literature [56], one finds reasonable qualitative agreement.

Let me finally point out that it is not entirely obvious to which actual momentum scale our results correspond. The transverse-momentum cutoff is x-dependent, $\Lambda^2(x) \simeq \Lambda^2 x(1-x)$, so, if we use the average x of (383), $\langle x \rangle \simeq 1/2$, a natural scale seems to be[13]

$$Q \equiv \Big[\langle x \rangle (1 - \langle x \rangle)\Big]^{1/2} \Lambda \simeq \Lambda/2 \simeq 600\,\text{MeV} . \tag{384}$$

Pion Distribution Amplitude. The pion distribution amplitude was originally introduced to describe hard exclusive processes involving pions [94,95], [92]. The pion formfactor at *large* Q^2, for example, is given by the following convolution formula,

$$F(Q^2) = \int_0^1 dx \int_0^1 dy\, \phi^*(x,Q)\, T_H(x,y;Q)\, \phi(y,Q) \Big[1 + O(1/Q)\Big] , \tag{385}$$

where Q denotes the *large* momentum transfer, T_H a 'hard scattering amplitude' and ϕ the pion distribution amplitude. While the amplitude T_H is the sum of all perturbative contributions to the scattering process, ϕ is nonperturbative in nature. The convolution formula is thus a prominent example where we see the principle of *factorization* into 'soft' and 'hard' physics at work.

The pion distribution amplitude is rather straightforwardly related to the light-cone wave function of the pion,

$$\begin{aligned} \phi(x,Q) &\sim \int \frac{dz^-}{4\pi}\, e^{ixP^+z^-/2} \langle 0 | \bar\psi(0) \gamma^+ \gamma_5 \psi(z^-, \mathbf{0}_\perp) | \pi(P^+) \rangle \\ &\sim \int \frac{d^2 k_\perp}{16\pi^3}\, \psi_{q\bar q}^{(Q)}(x, \mathbf{k}_\perp) . \end{aligned} \tag{386}$$

The normalization is fixed by demanding that ϕ integrates to unity.

Brodsky and Lepage have shown that ϕ obeys an evolution equation of the form [94,92],

$$Q \frac{\partial}{\partial Q} \phi(x,Q) = \int_0^1 dy\, V(x,y;Q)\, \phi(y,Q) . \tag{387}$$

where the evolution kernel V is determined by perturbative QCD. For $Q \to \infty$, (387) has the asymptotic solution

$$\phi_{\text{as}}(x) = 6\, x(1-x) , \tag{388}$$

(which is normalized to 1). The (nonasymptotic) pion distribution amplitude has been a rather controversial object. For a while people have tended to believe in a 'double-humped' shape of the amplitude (due to a factor $(1-2x)^2$), which was originally suggested by Chernyak and Zhitnitsky using QCD sum rules [33]. In 1995, however, the CLEO collaboration has published data

[13] I thank W. Schweiger for discussions on this point.

[131] that seemed to support an amplitude that is not too different from the asymptotic one [88]. Theoretical evidence for this fact has recently been reported in [10,11,121,122]. Belyaev and Johnson, for instance, have found two constraints which should be satisfied by the distribution amplitude [10],

$$\phi(x = 0.3) = 1 \pm 0.2 ,$$
$$\phi(x = 0.5) = 1.25 \pm 0.25 , \qquad (389)$$

which are consistent with an amplitude being close-to-asymptotic.

Last year, the experimental developments have culminated in a direct measurement of the distribution amplitude at Fermilab [4]. At a fairly low (i.e. nonasymptotic) momentum scale of $Q^2 \simeq 10$ GeV^2, one finds a distribution amplitude that is very close to the asymptotic one.

Let us see what we get in the NJL model. The distribution amplitude is given by

$$\phi_{\mathrm{NJL}}(x) = \frac{2\sqrt{3}}{f_\pi} \int \frac{d^2k_\perp}{16\pi^3} \psi_{2\uparrow\downarrow}(x, \mathbf{k}_\perp) = 6\,x(1-x) = \phi_{\mathrm{as}}(x) . \qquad (390)$$

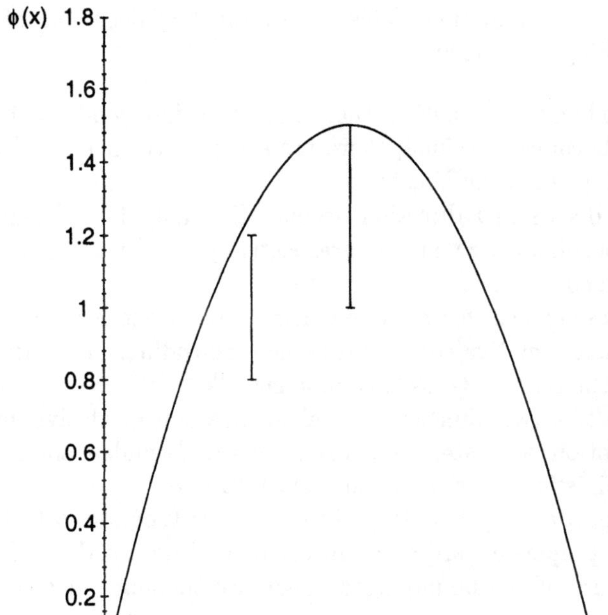

Fig. 11. *The (asymptotic) pion distribution amplitude. The vertical lines mark the constraints (389) of Belyaev and Johnson.*

Thus, in the large cutoff limit, the NJL distribution amplitude is exactly given by the asymptotic one! Upon comparing with (381), we see that ϕ_{NJL} coincides with the quark distribution f^v. This is accidental and stems from the fact that, due to the use of a step function in (356), $\psi_2 \sim |\psi_2|^2$.

In Fig. 11 we have displayed our distribution amplitude together with the constraints (389), represented by the vertical lines.

Our findings are thus consistent with the recent Fermilab experiment [4]. One should, however, be aware of the fact that our energy scale of $Q \simeq 0.6$ GeV from (384) is below the experimental one of $Q \simeq 3$ GeV. It is also somewhat lower than $Q \simeq 1$ GeV, which has been assumed by Belyaev and Johnson in their analysis of the distribution amplitude in terms of light-cone quark models [11].

6 Conclusions

In this lecture I have discussed an alternative approach to relativistic (quantum) physics based on Dirac's front form of dynamics. It makes use of the fact that for relativistic systems the choice of the time parameter is not unique. Our particular choice uses null-planes tangent to the light-cone as hypersurfaces of equal time. This apparently trivial change of coordinates has far-reaching consequences:

- The number of kinematical (i.e. interaction independent) Poincaré generators becomes maximal; there are seven of them instead of the usual six, among them the boosts.
- Lorentz boosts in z-direction become diagonal; the light-cone time and space coordinates, x^+ and x^-, respectively, do not get mixed but rather get rescaled.
- As a consequence, for many-particle systems one can introduce frame-independent relative coordinates, the longitudinal momentum fractions, x_i, and the relative transverse momenta, $\mathbf{k}_{\perp i}$.
- Because of a two-dimensional Galilei invariance, relative and center-of-mass motion separate. As a result, many formulae are reminiscent of nonrelativistic physics and thus very intuitive.
- This is particularly true for light-cone wave functions which, due to the last two properties, are boost invariant and do not depend on the total momentum of the bound state. They are therefore ideal tools to study relativistic particle systems.
- The last statement even holds for relativistic quantum field theory where one combines the unique properties of light-cone quantization with a Fock space picture. The central feature making this a reasonable idea is the triviality of the light-cone vacuum which accordingly is an eigenstate of the fully interacting Hamiltonian. This implies that the Fock operators create the *physical* particles from the ground state.

As expected, however, the principle of conservation of difficulties is at work so that there are problems to overcome. Of particular concern to us was the 'vacuum problem'. In instant-form quantum field theory, especially in QCD, many nonperturbative phenomena are attributed to the nontriviality of the vacuum which shows up via the appearance of condensates. These suggest that the instant-form vacuum is a complicated many-body state (like, e.g. the BCS ground state). In addition, many of these condensates signal the spontaneous (or anomalous) breakdown of a symmetry. The conceptual problem which arises at this point is to reconcile the existence of condensates with the triviality of the light-cone vacuum.

The idea put forward in these lectures is to reconstruct ground state properties from the particle spectrum. The latter is obtained by solving the light-cone Schrödinger equation for masses and wave functions of the associated bound states. For a relativistic quantum field theory, this, in principle, amounts to solving an infinite system of coupled integral equations for the amplitudes to find an ever increasing number of constituents in the bound state. Experience, however, shows that the *light-cone* amplitudes to find more than the valence quanta in the bound state tend to become rather small. Note that the same is *not* true within ordinary, that is, instant-form quantization.

A first explorative step towards solving a realistic light-cone Schrödinger equation was performed using an effective field theory, the NJL model. We have seen that, though we made a number of approximations, in particular by *enforcing* a constituent picture, a number of pionic observables are predicted with reasonable accuracy, among them the pion electromagnetic form factor, the pion charge and core radius and the pion valence structure function at low normalization scale. The pion distribution amplitude (in the chiral and large-cutoff limit) turns out to be asymptotic.

The results presented in these lectures provide some confidence that, also for real QCD, light-cone quantization may provide a road towards a reasonable constituent picture, in which hadrons are consistently described as bound states of a minimal number of constituents. How this hope can be turned into fact remains to be seen.

Acknowledgments

It is a pleasure to thank the organizers of the 39th Schladming winter school, in particular W. Schweiger and H. Latal, for inviting me to lecture and providing such a stimulating environment, both scientifically and socially.

References

1. E. Abdalla, M.C.B. Abdalla: Phys. Rep. **265**, 253 (1996)
2. S.R. Amendolia et al.: Nucl. Phys. B **277**, 168 (1986)
3. F. Antonuccio, S.J. Brodsky, S. Dalley: Phys. Lett. B **412**, 104 (1997)
4. D. Ashery: 'Diffractive Dissociation of High Momentum Pions', hep-ex/9910024 (1999)
5. T. Banks: TASI Lectures on Matrix Theory, hep-th/9911068 (1999)
6. T. Banks, W. Fischler, S. Shenker, L. Susskind: Phys. Rev. D **55**, 5112 (1997)
7. W.A Bardeen, R.B. Pearson, E. Rabinovici: Phys. Rev. D **21**, 1037 (1980)
8. A. Bassetto, G. Nardelli, R. Soldati: *Yang-Mills Theories in Algebraic Non-covariant Gauges: Canonical Quantization and Renormalization* (World Scientific, Singapore 1991)
9. A. Bassetto, F. Vian, L. Griguolo: 'Light-Front Vacuum and Instantons in Two Dimensions', hep-th/0004026 (2000)
10. V. Belyaev, M. Johnson: Mod. Phys. Lett. A **13**, 2909 (1998)
11. V. Belyaev, M. Johnson: Phys. Lett. B **423**, 379 (1998)
12. W. Bentz, T. Hama, T. Matsuki, K. Yazaki: Nucl. Phys. A **651**, 143 (1999)
13. V. Bernard, R. Brockmann, W. Weise: 'The Goldstone pion and the quark-antiquark pion (II). Pion size and decay', Regensburg preprint, TPR 85-02 (1985)
14. H. Bethe, E. Salpeter: *Quantum Mechanics of One- and Two-Electron Atoms* (Springer, Berlin, Heidelberg, New York 1957)
15. J. Bjorken, S. Drell: *Relativistic Quantum Mechanics* (McGraw-Hill, 1964)
16. A.H. Blin, B. Hiller, M. Schaden: Z. Phys. A **331**, 75 (1988)
17. N. Bogolubov, A. Logunov, A. Oksak, I. Todorov: *General Principles of Quantum Field Theory* (Kluwer Academic, 1990)
18. N. Bogolubov, A. Logunov, I. Todorov: *Introduction to Axiomatic Field Theory* (Benjamin, New York 1975)
19. A. Borderies, P. Grange, E. Werner: Phys. Lett. B **319**, 490 (1993)
20. A. Borderies, P. Grange, E. Werner: Phys. Lett. B **345**, 458 (1995)
21. S. Brodsky, C.-R. Ji, M. Sawicki: Phys. Rev. D **32**, 1530 (1985)
22. S. Brodsky, G. Lepage: 'Exclusive Processes in Quantum Chromodynamics'. In: *Perturbative Quantum Chromodynamics*, ed. by A.H. Mueller (World Scientific, 1989)
23. S. Brodsky, J.R. Roskies, R. Suaya: Phys. Rev. D **8**, 4574 (1973)
24. M. Burkardt: Phys. Rev. D **54**, 2913 (1996)
25. M. Burkardt: Phys. Rev. D **57**, 1136 (1998)
26. C. Callan, N. Coote, D. Gross: Phys. Rev. D **13**, 1649 (1976)
27. A. Casher: Phys. Rev. D **14**, 452 (1976)
28. A. Casher, S. Noskowicz, L. Susskind: Nucl. Phys. B **32**, 75 (1971)
29. A. Casher, L. Susskind: Phys. Rev. D **9**, 436 (1974)
30. D. Chakrabarti, A. Mukherjee, R. Kundu, A. Harindranath: Phys.Lett. B **480**, 409 (1999)
31. S. Chakrabarty, K. Gupta, N.N. Singh, A.N. Mitra: Prog. Part. Nucl. Phys. **22**, 43 (1989)
32. S.-J. Chang, R. G. Root, T.-M. Yan: Phys. Rev. D **7**, 1133 (1973)
33. V. Chernyak, A. Zhitnitsky: Phys. Rep. **112**, 173 (1984)
34. P. Chung, F. Coester, W. Polyzou: Phys. Lett. B **205**, 545 (1988)

35. F. Coester: Prog. Part. Nucl. Phys. **29**, 1 (1992)
36. F. Coester, W. Polyzou: Phys. Rev. D **26**, 1348 (1982)
37. S. Coleman: Comm. Math. Phys. **31**, 259 (1973)
38. R. Courant, D. Hilbert: *Methods of Mathematical Physics* (Interscience, New York 1962)
39. S. Dancoff: Phys. Rev. **78**, 382 (1950)
40. B. DeWitt: 'The space-time approach to quantum field theory'. In: *Relativity, Groups and Topology, II, Les Houches 1983, Session XL*, ed. by B.S. DeWitt, R. Stora (North-Holland, Amsterdam 1984)
41. D. Diakonov: 'Chiral Symmetry Breaking by Instantons'. In: *Selected Topics in Nonperturbative QCD, "Enrico Fermi", Course CXXX, Varenna, Italy, 1995*, ed. by A. Di Giacomo, D. Diakonov, (Proceedings International School of Physics IOS Press, Amsterdam 1996)
42. C. Dietmaier, T. Heinzl, M. Schaden, E. Werner: Z. Phys. A **334**, 220 (1989)
43. P.A.M. Dirac: Rev. Mod. Phys. **21**, 392 (1949)
44. P.A.M. Dirac: *Lectures on quantum mechanics* (Benjamin, New York 1964)
45. G. Domokos: 'Introduction to the Characteristic Initial-Value Problem in Quantum Field Theory'. In: *Boulder Lectures, Vol. XIV*, ed. by A.O. Barut, W.E. Brittin (Colorado University Press, Boulder 1972)
46. H. Dosch, S. Narison: Phys. Lett. B **417**, 173 (1998)
47. S. Drell, T.-M. Yan: Phys. Rev. Lett. **24**, 181 (1970)
48. Z. Dziembowski: Phys.Rev. D **37**, 778 (1988)
49. L. Faddeev, R. Jackiw: Phys. Rev. Lett. **60**, 1692 (1988)
50. V. Fock: Z. Phys. **98**, 145 (1935)
51. S. Fubini, A. Hanson, R. Jackiw: Phys. Rev. D **7**, 1732 (1973)
52. M. Fuda: Ann. Phys. (N.Y.) **197**, 265 (1990)
53. I. Gelfand, G. Shilov: *Generalized Functions* (Academic Press, New York 1964)
54. M. Gell-Mann, R. J. Oakes, B. Renner: Phys. Rev. **175**, 2195 (1968)
55. D. Gitman, I. Tyutin: *Quantization of Fields with Constraints* (Springer, Berlin, Heidelberg, New York 1990)
56. M. Glück, E. Reya, I. Schienbein: Eur. Phys. J. C **10**, 313 (1999)
57. J. Goldstone: Nuovo Cim. **19**, 154 (1961)
58. M. Green, J. Schwarz, E. Witten: *Superstring Theory* (Cambridge University Press, Cambridge 1987)
59. V.N. Gribov: Nucl. Phys. B **139**, 1 (1978)
60. F. Gross: 'Relativistic Nuclear Physics with the Spectator Model'. In: *Nuclear and Particle Physics on the Light-Cone, LAMPF Workshop, Los Alamos*, ed. by M.B. Johnson, L.S. Kisslinger (World Scientific, Singapore 1989)
61. A. Hanson, T. Regge, C. Teitelboim: *Constrained Hamiltonian Systems* (Accademia Nazionale dei Lincei, Rome 1976)
62. K. Harada, T. Heinzl, C. Stern: Phys. Rev. D **57**, 2460 (1998)
63. K. Harada, T. Sugihara, M. Taniguchi, M. Yahiro: Phys. Rev. D **49**, 4226 (1994)
64. T. Hatsuda, T. Kunihiro: Phys. Rep. **247**, 221 (1994)
65. T. Heinzl: 'Hamiltonian Formulations of Yang-Mills Quantum Theory and the Gribov Problem', hep-th/9604018 (1996)
66. T. Heinzl: Phys. Lett. B **388**, 129 (1996)
67. T. Heinzl: 'Light-Cone Dynamics of Particles and Fields', hep-th/9812190 (1998)

68. T. Heinzl, S. Krusche, S. Simbürger, E. Werner: Z. Phys. C **56**, 415 (1992)
69. T. Heinzl, N. Scheu, H. Kröger: 'Loss of Causality in Discretized Light-Cone Quantization', hep-th/9908173 (1999)
70. T. Heinzl, C. Stern, E. Werner, B. Zellermann: Z. Phys. C **72**, 353 (1996)
71. T. Heinzl, E. Werner: Z. Phys. C **62**, 521 (1994)
72. S. Hellerman, J. Polchinski: Phys. Rev. D **59**, 125002 (1999)
73. B.R. Holstein: 'Chiral Perturbation Theory: A Primer', hep-ph/9510344 (1995)
74. B.R. Holstein: Am. J. Phys. **66**, 507 (1998)
75. K. Hornbostel: Phys. Rev. D **45**, 3781 (1992)
76. K. Hornbostel, S. J. Brodsky, H.-C. Pauli: Phys. Rev. D **37**, 2363 (1988)
77. K. Itakura, S. Maedan: Phys. Rev. D **61**, 045009 (2000)
78. R. Jackiw: 'Functional Representations for Quantized Fields'. In: *Conformal Field Theory, Anomalies, and Superstrings, First Asia Pacific Workshop on High Energy Physics, Singapore 1987*, ed. by C.K. Chew et al. (World Scientific, Singapore 1987)
79. R. Jackiw: '(Constrained) Quantisation without Tears', hep-th/9306075 (1993)
80. X. Jaen, A. Molina, J. Llosa: 'Front Form Predictive Mechanics Non-Interaction Theorem' (LNP 212, Springer, Berlin, Heidelberg, New York 1984)
81. W. Jaus: Phys. Rev. D **41**, 3394 (1990)
82. C.-R. Ji, P. Chung, S. Cotanch: Phys. Rev. D **45**, 4214 (1992)
83. C.-R. Ji, S. Cotanch: Phys. Rev. D **41**, 2319 (1990)
84. P. Jordan, W. Pauli: Z. Phys. **47**, 151 (1928)
85. D. Kaplan: 'Effective field theories', nucl-th/9506035 (1995)
86. C. Kiefer, A. Wipf: Ann. Phys. (N.Y.) **236**, 241 (1994)
87. S.P. Klevansky: Rev. Mod. Phys. **64**, 649 (1992)
88. P. Kroll, M. Raulfs: Phys. Lett. B **387**, 848 (1996)
89. M. Lavelle, E. Werner, S. Glazek: Few Body Systems, Suppl. **2**, 519 (1987)
90. G. Leibbrandt: Phys. Rev. D **29**, 1699 (1984)
91. F. Lenz, M. Thies, K. Yazaki, S. Levit: Ann. Phys. (N.Y.) **208**, 1 (1991)
92. G. Lepage, S. Brodsky: Phys. Rev. D **22**, 2157 (1980)
93. G. Lepage, S. Brodsky, T. Huang, P. Mackenzie: 'Hadronic Wave Functions in QCD'. In: *Proceedings of the Banff Summer Institute, 1981*
94. G.P. Lepage, S. J. Brodsky: Phys. Lett. B **87**, 359 (1979)
95. G.P. Lepage, S. J. Brodsky: Phys. Rev. Lett. **43**, 545 (1979)
96. H. Leutwyler: Nuovo Cim. **37**, 556 (1965)
97. H. Leutwyler: Phys. Lett. B **48**, 45 (1974)
98. H. Leutwyler: Nucl. Phys. B **76**, 413 (1974)
99. H. Leutwyler, J. Stern: Ann. Phys. (N.Y.) **112**, 94 (1978)
100. H.H. Liu, D. E. Soper: Phys. Rev. D **48**, 1841 (1993)
101. W. Lucha, F. Schöberl, D. Gromes: Phys. Rep. **200**, 127 (1991)
102. S. Mandelstam: Nucl. Phys. B **213**, 149 (1983)
103. A. Manohar: 'Effective field theories'. In: *Perturbative and Nonperturbative Aspects of Quantum Field Theory*, ed. by H. Latal, W. Schweiger (LNP 479, Springer, Berlin, Heidelberg, New York 1997)
104. A. Manohar, H. Georgi: Nucl. Phys. B **234**, 189 (1984)
105. P. Maris, P.C. Tandy: Phys. Rev. C **61**, 045202 (2000)

106. T. Maskawa, K. Yamawaki: Prog. Theor. Phys. **56**, 270 (1976)
107. H. Melosh: Phys.Rev. D **9**, 1095 (1974)
108. C. Michael, F. Payne: Z. Phys. C **12**, 145 (1982)
109. S. Mukherjee, R. Nag, S. Sanyal, T. Morii, J. Morishita, M. Tsuge: Phys. Rep. **231**, 201 (1993)
110. N. Nakanishi, K. Yamawaki: Nucl. Phys. B **122**, 15 (1977)
111. Y. Nambu, G. Jona-Lasinio: Phys. Rev. **122**, 345 (1961)
112. J. Namysłowski: Prog. Part. Nucl. Phys. **14**, 49 (1985)
113. M. Neubert: Phys. Rep. **245**, 259 (1994)
114. R. Neville, F. Rohrlich: Nuovo Cim. **1A**, 625 (1971)
115. T. Newton, E. Wigner: Rev. Mod. Phys. **21**, 400 (1949)
116. L. Parker, G. Schmieg: Am. J. Phys. **38**, 218 (1970)
117. H.-C. Pauli, S. J. Brodsky: Phys. Rev. D **32**, 1993 (1985)
118. H.-C. Pauli, S. J. Brodsky: Phys. Rev. D **32**, 2001 (1985)
119. R. Peierls: Proc. Roy. Soc. A **214**, 143 (1952)
120. R.J. Perry: 'Light-Front Quantum Chromodynamics', nucl-th/9901080 (1999)
121. V.Y. Petrov, P. Pobylitsa: 'Pion Wave Function from the Instanton Vacuum Model', hep-ph/9712203 (1997)
122. V.Y. Petrov, M.V. Polyakov, R. Ruskov, C. Weiss, K. Goeke: Phys. Rev. D **59**, 114018 (1999)
123. A Polyakov.: *Gauge Fields and Strings* (Harwood Academic, Chur 1987)
124. E. Prokhvatilov, V. Franke: Sov. J. Nucl. Phys. **47**, 559 (1988)
125. E. Prokhvatilov, V. Franke: Sov. J. Nucl. Phys. **49**, 688 (1989)
126. A.V. Radyushkin: Acta Phys. Polon. B **26**, 2067 (1995)
127. C.D. Roberts, A. G. Williams: Prog. Part. Nucl. Phys. **33**, 477 (1994)
128. C.D. Roberts, S.M. Schmidt: 'Dyson-Schwinger Equations: Density, Temperature and Continuum Strong QCD', nucl-th/0005064 (2000)
129. F. Rohrlich: Acta Physica Austriaca, Suppl. VIII, 277 (1971)
130. S. Salmons, P. Grange, E. Werner: Phys. Rev. D **60**, 067701 (1999)
131. V. Savinov: 'A Measurement of the Form Factors of Light Pseudoscalar Mesons at a Large Momentum Transfer', hep-ex/9507005 (1995)
132. Ç. Şavklı, F. Tabakin: Nucl. Phys. A **628**, 645 (1998)
133. N. Scheu: On the Computation of Structure Functions and Mass Spectra in a Relativistic Hamiltonian Formalism: A Lattice Point of View. Ph.D. Thesis, Université Laval, Québec (1997), hep-th/9804190
134. J. Schwinger: Phys. Rev. **75**, 651 (1949)
135. J. Schwinger: Phys. Rev. **82**, 914 (1951)
136. J. Schwinger: Phys. Rev. **91**, 713 (1953)
137. J. Schwinger: Phil. Mag. **44**, 1171 (1953)
138. M.A. Shifman, A. I. Vainshtein, V. I. Zakharov: Nucl. Phys. B **147**, 385 (1979)
139. T. Shigetani, K. Suzuki, H. Toki: Phys. Lett. B **308**, 383 (1993)
140. S. Sokolov, A. Shatnii: Theor. Math. Phys. **37**, 1029 (1979)
141. P.P. Srivastava, S.J. Brodsky: Phys. Rev. D **61**, 025013 (2000)
142. C. Stern: Chirale Symmetriebrechung und gebundene Zustände in Lichtkegelquantenfeldtheorien. Ph.D. Thesis, Regensburg (1999)
143. R. Streater, A. Wightman: *PCT, Spin and Statistics, and All That* (Benjamin, New York 1963)
144. K. Sundermeyer: *Constrained Dynamics* (Springer, Berlin, Heidelberg, New York 1982)

145. L. Susskind: 'Another Conjecture about M(atrix) Theory', hep-th/9704080 (1997)
146. G. 't Hooft: Nucl. Phys. B **75**, 461 (1974)
147. G. 't Hooft: 'Gauge Theory for Strong Interactions'. In: *New Phenomena in Subnuclear Physics, Part A, International School of Subnuclear Physics, Erice, Italy, 1975*, ed. by A. Zichichi (Plenum, New York 1975)
148. I. Tamm: J. Phys. (Moscow) **9**, 449 (1945)
149. M. Terent'ev: Sov. J. Nucl. Phys. **24**, 106 (1976)
150. E. Tomboulis: Phys. Rev. D **8**, 2736 (1973)
151. U. Vogl, W. Weise: Prog. Part. Nucl. Phys. **27**, 195 (1991)
152. W. Weise: 'Quarks, Chiral Symmetry and Dynamics of Nuclear Constituents'. In: *International Review of Nuclear Physics, Vol.1, Quarks and Nuclei*, ed. by W. Weise (World Scientific, Singapore 1984)
153. K.G. Wilson, T.S. Walhout, A. Harindranath, W.-M. Zhang, R.J. Perry, S.D. Glazek: Phys. Rev. D **49**, 6720 (1994)
154. E. Witten: Nucl. Phys. B **145**, 110 (1978)
155. R.S. Wittman: 'Symmetry Breaking in the $(\phi^4)_2$ Theory and the Light-Front Vacuum'. In: *Nuclear and Particle Physics on the Light-Cone, LAMPF Workshop, Los Alamos, 1988*, ed. by M.B. Johnson, L.S. Kisslinger (World Scientific, Singapore 1989)
156. T.T. Wu: Phys. Lett. B **71**, 142 (1977)
157. F. Yndurain: *Relativistic Quantum Mechanics and Introduction to Field Theory* (Springer, Berlin, Heidelberg, New York 1996)
158. A. Zhitnitsky: Sov. J. Nucl. Phys. **43**, 999 (1986)
159. J. Zinn-Justin: *Quantum Field Theory and Critical Phenomena* (Clarendon Press, Oxford 1996)

Quantization of Constrained Systems

John R. Klauder

University of Florida, Departments of Physics and Mathematics,
Gainesville, FL 32611, USA

Abstract. The present article is primarily a review of the projection-operator approach to quantize systems with constraints. We study the quantization of systems with general first- and second-class constraints from the point of view of coherent-state, phase-space path integration, and show that all such cases may be treated, within the original classical phase space, by using suitable path-integral measures for the Lagrange multipliers which ensure that the quantum system satisfies the appropriate quantum constraint conditions. Unlike conventional methods, our procedures involve no δ-functionals of the classical constraints, no need for dynamical gauge fixing of first-class constraints nor any average thereover, no need to eliminate second-class constraints, no potentially ambiguous determinants, as well as no need to add auxiliary dynamical variables expanding the phase space beyond its original classical formulation, including no ghosts. Besides several pedagogical examples, we also study: (i) the quantization procedure for reparameterization invariant models, (ii) systems for which the original set of Lagrange multipliers are elevated to the status of dynamical variables and used to define an extended dynamical system which is completed with the addition of suitable conjugates and new sets of constraints and their associated Lagrange multipliers, (iii) special examples of alternative but equivalent formulations of given first-class constraints, as well as (iv) a comparison of both regular and irregular constraints.

1 Introduction

1.1 Initial Comments

The quantization of systems with constraints is important conceptually as well as practically. Principal techniques for the quantization of such systems involve conventional operator techniques [10], path integral techniques in terms of the original phase space variables [11], extended operator techniques involving ghost variables in addition to the original variables and extended path integral techniques also including ghost fields (see, e.g., [14,15,21]). However, these standard approaches are generally not unambiguous and may exhibit certain difficulties in application. A recent review [44] carefully analyzes these traditional methods and details their weaknesses as well as their strengths.

Canonical quantization generally requires the use of Cartesian coordinates and not more general coordinates [9]. Therefore, whenever we consider a dynamical system without any constraints whatsoever, we assume that the phase space of the unconstrained system is flat and admits a standard quantization of its canonical variables either in an operator form or in an equivalent path integral form. Next, suppose constraints exist, which, for the sake of discussion, we choose as a closed set of first-class constraints; extensions to treat more general constraints are presented in later sections. Whenever there are constraints the original set of variables is no longer composed solely of physical variables but now contains some unphysical variables as well. While such variables cause little concern from a classical standpoint, they are viewed as highly unwelcome from a quantum standpoint inasmuch as one generally wants to quantize only physical variables. Thus it is often deemed necessary to eliminate the unphysical variables leaving only the true physical degrees of freedom. Quantization of the true degrees of freedom is supposed to proceed as in the initial step. In the general case, however, a quantization of the remaining degrees of freedom is not straightforward or perhaps not even possible because the physical (reduced) phase space is non-Euclidean meaning that an obstruction has arisen where none existed before. An obstruction generally precludes the existence of self-adjoint (observable!) canonical operators satisfying the canonical commutation relations. In path integral treatments, such obstructions arise from the introduction of delta functionals that enforce the classical constraints and the concomitant need to introduce subsidiary delta functionals to select a compatible dynamical gauge in order to introduce a canonical symplectic structure on the physical phase space that generally is not flat. These are fundamental problems that seem difficult to overcome.

This article reviews a middle ground in the quantization procedure of systems with constraints which may be called the *projection-operator, coherent-state approach.* Briefly stated, quantization of the original, unconstrained variables proceeds without obstruction or ambiguity, while constraints are enforced by means of a well-chosen projection operator projecting the original Hilbert space onto the physical Hilbert subspace. This conservative framework is presented in the form of a phase-space path integral with the help of coherent states (which, while convenient, are not necessary). The difference between the present approach and other functional integral methods may be attributed to an alternative choice for the integration measure for the Lagrange multiplier variables. The present approach may be traced from [27]. In addition, some aspects of the projection operator approach have been presented in unpublished work of Shabanov [42]; see also [43].

1.2 Classical Background

For our initial discussion, let us briefly review the classical theory of constraints. Let $\{p_j, q^j\}$, $1 \leq j \leq J$, denote a set of dynamical variables, $\{\lambda^a\}$, $1 \leq a \leq A$, a set of Lagrange multipliers, and $\{\phi_a(p,q)\}$ a set of constraints. Then the dynamics of a constrained system may be summarized in the form of an action principle by means of the classical action (summation implied)

$$I = \int [p_j \dot{q}^j - H(p,q) - \lambda^a \phi_a(p,q)]\, dt \,. \tag{1}$$

The resultant equations that arise from the action read

$$\begin{aligned}
\dot{q}^j &= \frac{\partial H(p,q)}{\partial p_j} + \lambda^a \frac{\partial \phi_a(p,q)}{\partial p_j} \equiv \{q^j, H\} + \lambda^a \{q^j, \phi_a\} \,, \\
\dot{p}_j &= -\frac{\partial H(p,q)}{\partial q^j} - \lambda^a \frac{\partial \phi_a(p,q)}{\partial q^j} \equiv \{p_j, H\} + \lambda^a \{p_j, \phi_a\} \,, \\
\phi_a&(p,q) = 0 \,,
\end{aligned} \tag{2}$$

where $\{\cdot,\cdot\}$ denotes the Poisson bracket. The set of conditions $\{\phi_a(p,q) = 0\}$ defines the *constraint hypersurface.* If the constraints satisfy

$$\{\phi_a(p,q), \phi_b(p,q)\} = c_{ab}{}^c \, \phi_c(p,q) \,, \tag{3}$$
$$\{\phi_a(p,q), H(p,q)\} = h_a{}^b \, \phi_b(p,q) \,, \tag{4}$$

then we are dealing with a system of first-class constraints. If the coefficients $c_{ab}{}^c$ and $h_a{}^b$ are constants, then it is a closed system of first-class constraints; if they are suitable functions of the variables p, q, then it is called an open first-class constraint system. If (3) fails, or (3) and (4) fail, then the constraints are said to be second class (see below).

For first-class constraints it is sufficient to impose the constraints at the initial time inasmuch as the equations of motion will ensure that the constraints are fulfilled at all future times. Such an initial imposition of the constraints is called an *initial value equation.* Furthermore, the Lagrange multipliers are not determined by the equations of motion; rather the solutions of the equations of motion depend on them. By specifying the Lagrange multipliers, the solution can be forced to satisfy an additional ("gauge") condition. Observable quantities are gauge invariant and, hence, do not depend on the gauge arbitrariness. For second-class constraints, on the other hand, the Lagrange multipliers are determined by the equations of motion in such a way that the constraints are satisfied for all time.

In the remainder of this section we review standard quantization procedures for systems with closed first-class constraints, both of the operator and path integral variety, pointing out some problems in each approach.

1.3 Quantization First: Standard Operator Quantization

For a system of closed first-class constraints we assume (with $\hbar = 1$) that

$$[\Phi_a(P,Q), \Phi_b(P,Q)] = i c_{ab}{}^c \Phi_c(P,Q) \,, \tag{5}$$
$$[\Phi_a(P,Q), \mathcal{H}(P,Q)] = i h_a{}^b \Phi_b(P,Q) \,, \tag{6}$$

where Φ_a and $\mathcal{H}$ denote self-adjoint constraint and Hamiltonian operators, respectively. Following [10], we adopt the quantization prescription given by

$$i\dot{W}(P,Q) = [W(P,Q), \mathcal{H}(P,Q)] \tag{7}$$

where W denotes a general function of the kinematical operators $\{Q^j\}$ and $\{P_j\}$ which are taken as a self-adjoint, irreducible representation of the commutation rules $[Q^j, P_k] = i\delta^j_k \mathbb{1}$, with all other commutators vanishing. The equations of motion hold for all time t, say $0 < t < T$. On the other hand, the conditions

$$\Phi_a(P,Q)|\psi\rangle_{\mathrm{phys}} = 0 \tag{8}$$

to select the physical Hilbert space are imposed only at time $t = 0$ as the analog of the initial value equation; the quantum equations of motion ensure that the constraint conditions are fulfilled for all time.

The procedure of Dirac has potential difficulties if zero lies in the continuous spectrum of the constraint operators for in that case there are no normalizable solutions of the constraint condition. We face the same problem, of course, and our resolution is discussed below.

1.4 Reduction First: Standard Path Integral Quantization

Faddeev [11] has given a path integral formulation in the case of closed first-class constraint systems as follows. The formal path integral

$$\begin{aligned}&\int \exp\{i\textstyle\int_0^T [p_j\dot{q}^j - H(p,q) - \lambda^a\phi_a(p,q)]\,dt\}\,\mathcal{D}p\,\mathcal{D}q\,\mathcal{D}\lambda \\ &\qquad = \int \exp\{i\textstyle\int_0^T [p_j\dot{q}^j - H(p,q)]\,dt\}\,\delta\{\phi(p,q)\}\,\mathcal{D}p\,\mathcal{D}q\end{aligned} \tag{9}$$

may well encounter divergences in the remaining integrals. Therefore, subsidiary conditions in the form $\chi^a(p,q) = 0$, $1 \le a \le A$, are imposed picking out (ideally) one gauge equivalent point per gauge orbit, and in addition a factor in the form of the Faddeev-Popov determinant is introduced to formally preserve canonical covariance. The result is the path integral

$$\int \exp\{i\textstyle\int_0^T [p_j\dot{q}^j - H(p,q)]\,dt\}\,\delta\{\chi(p,q)\}\det(\{\chi^a,\phi_b\})\delta\{\phi(p,q)\}\,\mathcal{D}p\,\mathcal{D}q\,. \tag{10}$$

This result may also be expressed as

$$\int \exp\{i\textstyle\int_0^T [p^*_j\dot{q}^{*j} - H^*(p^*,q^*)]\,dt\}\,\mathcal{D}p^*\,\mathcal{D}q^*\,, \tag{11}$$

namely, as a path integral over a reduced phase space in which the δ-functionals have been used to eliminate $2A$ integration variables.

The final expression generally involves an integration over a non-Euclidean phase space for which the conventional definition of the path integral is typically ill defined. Thus this widely used prescription is not without its difficulties.

1.5 Quantization First $\not\equiv$ Reduction First

The two schemes illustrated in the preceding sections are different in principle. In the initial case, one quantizes *first* and reduces *second;* in the latter case, one reduces *first* and quantizes *second.* For certain systems the results of these different procedures are the same, but that is not universally the case, as we now proceed to illustrate.

Let us consider the example of a single degree of freedom specified by the classical action

$$I = \int [p\dot{q} - \lambda(p^2 + q^4 - E)]\, dt \, . \tag{12}$$

Observe that the classical Hamiltonian vanishes and there is a single constraint. The question we pose is: For what values of E, $E > 0$, is the quantum theory nontrivial?

On the one hand, according to the procedure of Dirac, the physical Hilbert space is either empty or one-dimensional, spanned by the nonvanishing eigenvector $|\psi_n\rangle$ that satisfies

$$(P^2 + Q^4)|\psi_n\rangle = E_n\, |\psi_n\rangle \, , \tag{13}$$

for E_n one of the purely discrete eigenvalues for the "Hamiltonian" $P^2 + Q^4$.

On the other hand, the procedure of Faddeev leads initially to

$$\int e^{i\int p\, dq}\, \delta\{p^2 + q^4 - E\}\, \mathcal{D}p\, \mathcal{D}q \, . \tag{14}$$

Next, we fix a gauge, e.g., $p = 0$, in which case the reduced phase space propagator is given by

$$\begin{aligned} \int e^{i\int p\, dq}\, \delta\{p^2 + q^4 - E\}\, \Pi(4q^3)\, \delta\{p\}\, \mathcal{D}p\, \mathcal{D}q & \\ = 0 \, , & \end{aligned} \tag{15}$$

which vanishes due to cancellation between the term with $q > 0$ and the term with $q < 0$. Note that the symbol Π denotes a formal multiplication over all time points. An alternative evaluation may be given if we allow only the term with $q > 0$, which is achieved by instead using

$$\begin{aligned} & \int e^{i\int p\, dq}\, \delta\{p^2 + q^4 - E\}\, \delta\{p\}\, \mathcal{D}p\, \mathcal{D}q^4 \\ & = \int \delta\{q^4 - E\}\, \mathcal{D}q^4 \\ & = 1 \, . \end{aligned} \tag{16}$$

Either of these choices imposes *no* restriction on E whatsoever. Ignoring the nonphysical nature of the variables involved, one might possibly impose the condition

$$\oint p\, dq = 2\pi n \, , \tag{17}$$

leading to a Bohr-Sommerfeld spectrum, which for this problem is incorrect. (The reader is encouraged to examine alternative choices of gauge.)
Remark: It is instructive in this example to note that the Faddeev-Popov determinant $\Delta = \Pi(4q^3)$ and the reduced phase space is the single point $(p,q) = (0, E^{1/4})$. The point $(p,q) = (0, -E^{1/4})$ corresponds to a Gribov copy.

Clearly, in this case, reduction before quantization has led to the wrong result. Some workers may assert that such errors are merely "order $\hbar$ corrections". Although true, this argument cannot be used to defend the general procedure since the role of a quantization procedure, after all, should be to determine the *correct* spectrum for a specific problem, not a spectrum that is potentially incorrect even in its *leading order*. Examples of other work which arrive at the same conclusion are given in [7,39,1,40].

1.6 Outline of the Remaining Sections

In the following section, Sect. 2, we present an overview of the *projection operator approach* to constrained system quantization with an emphasis on coherent-state representations. Section 3 deals with coherent-state path integrals without gauge fixing for closed first-class constrained systems. Extensions to general constraints such as open first-class or second-class systems are the subject of Sect. 4. Section 5 is devoted to selected examples of first-class systems, while Sect. 6 concentrates on two rather special applications. Finally, in Sect. 7 we comment on some other applications of the projection operator approach that have not been discussed in this paper.

2 Overview of the Projection Operator Approach to Constrained System Quantization

2.1 Coherent States

Canonical quantization is consistent only for Cartesian phase space coordinates [9], and we assume that our original and unconstrained set of classical dynamical variables fulfill that condition. Then, for each classical coordinate q^j and momentum p_j, $1 \leq j \leq J$, we may introduce associated self-adjoint canonical operators Q^j and P_j, acting in a separable Hilbert space $\mathfrak{H}$, and which satisfy, in units where $\hbar = 1$, the canonical commutation relations $[Q^j, P_k] = i\delta^j_k\,\mathbb{1}$, with all other commutation relations vanishing. With the fiducial vector $|0\rangle \in \mathfrak{H}$ a suitable normalized state – typically the ground state of a (unit-frequency) harmonic oscillator (but not always!) – we introduce the *canonical coherent states* (see, e.g., [33,32])

$$|p,q\rangle \equiv e^{-iq^jP_j}\, e^{ip_jQ^j}\,|0\rangle\,, \tag{18}$$

for all $(p,q) \in \mathbb{R}^{2J}$, where $p = \{p_j\}$ and $q = \{q^j\}$. These states admit a resolution of unity in the form [37]

$$\mathbb{1} = \int |p,q\rangle\langle p,q|\, d\mu(p,q)\,, \qquad d\mu(p,q) \equiv d^J p\, d^J q/(2\pi)^J\,, \tag{19}$$

integrated over $\mathbb{R}^{2J}$.

The unit operator resides in the Hilbert space $\mathfrak{H}$ of the unconstrained system. We may conveniently represent this Hilbert space as follows. We first introduce the *reproducing kernel* $\langle p'',q''|p',q'\rangle$ as the overlap matrix element between any two coherent states. This expression is a bounded, continuous function that characterizes a (reproducing kernel Hilbert space) representation of $\mathfrak{H}$ appropriate to the unconstrained system as follows. A dense set of vectors in the associated functional Hilbert space is given by vectors of the form

$$\psi(p,q) \equiv \langle p,q|\psi\rangle = \sum_{l=1}^{L} \alpha_l\, \langle p,q|p_{(l)},q_{(l)}\rangle\,, \tag{20}$$

for arbitrary sets $\{\alpha_l\}$ and $\{p_{(l)},q_{(l)}\}$ with $L<\infty$. The inner product of two such vectors is given by

$$(\psi,\xi) \equiv \sum_{l,m=1}^{L,M} \alpha_l^*\, \beta_m\, \langle p_{(l)},q_{(l)}|\overline{p}_{(m)},\overline{q}_{(m)}\rangle \tag{21}$$

$$= \int \psi(p,q)^*\xi(p,q)\, d\mu(p,q)\,, \tag{22}$$

where ξ is a second function defined in a manner analogous to ψ. A general vector in the functional Hilbert space is defined by a Catchy sequence of such vectors, and all such vectors are given by bounded, continuous functions. The first form of the inner product applies in general only to vectors in the dense set, while the second form of the inner product holds for arbitrary vectors in the Hilbert space. We shall have more to say below regarding reproducing kernels and reproducing kernel Hilbert spaces.

2.2 Constraints

Now suppose we introduce constraints into the quantum theory [27]. In particular, we assume that $\mathbb{E}$ denotes a *projection operator onto the constraint subspace*, i.e., the subspace on which the quantum constraints are satisfied (in a sense to be defined below), and which is called the physical Hilbert space $\mathfrak{H}_{\rm phys} \equiv \mathbb{E}\mathfrak{H}$. Later we shall discuss examples of $\mathbb{E}$. Hence, if $|\psi\rangle \in \mathfrak{H}$ denotes a general vector in the original (unconstrained) Hilbert space, the vector $\mathbb{E}|\psi\rangle \in \mathfrak{H}_{\rm phys}$ represents its component within the physical subspace.

As a Hilbert space, the physical subspace also admits a functional representation by means of a reproducing kernel which may be taken as $\langle p'', q''|\mathbf{E}|p', q'\rangle$. In the same manner as before, it follows that a dense set of vectors in $\mathfrak{H}_{\text{phys}}$ is given by functions of the form

$$\psi(p,q) \equiv \langle p,q|\mathbf{E}|\psi\rangle = \sum_{l=1}^{L} \alpha_l \, \langle p,q|\mathbf{E}|p_{(l)}, q_{(l)}\rangle \ , \tag{23}$$

for arbitrary sets $\{\alpha_l\}$ and $\{p_{(l)}, q_{(l)}\}$ with $L < \infty$. The inner product of two such vectors is given by

$$\begin{aligned}(\psi,\xi) &\equiv \sum_{l,m=1}^{L,M} \alpha_l^* \, \beta_m \, \langle p_{(l)}, q_{(l)}|\mathbf{E}|\overline{p}_{(m)}, \overline{q}_{(m)}\rangle \\ &= \int \psi(p,q)^* \xi(p,q) \, d\mu(p,q) \ . \end{aligned}\tag{24}$$

Again, a general vector in the functional Hilbert space is defined by means of a Catchy sequence, and all such vectors are given by bounded, continuous functions. Note well, in the case illustrated, that even though $\mathbf{E}\mathfrak{H} \subset \mathfrak{H}$, the functional representation of the unconstrained and the constrained Hilbert spaces are *identical*, namely by functions of $(p,q) \in \mathbb{R}^{2J}$, and the form of the inner product is *identical* in the two cases. This situation holds even if $\mathfrak{H}_{\text{phys}}$ is one dimensional!

The relation between the self-adjoint constraint operators Φ_a, $1 \leq a \leq A$, $A < \infty$, and the projection operator $\mathbf{E}$ may take several different forms. Unless otherwise specified, we shall assume that $\Sigma_a \Phi_a^2$ is self adjoint and that

$$\mathbf{E} = \mathbf{E}(\Sigma_a \Phi_a^2 \leq \delta(\hbar)^2) \ , \tag{25}$$

where $\delta = \delta(\hbar)$ (*not* a Dirac δ-function!) is a reauthorization parameter which is chosen in accord with rules to be discussed below.

2.3 Dynamics for First-Class Systems

Suppose further that the Hamiltonian $\mathcal{H}$ respects the first-class character of the constraints. It follows in this case that $[\mathbf{E}, \mathcal{H}] = 0$ or stated otherwise that

$$e^{-i\mathcal{H}t}\mathbf{E} \equiv \mathbf{E}\, e^{-i\mathcal{H}t}\mathbf{E} \equiv \mathbf{E}\, e^{-i(\mathbf{E}\mathcal{H}\mathbf{E})\,t}\mathbf{E} \, . \tag{26}$$

Dynamics in the physical subspace is then fully determined by the propagator on $\mathfrak{H}_{\text{phys}}$, which is given in the relevant functional representation by

$$\langle p'', q''|e^{-i\mathcal{H}t}\mathbf{E}|p', q'\rangle \ . \tag{27}$$

In (27) we have achieved a fully gauge invariant propagator without having to reduce the range or even the number of the original classical variables nor change the original form of the inner product on the functional Hilbert space representation. Any observable $\mathcal{O}$ – $\mathcal{H}$ included – satisfies $[\mathbb{E}, \mathcal{O}] = 0$, and relations similar to (26) follow with $\mathcal{H}$ replaced by $\mathcal{O}$.

2.4 Zero in the Continuous Spectrum

The foregoing scenario has assumed that the appropriate $\mathfrak{H}_{\rm phys}$ is given by means of a projection operator $\mathbb{E}$ acting on the original Hilbert space. This situation holds true whenever the set of quantum constraints admits zero as a common point in their discrete spectrum; in that case $\mathbb{E}$ defines the subspace where the constraints all vanish. That situation may not always hold true, but even in case zero lies in the continuous spectrum for some or all of the constraints, a suitable result may generally be given by matrix elements of a sequence of rescaled projection operators, say $c_\delta\,\mathbb{E}$, $c_\delta > 0$, as $\delta \to 0$. Specifically, we consider the limit of a sequence of reproducing kernels $c_\delta\,\langle p'', q''|\mathbb{E}|p', q'\rangle$, which – if the limit is a nonvanishing continuous function – defines a new reproducing kernel, and thereby a new reproducing kernel Hilbert space, within which the appropriate constraints are fulfilled. In such a limit certain variables may cease to be relevant and as a consequence the local integral representation of the inner product, if any, may require modification. On the other hand, the definition of the inner product by sums involving the reproducing kernel will always hold. We refer to the result of such a limiting operation as a *reduction of the reproducing kernel.* A simple example should help clarify what we mean by a reduction of the reproducing kernel.

Consider the example

$$\langle p'', q''|\mathbb{E}|p', q'\rangle$$
$$= \pi^{-1/2} \int_{-\delta}^{\delta} \exp[-\tfrac{1}{2}(k-p'')^2 + ik(q''-q') - \tfrac{1}{2}(k-p')^2]\,dk\,, \quad (28)$$

where $\mathbb{E} = \mathbb{E}(P^2 \leq \delta^2)$, which defines a reproducing kernel for any $\delta > 0$ that corresponds to an infinite dimensional Hilbert space. (If $\delta = \infty$ the result is the usual canonical coherent state overlap and characterizes the unconstrained Hilbert space.) If we take the limit of the expression as it stands as $\delta \to 0$, the result will vanish. What we need to do is extract the *germ* of the projection operator as we let δ go to zero. Therefore, let us first multiply this expression by $\pi^{1/2}/(2\delta)$ (c_δ in this case) and take the limit $\delta \to 0$. The result is the expression

$$\mathcal{K}(p''; p') = e^{-\frac{1}{2}(p''^2 + p'^2)}\,, \quad (29)$$

which has become a reproducing kernel that characterizes a *one*-dimensional Hilbert space with every functional representative proportional to $\chi_o(p) \equiv$

$\exp(-p^2/2)$. This one-dimensional Hilbert space representation also admits a local integral representation for the inner product given by

$$(\chi,\chi) = \int |\chi(p)|^2 \, dp/\sqrt{\pi} \,. \tag{30}$$

In the present case, it is clear that one may reduce the reproducing kernel even further by choosing $p = c$, an arbitrary but fixed constant. This kind of reduction – in which the latter reproducing kernel Hilbert space is equivalent to the former reproducing kernel Hilbert space – is analogous to choosing a gauge in the classical theory. We shall see another example of this latter kind of reduction later.

The example presently under discussion is also an important one inasmuch as it illustrates how a constraint operator with its zero lying in the continuous spectrum is dealt with in the coherent-state, projection-operator approach. Some other approaches to deal with the problem of zero in the continuous spectrum may be traced from [46,19,20,22,36].

2.5 Alternative View of Continuous Zeros

If $\delta \ll 1$ in (28), then it may be approximately evaluated as

$$\begin{aligned} &\langle p'', q''|\mathbf{E}|p', q'\rangle \\ &\quad = \pi^{-1/2}\delta \, e^{-\frac{1}{2}(p''^2+p'^2)} \frac{\sin[\delta(q''-q')]}{\delta(q''-q')} + O(\delta^2) \,. \end{aligned} \tag{31}$$

When $\delta = 10^{-1000}$, or some other extremely tiny factor, it is clear that for all practical purposes it is sufficient to accept just the first term in (31), ignoring the term $O(\delta^2)$, as the "reduced" reproducing kernel. The resultant expression is indeed a proper reproducing kernel for which inner products are given with the full set of integration variables and the normal integration range. So long as q values are "normal sized", e.g., $|q| < 10^{500}$ in the present case, there is no practical distinction between the space of functions generated by (29) and that generated by (31). In other words, if δ is chosen extremely close to zero, but still positive, it is not actually necessary to take the limit $\delta \to 0$ in order to do practical calculations. Even though this is the case, we shall for the most part in the examples we study take a full reduction by first rescaling the reproducing kernel (by an appropriate factor c_δ) and then taking the limit $\delta \to 0$.

3 Coherent State Path Integrals Without Gauge Fixing

As introduced above, canonical coherent states may be defined by the relation

$$|p,q\rangle \equiv e^{-iq^j P_j} \, e^{ip_j Q^j} \, |0\rangle \,, \tag{32}$$

for all (p,q), where the financial vector $|0\rangle$ traditionally denotes a normalized, unit frequency, harmonic oscillator ground state, and the coherent states admit a resolution of unity in the form

$$\mathbb{1} = \int |p,q\rangle\langle p,q|\, d\mu(p,q)\,, \qquad\qquad d\mu(p,q) \equiv d^J p\, d^J q/(2\pi)^J\,, \tag{33}$$

where the integration is over $\mathbb{R}^{2J}$. Note that the integration domain and the form of the measure are unique.

Based on such coherent states, we introduce the upper symbol for a general operator $\mathcal{H}(P,Q)$,

$$H(p,q) \equiv \langle p,q|\mathcal{H}(P,Q)|p,q\rangle = \langle p,q| : H(P,Q) : |p,q\rangle \tag{34}$$

which is related to the normal-ordered form as shown. (N.B. Some workers would call $H(p,q)$ the lower symbol.) If $\mathcal{H}(P,Q)$ denotes the quantum Hamiltonian, then we shall adopt $H(p,q)$ as the classical Hamiltonian. We also note that an important one-form generated by the coherent states is given by $i\langle p,q|d|p,q\rangle = p_j\, dq^j$.

Using these quantities, and the time ordering operator T, the coherent state path integral for the propagator generated by the time-dependent Hamiltonian $\mathcal{H}(P,Q) + \lambda^a(t)\Phi_a(P,Q)$ is readily given by

$$\begin{aligned}
&\langle p'',q''|\mathsf{T}e^{-i\int_0^T [\mathcal{H}(P,Q)+\lambda^a(t)\Phi_a(P,Q)]\,dt}|p',q'\rangle \\
&= \lim_{\epsilon\to 0}\int \prod_{l=0}^{N}\langle p_{l+1},q_{l+1}|e^{-i\epsilon(\mathcal{H}+\lambda_l^a\Phi_a)}\,|p_l,q_l\rangle \prod_{l=1}^{N} d\mu(p_l,q_l) \\
&= \int \exp\{i\textstyle\int[i\langle p,q|(d/dt)|p,q\rangle - \langle p,q|\mathcal{H}+\lambda^a(t)\Phi_a|p,q\rangle]\,dt\}\,\mathcal{D}\mu(p,q) \\
&= \mathcal{M}\int \exp\{i\textstyle\int[p_j\dot{q}^j - H(p,q) - \lambda^a(t)\phi_a(p,q)]\,dt\}\,\mathcal{D}p\,\mathcal{D}q\,.
\end{aligned} \tag{35}$$

Here, in the second line, we have set $\epsilon \equiv T/(N+1)$, made a Trotter- product like approximation to the evolution operator, repeatedly inserted the resolution of unity, and set $p_{N+1}, q_{N+1} = p'', q''$ and $p_0, q_0 = p', q'$. In the third and fourth lines we have formally interchanged the continuum limit and the integrations, and written for the integrand the form it would assume for continuous and differentiable paths ($\mathcal{M}$ denotes a formal normalization constant). The result evidently depends on the chosen form of the functions $\{\lambda^a(t)\}$.

3.1 Enforcing the Quantum Constraints

Let us next introduce the quantum analog of the initial value equation. For simplicity we assume that the constraint operators form a compact group; more general situations are dealt with below. In that case

$$\mathbb{E} \equiv \int e^{-i\xi^a\Phi_a(P,Q)}\,\delta\xi = \mathbb{E}(\Phi_a = 0,\ 1 \le a \le A) = \mathbb{E}(\Sigma_a\Phi_a^2 = 0) \tag{36}$$

defines a *projection operator* onto the subspace for which $\Phi_a = 0$ provided that $\delta\xi$ denotes the normalized, $\int \delta\xi = 1$, group invariant measure. It follows from (36) that

$$e^{-i\tau^a \Phi_a}\mathbb{E} = \mathbb{E}\ . \tag{37}$$

We now project the propagator (35) onto the quantum constraint subspace which leads to the following set of relations

$$\begin{aligned} &\int \langle p'', q''|\mathsf{T}e^{-i\int[\mathcal{H}+\lambda^a(t)\Phi_a]\,dt}\,|\overline{p}', \overline{q}'\rangle\langle\overline{p}', \overline{q}'|\mathbb{E}\,|p', q'\rangle\, d\mu(\overline{p}', \overline{q}') \\ &\quad = \langle p'', q''|\mathsf{T}e^{-i\int[\mathcal{H}+\lambda^a(t)\Phi_a]\,dt}\,\mathbb{E}\,|p', q'\rangle \\ &\quad = \lim\,\langle p'', q''|[\overleftarrow{\prod_l}(e^{-i\epsilon\mathcal{H}}e^{-i\epsilon\lambda^a_l\Phi_a})]\,\mathbb{E}\,|p', q'\rangle \\ &\quad = \langle p'', q''|e^{-iT\mathcal{H}}e^{-i\tau^a\Phi_a}\,\mathbb{E}\,|p', q'\rangle \\ &\quad = \langle p'', q''|e^{-iT\mathcal{H}}\,\mathbb{E}\,|p', q'\rangle\ , \end{aligned} \tag{38}$$

where τ^a incorporates the functions λ^a as well as the structure parameters ${c_{ab}}^c$ and ${h_a}^b$. Alternatively, this expression has the formal path integral representation

$$\int \exp\{i\textstyle\int[p_j\dot{q}^j - H(p,q) - \lambda^a(t)\phi_a(p,q)]\,dt - i\xi^a\phi_a(p', q')\}\,\mathcal{D}\mu(p,q)\,\delta\xi\,. \tag{39}$$

On comparing (35) and (39), we observe that *after projection onto the quantum constraint subspace the propagator is entirely independent of the choice of the Lagrange multiplier functions. In other words, the projected propagator is gauge invariant.*

We may also express the physical (projected) propagator in a more general form, namely,

$$\begin{aligned} &\int \exp\{i\textstyle\int[p_j\dot{q}^j - H(p,q) - \lambda^a(t)\phi_a(p,q)]\,dt\}\,\mathcal{D}\mu(p,q)\,\mathcal{D}C(\lambda) \\ &\quad = \langle p'', q''|e^{-iT\mathcal{H}}\,\mathbb{E}\,|p', q'\rangle \end{aligned} \tag{40}$$

provided that $\int \mathcal{D}C(\lambda) = 1$ and that such an average over the functions $\{\lambda^a(t)\}$ introduces (at least) one factor $\mathbb{E}$.

3.2 Reproducing Kernel Hilbert Spaces

The coherent-state matrix elements of $\mathbb{E}$ define a fundamental kernel

$$\mathcal{K}(p'', q''; p', q') \equiv \langle p'', q''|\mathbb{E}|p', q'\rangle\ , \tag{41}$$

which is a bounded, continuous function for any projection operator $\mathbf{E}$, including the unit operator. It follows that $\mathcal{K}(p'',q'';p',q')^* = \mathcal{K}(p',q';p'',q'')$ as well as

$$\sum_{k,l=1}^{K} \alpha_k^* \alpha_l \mathcal{K}(p_k,q_k;p_l,q_l) \geq 0 \tag{42}$$

for all sets $\{\alpha_k\}$, $\{(p_k,q_k)\}$, and all $K < \infty$. The last relation is an automatic consequence of the complex conjugate property and the fact that

$$\mathcal{K}(p'',q'';p',q') = \int \mathcal{K}(p'',q'';p,q)\,\mathcal{K}(p,q;p',q')\,d\mu(p,q) \tag{43}$$

holds in virtue of the coherent state resolution of unity and the properties of $\mathbf{E}$. As noted earlier, the function $\mathcal{K}$ is called the *reproducing kernel* and the Hilbert space it engenders is termed a *reproducing kernel Hilbert space* [2,3,38]. A dense set of elements in the Hilbert space is given by functions of the form

$$\psi(p,q) = \sum_{k=1}^{K} \alpha_k \mathcal{K}(p,q;p_k,q_k)\ , \tag{44}$$

and the inner product of this function has two equivalent forms given by

$$(\psi,\psi) = \sum_{k,l=1}^{K} \alpha_k^* \alpha_l \mathcal{K}(p_k,q_k;p_l,q_l) \tag{45}$$

$$= \int \psi(p,q)^* \psi(p,q)\,d\mu(p,q)\ . \tag{46}$$

The inner product of two distinct functions may be determined by polarization of the norm squared [41]. Clearly, the entire Hilbert space is characterized by the reproducing kernel $\mathcal{K}$. Change the kernel $\mathcal{K}$ and one changes the representation of the Hilbert space. Following a suitable limit of the kernel $\mathcal{K}$, it is even possible to change the *dimension* of the Hilbert space, as already illustrated earlier.

3.3 Reduction of the Reproducing Kernel

Suppose the reproducing kernel depends on a number of variables and additional parameters. We can generate new reproducing kernels from a given kernel by a variety of means. For example, the expressions

$$\mathcal{K}_1(p'';p') = \mathcal{K}(p'',c;p',c)\ , \tag{47}$$

$$\mathcal{K}_2(p'';p') = \int f(q'')^* f(q')\mathcal{K}(p'',q'';p',q')\,dq''\,dq'\ , \tag{48}$$

$$\mathcal{K}_3(p'',q'';p',q') = \lim \mathcal{K}(p'',q'';p',q') \tag{49}$$

each generate a new reproducing kernel provided the resultant function remains continuous. In general, however, the inner product in the Hilbert space generated by the new reproducing kernel is only given by an analog of (21) and not by (22), although frequently some sort of local integral representation for the inner product may exist.

Let us offer an example of the reduction of a reproducing kernel that is a slight generalization of the earlier example. Let the expression

$$\langle p'', q''|\mathbb{E}|p', q'\rangle \equiv \\ \pi^{-J/2}\int_{-\delta}^{\delta}\cdots\int_{-\delta}^{\delta}\exp[-\tfrac{1}{2}(k-p'')^2+ik\cdot(q''-q')-\tfrac{1}{2}(k-p')^2]\,d^Jk \quad (50)$$

denote a reproducing kernel for any $\delta > 0$. In the present case it follows that $\mathbb{E} \equiv \Pi_{j=1}^{J}\mathbb{E}(-\delta \leq P_j \leq \delta)$. When $\delta \to 0$, then (50) vanishes. However, if we first multiply by δ^{-J} – or more conveniently by $\pi^{J/2}(2\delta)^{-J}$ – before taking the limit, the result becomes

$$\lim_{\delta\to 0}\pi^{J/2}(2\delta)^{-J}\langle p'', q''|\mathbb{E}|p', q'\rangle = \exp(-\tfrac{1}{2}p''^2)\exp(-\tfrac{1}{2}p'^2)\,, \quad (51)$$

which is continuous and therefore denotes the reproducing kernel for some Hilbert space. Note that the classical variables q'' and q' have disappeared, which on reference to (32) implies that all "$P_j = 0$". In the present example, the resultant Hilbert space is one dimensional, and the inner product may be given either by a sum as in (21) involving the p variables alone or by a local integral representation now using the measure $\pi^{-J/2}\,d^Jp$, namely,

$$(\chi, \chi) = \textstyle\int|\chi(p)|^2\pi^{-J/2}\,d^Jp\,. \quad (52)$$

This example illustrates the case where the constraints are "$P_j = 0$", for all j, a situation where zero lies in the continuous spectrum.

We may also use this example to illustrate how *several* constraints may be replaced by a *single* constraint. The several constraints "$P_j = 0$", for all j, were first approximated by the regularized constraints $P_j^2 \leq \delta^2$, $\delta > 0$, for all j. Alternatively, we may also regularize the constraints in the form $\Sigma_j P_j^2 \leq \delta^2$. Furthermore, if we use $\mathbb{E} = \mathbb{E}(\Sigma_j P_j^2 \leq \delta^2)$, then it is clear that a new prefactor, also proportional to δ^{-J}, can be chosen so that (51) again emerges as $\delta \to 0$.

3.4 Single Regularized Constraints

Clearly, the *set* of real classical constraints $\phi_a = 0$, $1 \leq a \leq A$, is equivalent to the *single* classical constraint $\Sigma_a\phi_a^2 = 0$. Likewise, the set of (idealized) quantum constraints "$\Phi_a|\psi\rangle_{\rm phys} = 0$", $1 \leq a \leq A$, where each Φ_a is self adjoint, is equivalent to the single (idealized) quantum constraint "$\Sigma_a\Phi_a^2|\psi\rangle_{\rm phys} = 0$", where we further assume that $\Sigma_a\Phi_a^2$ is a self-adjoint operator. In general, however, the only solution of the idealized quantum constraint is the zero vector, $|\psi\rangle_{\text{phys}} = 0$.

To overcome this difficulty, we relax the idealized quantum constraint and instead generally adopt the regularized form of the constraint given by $|\psi\rangle_{\text{phys}} \in \mathcal{H}_{\text{phys}} \equiv \mathbb{E}\,\mathcal{H}$, where

$$\mathbb{E} = \mathbb{E}(\Sigma_a \Phi_a^2 \le \delta(\hbar)^2) . \tag{53}$$

Here $\delta(\hbar)$ is a reauthorization parameter and the inequality means that in a spectral resolution of $\Sigma_a \Phi_a^2 \equiv \int_0^\infty \lambda\, dE(\lambda)$ that

$$\mathbb{E} \equiv \int_0^{\delta(\hbar)^2} dE(\lambda) = E(\delta(\hbar)^2) . \tag{54}$$

Let us examine three basic examples.

First, let zero be in the discrete spectrum of $\Sigma_a \Phi_a^2$. Then, it follows that there exists a $\delta_1(\hbar)^2$ such that for all $\delta(\hbar)^2$, $0 < \delta(\hbar)^2 < \delta_1(\hbar)^2$, then $\mathbb{E}(\Sigma_a \Phi_a^2 \le \delta(\hbar)^2) = \mathbb{E}(\Sigma_a \Phi_a^2 = 0)$.

Second, if $\Sigma_a \Phi_a^2$ has its zero in the continuum, then $\mathbb{E}(\Sigma_a \Phi_a^2 \le \delta^2)$ is infinite dimensional for all $\delta > 0$, but $\mathbb{E}$ vanishes weakly as $\delta \to 0$. For such cases we consider $c_\delta \mathbb{E}$ and choose the sequence c_δ to weakly extract the *germ* of $\mathbb{E}$ as $\delta \to 0$, just as in the examples illustrated above.

Third, in a case to be studied later, suppose that zero is *not* in the spectrum of the operator $\Sigma_a \Phi_a^2$. Since $\Sigma_a \phi_a^2 = 0$ classically, it follows that spectral values of $\Sigma_a \Phi_a^2$ are $o(\hbar^0)$ close to zero. A relevant example discussed later is where $\Phi_1 = P$ and $\Phi_2 = Q$. Then $\mathbb{E}(P^2 + Q^2 \le \hbar) = |0\rangle\langle 0|$ is a one- dimensional projection operator onto the harmonic oscillator ground state $|0\rangle$. Observe in this case that $\delta(\hbar)^2 = \hbar$, which vanishes when $\hbar \to 0$; note also that we cannot reduce this parameter further since $\mathbb{E}(P^2 + Q^2 < \hbar) \equiv 0$. Thus, in some cases, whether we use "$\le$" or "$<$" in the inequality defining the projection operator can make a real difference.

The three types of examples discussed above illustrate three qualitatively different behaviors possible for the projection operator $\mathbb{E}$. As we proceed, we shall find the use of a single regularized constraint will be an important unifying principle in treating the most general multiple constraint situation imaginable.

3.5 Basic First-Class Constraint Example

Consider the system with two degrees of freedom, a vanishing Hamiltonian, and a single constraint, characterized by the action

$$I = \int [\tfrac{1}{2}(p_1 \dot{q}_1 - q_1 \dot{p}_1 + p_2 \dot{q}_2 - q_2 \dot{p}_2) - \lambda(q_2 p_1 - p_2 q_1)]\, dt , \tag{55}$$

where for notational convenience we have lowered the index on the q variables. Note that we have chosen a different form for the kinematic part of the action which amounts to a change of phase for the coherent states, and in particular a factor of $e^{ipq/2}$ has been introduced on the right side of (18), or, equivalently, both generators appear in the same exponent. It follows that

$$\begin{aligned}\mathcal{M}\int \exp\{i{\textstyle\int}[\tfrac{1}{2}(p_1\dot{q}_1 - q_1\dot{p}_1 + p_2\dot{q}_2 - q_2\dot{p}_2) - \lambda(q_2p_1 - p_2q_1)]\,dt\}\\ \times \mathcal{D}p\,\mathcal{D}q\,\mathcal{D}C(\lambda)\\ = \langle p'', q''|\mathbb{E}\,|p', q'\rangle\,,\end{aligned} \tag{56}$$

where we choose

$$\mathbb{E} = (2\pi)^{-1}\int_0^{2\pi} e^{-i\xi(Q_2P_1 - P_2Q_1)}\,d\xi = \mathbb{E}\,(L_3 = 0)\,. \tag{57}$$

Based on the fact [33] that

$$\begin{aligned}\langle p'', q''|p', q'\rangle = \exp(-\tfrac{1}{2}|z_1''|^2 - \tfrac{1}{2}|z_2''|^2 - \tfrac{1}{2}|z_1'|^2 - \tfrac{1}{2}|z_2'|^2)\\ \times \exp(z_1''^{*}z_1' + z_2''^{*}z_2')\,,\end{aligned} \tag{58}$$

where $z_1' \equiv (q_1' + ip_1')/\sqrt{2}$, etc., it is straightforward to show that

$$\begin{aligned}\langle p'', q''|\mathbb{E}\,|p', q'\rangle = \exp(-\tfrac{1}{2}|z_1''|^2 - \tfrac{1}{2}|z_2''|^2 - \tfrac{1}{2}|z_1'|^2 - \tfrac{1}{2}|z_2'|^2)\\ \times I_0(\,(z_1''^{*2} + z_2''^{*2})^{1/2}(z_1'^2 + z_2'^2)^{1/2}\,)\,,\end{aligned} \tag{59}$$

with I_0 a standard Bessel function. We emphasize again that although the Hilbert space has been strictly reduced by the introduction of $\mathbb{E}$, the reproducing kernel (59) leads to a reproducing kernel Hilbert space with an inner product having the same number of integration variables and domain of integration as in the unconstrained case.

4 Application to General Constraints

4.1 Classical Considerations

When dealing with a general constraint situation it will typically happen that the self-consistency of the equations of motion may determine some or all of the Lagrange multipliers in order for the system to remain on the classical constraint hypersurface. For example, if the Hamiltonian attempts to force points initially lying on the constraint hypersurface to leave that hypersurface, then the Lagrange multipliers must supply the necessary forces for the system to remain on the constraint hypersurface.

We may elaborate on this situation as follows. Since $\phi_a(p,q) = 0$ for all a defines the constraint hypersurface, it is also necessary, for all a, that

$$\dot{\phi}_a(p,q) \equiv \{\phi_a(p,q), H(p,q)\} + \lambda^b(t)\{\phi_a(p,q), \phi_b(p,q)\} \equiv 0 \tag{60}$$

also holds on the constraint hypersurface. If the Poisson brackets fulfill the conditions given in (3) and (4), then it follows that $\dot{\phi}_a(p,q) \equiv 0$ on the constraint hypersurface for any choice of the Lagrange multipliers $\{\lambda^a(t)\}$. This is the case for first-class constraints, and to obtain specific solutions to the dynamical equations it is necessary to specify some choice of the Lagrange multipliers, i.e., to select a gauge. However, if (3), or (3) and (4) do *not* hold on the constraint hypersurface, the situation changes. For example, let us first assume that (4) holds but that

$$\Delta_{ab}(p,q) \equiv \{\phi_a(p,q), \phi_b(p,q)\} \tag{61}$$

is a *non*singular matrix on the constraint hypersurface. In this case it follows that we must choose $\lambda^a(t) \equiv 0$ for all a to satisfy (60). More generally, we must choose

$$\lambda^a(t) \equiv -(\Delta^{-1}(p,q))^{ab} \{\phi_b(p,q), H(p,q)\} \tag{62}$$

in order that (60) will be satisfied. When the Lagrange multipliers are not arbitrary but rather must be specifically chosen in order to keep the system on the constraint hypersurface, then we say that we deal with second-class constraints. Of course, there are also intermediate situations where part of the constraints are first class while some are second class; in this case the matrix $\Delta_{ab}(p,q)$ would be singular but would have a nonzero rank on the constraint hypersurface.

Remark: It is useful to also imagine solving the differential equation (60) as a computer might do it, namely, by an iteration procedure. In particular, we could imagine evolving by a small time step ϵ by the first (Hamiltonian) term, then using the second (constraint) term to choose λ^a at that moment to force the system back onto the constraint hypersurface, and afterwards continuing this procedure over and over. A proper solution can be obtained this way by taking the limit of these approximate solutions as $\epsilon \to 0$. An analogue of this procedure will be used in our quantum discussion.

There is also a third situation that may arise, namely constraints that are *first* class from a classical point of view but are *second* class quantum mechanically. Such constraints would arise if

$$\Delta_{ab}(p,q) = Y_{ab}{}^c(p,q)\,\phi_c(p,q)\ , \tag{63}$$

where, for the sake of convenience, we assume that the quantities $Y_{ab}{}^c(p,q)$ are all uniformly bounded away from zero and infinity, i.e., $0 < C \leq Y_{ab}{}^c(p,q) \leq D < \infty$. In that case $\Delta_{ab}(p,q)$ would vanish on the constraint hypersurface classically. Quantum mechanically, the expression for the commutator is proportional to $\hbar$ and may be taken as

$$i[\Phi_a(P,Q), \Phi_b(P,Q)] = \tfrac{1}{2}[Y_{ab}{}^c(P,Q)\Phi_c(P,Q) + \Phi_c(P,Q)Y_{ab}{}^c(P,Q)]\,. \tag{64}$$

If we assume that "$\Phi_a(P,Q)|\psi\rangle_{\rm phys} = 0$", then self-consistency requires that "$[\Phi_c(P,Q), Y_{ab}{}^c(P,Q)]|\psi\rangle_{\rm phys} = 0$", an expression which is now proportional to $\hbar^2$. If this expression vanishes it causes no problem; if it does not vanish one says that there is a "factor ordering problem" or an "anomaly". As Jackiw has often stressed, it would be preferable to call an anomaly "quantum mechanical symmetry breaking", a phrase which more accurately describes what it is and what it does. Whatever it is called, the resultant quantum constraints are second class even though they were classically first class. As is well known, gravity falls into just this category.

In this section we take up the quantization of these more general situations involving both first and second class constraints [27].

4.2 Quantum Considerations

As in previous sections, we let $\mathbb{E}$ denote the projection operator onto the quantum constraint subspace. Motivated by the classical comments given above we consider the quantity

$$\lim \langle p'', q''|\mathbb{E}\, e^{-i\epsilon\mathcal{H}}\mathbb{E}\, e^{-i\epsilon\mathcal{H}} \cdots \mathbb{E}\, e^{-i\epsilon\mathcal{H}}\mathbb{E}\,|p', q'\rangle \tag{65}$$

where the limit, as usual, is for $\epsilon \to 0$. The physics behind this expression is as follows. Reading from right to left we first impose the quantum initial value equation, and then propagate for a small amount of time (ϵ). Next we recognize that the system may have left the quantum constraint subspace, and so we project it back onto that subspace, and so on over and over. In the limit that $\epsilon \to 0$ the system remains within the quantum constraint subspace and (65) actually leads to

$$\langle p'', q''|\mathbb{E}\, e^{-iT(\mathbb{E}\,\mathcal{H}\mathbb{E})}\mathbb{E}\,|p', q'\rangle\ , \tag{66}$$

which clearly illustrates temporal evolution entirely within the quantum constraint subspace. If we assume that $\mathbb{E}\,\mathcal{H}\mathbb{E}$ is a self-adjoint operator, then we conclude that (66) describes a unitary time evolution within the quantum constraint subspace.

The expression (65) may be developed in two additional and alternative ways. First, we repeatedly insert the resolution of unity in such a way that (65) becomes

$$\lim \int \prod_{l=0}^{N} \langle p_{l+1}, q_{l+1}|\mathbb{E}\, e^{-i\epsilon\mathcal{H}}\mathbb{E}\,|p_l, q_l\rangle \prod_{l=1}^{N} d\mu(p_l, q_l)\ . \tag{67}$$

We wish to turn this expression into a formal path integral, but the procedure used previously relied on the use of unit vectors, and the vectors $\mathbb{E}\,|p,q\rangle$ are generally not unit vectors. Thus, let us rescale the factors in the integrand introducing

$$|p, q\rangle\rangle \equiv \mathbb{E}\,|p, q\rangle / \|\mathbb{E}\,|p, q\rangle\| \tag{68}$$

which are unit vectors. If we let $M'' \equiv \|\mathbf{E}\,|p'',q''\rangle\|$, $M' \equiv \|\mathbf{E}\,|p',q'\rangle\|$, and observe that $\|\mathbf{E}\,|p,q\rangle\|^2 = \langle p,q|\mathbf{E}\,|p,q\rangle$, it follows that (67) may be rewritten as

$$M''M' \lim \int \prod_{l=0}^{N} \langle\!\langle p_{l+1},q_{l+1}|e^{-i\epsilon\mathcal{H}}|p_l,q_l\rangle\!\rangle \prod_{l=1}^{N} \langle p_l,q_l|\mathbf{E}\,|p_l,q_l\rangle\, d\mu(p_l,q_l)\,. \quad (69)$$

This expression is represented by the formal path integral

$$M''M' \int \exp\{i{\textstyle\int}[i\langle\!\langle p,q|(d/dt)|p,q\rangle\!\rangle - \langle\!\langle p,q|\mathcal{H}\,p,q\rangle\!\rangle]\,dt\}\,\mathcal{D}_E\mu(p,q)\,, \quad (70)$$

where the new formal measure for the path integral is defined in an evident fashion from its lattice prescription. We can also reexpress this formal path integral in terms of the original bra and ket vectors in the form

$$\begin{aligned} M''M' \int \exp\{i{\textstyle\int}[i\langle p,q|\mathbf{E}\,(d/dt)\mathbf{E}\,|p,q\rangle/\langle p,q|\mathbf{E}\,|p,q\rangle \\ -\langle p,q|\mathbf{E}\,\mathcal{H}\mathbf{E}\,|p,q\rangle/\langle p,q|\mathbf{E}\,|p,q\rangle]\,dt\}\,\mathcal{D}_E\mu(p,q)\,. \end{aligned} \quad (71)$$

This last relation concludes our second route of calculation beginning with (65).

The third relation we wish to derive uses an integral representation for the projection operator $\mathbf{E}$ generally given by

$$\mathbf{E} = {\textstyle\int} e^{-i\xi^a\Phi_a(P,Q)}\, f(\xi)\,\delta\xi \quad (72)$$

for a suitable function f. Thus we rewrite (65) in the form

$$\begin{aligned} \lim \int \langle p'',q''|e^{-i\epsilon\lambda_N^a\Phi_a}e^{-i\epsilon\mathcal{H}}e^{-i\epsilon\lambda_{N-1}^a\Phi_a}e^{-i\epsilon\mathcal{H}}\cdots e^{-i\epsilon\lambda_1^a\Phi_a}e^{-i\epsilon\mathcal{H}}e^{-i\epsilon\lambda_0^a\Phi_a}|p',q'\rangle \\ \times f(\epsilon\lambda_N)\cdots f(\epsilon\lambda_0)\,\delta\epsilon\lambda_N\cdots\delta\epsilon\lambda_0\,. \end{aligned} \quad (73)$$

Next we insert the coherent-state resolution of unity at appropriate places to find that (73) may also be given by

$$\begin{aligned} \lim \int \langle p_{N+1},q_{N+1}|e^{-i\epsilon\lambda_N^a\Phi_a}|p_N,q_N\rangle \prod_{l=0}^{N-1} \langle p_{l+1},q_{l+1}|e^{-i\epsilon\mathcal{H}}e^{-i\epsilon\lambda_l^a\Phi_a}|p_l,q_l\rangle \\ \times[\prod_{l=1}^{N} d\mu(p_l,q_l)\,f(\epsilon\lambda_l)\,\delta\epsilon\lambda_l]\,f(\epsilon\,l_0)\,\delta\epsilon\lambda_0\,. \end{aligned} \quad (74)$$

Following the normal pattern, this last expression may readily be turned into a formal coherent-state path integral given by

$$\int \exp\{i{\textstyle\int}[p_j\dot{q}^j - H(p,q) - \lambda^a(t)\phi_a(p,q)]\,dt\}\,\mathcal{D}\mu(p,q)\mathcal{D}E(\lambda)\,, \quad (75)$$

where $E(\lambda)$ is a measure designed so as to insert the projection operator $\mathbb{E}$ at every time slice. This usage of the Lagrange multipliers to ensure that the quantum system remains within the quantum constraint subspace is similar to their usage in the classical theory to ensure that the system remains on the classical constraint hypersurface. On the other hand, it is also possible to use the measure $E(\lambda)$ in the case of closed *first*-class constraints as well; this would be just one of the acceptable choices for the measure $C(\lambda)$ designed to put at least one projection operator $\mathbb{E}$ into the propagator.

In summary, we have established the equality of the three expressions

$$\begin{aligned}
&\langle p'',q''|\mathbb{E}\, e^{-iT(\mathbb{E}\mathcal{H}\mathbb{E})}\mathbb{E}\,|p',q'\rangle \\
&\quad = M''M'\int \exp\{i{\textstyle\int}[i\langle p,q|\mathbb{E}\,(d/dt)\mathbb{E}\,|p,q\rangle/\langle p,q|\mathbb{E}\,|p,q\rangle \\
&\qquad\qquad -\langle p,q|\mathbb{E}\,\mathcal{H}\mathbb{E}\,|p,q\rangle/\langle p,q|\mathbb{E}\,|p,q\rangle]\,dt\}\,\mathcal{D}_E\mu(p,q) \\
&\quad = \int \exp\{i{\textstyle\int}[p_j\dot{q}^j - H(p,q) - \lambda^a(t)\phi_a(p,q)]\,dt\}\,\mathcal{D}\mu(p,q)\mathcal{D}E(\lambda)\,. \qquad (76)
\end{aligned}$$

This concludes our initial derivation of path integral formulas for general constraints. Observe that we have not introduced any δ-functionals, nor, in the middle expression, reduced the number of integration variables or the limits of integration in any way even though in that expression the integral over the Lagrange multipliers has been carried out.

4.3 Universal Procedure to Generate Single Regularized Constraints

The preceding section developed a functional integral approach suitable for a general set of constraints, but it had one weak point, namely, it required prior knowledge of the constraints themselves in order to choose $f(\xi)$ in (72) so as to construct the appropriate projection operator. Is there any way to construct $\mathbb{E}$ *without* prior knowledge of the form the constraints will take? The answer is *yes!*

We first observe that the evolution operator appearing in (35) may be written in the form of a lattice limit given by

$$\lim_{\epsilon\to 0}\overleftarrow{\prod_{1\le n\le N}}\left[\mathsf{T}e^{-i\int_{(n-1)\epsilon}^{n\epsilon}\mathcal{H}(t)\,dt}\right]\left[\mathsf{T}e^{-i\int_{(n-1)\epsilon}^{n\epsilon}\lambda^a(t)\Phi_a\,dt}\right], \qquad (77)$$

where $\epsilon\equiv T/N$ and the directed product (symbol $\longleftarrow$) also respects the time ordering. Thus, this expression is simply an alternating sequence of short-time evolutions, first by $\lambda^a(t)\Phi_a$, second by $\mathcal{H}(t)$, a pattern which is then repeated $N-1$ more times. The validity of this Trotter-product form follows whenever $\mathcal{H}(t)^2+\Phi_a\delta^{ab}\Phi_b$ is essentially self adjoint for all t, $0\le t\le T$. As a slight generalization, we shall assume that $\mathcal{H}(t)^2+\Phi_a M^{ab}\Phi_b$ is essentially self adjoint for all t, $0\le t\le T$. Here the real, symmetric coefficients M^{ab} $(=M^{ba})$

are the elements of a positive-definite matrix, i.e., $\{M^{ab}\} > 0$. For a finite number of constraints, $A < \infty$, it is sufficient to assume that $M^{ab} = \delta^{ab}$. Other choices for M^{ab} may be relevant when $A = \infty$. (We do not explicitly consider the case $A = \infty$ in this article; for some examples see [28].)

With all this in mind, we shall explain the construction of a formal integration procedure [29] whereby

$$\int \mathsf{T}e^{-i\int_{(n-1)\epsilon}^{n\epsilon} \lambda^a(t)\Phi_a\,dt}\,\mathcal{D}R(\lambda) = \mathbb{E}(\Phi_a M^{ab}\Phi_b \le \delta(\hbar)^2)\,, \tag{78}$$

and for which the integral represented by $\int \cdots \mathcal{D}R(\lambda)$ is independent of the set of operators $\{\Phi_a\}$ and the Hamiltonian operator $\mathcal{H}(t)$ for all t. First, introduce a formal Gaussian measure $\mathcal{D}S_{\gamma_n}(\lambda)$ such that

$$\begin{aligned}&\int \mathsf{T}e^{-i\int_{(n-1)\epsilon}^{n\epsilon} \lambda^a(t)\Phi_a\,dt}\,\mathcal{D}S_{\gamma_n}(\lambda)\\ &= \mathcal{N}\int \mathsf{T}e^{-i\int_{(n-1)\epsilon}^{n\epsilon} \lambda^a(t)\Phi_a\,dt}\,e^{(i/4\gamma_n)\int_{(n-1)\epsilon}^{n\epsilon} \lambda^a(t)(M^{-1})_{ab}\lambda^b(t)\,dt}\,\Pi_a\mathcal{D}\lambda^a\\ &= e^{-i\epsilon\gamma_n(\Phi_a M^{ab}\Phi_b)}\,.\end{aligned} \tag{79}$$

The second and last step in the construction involves an integration over γ_n given by

$$\begin{aligned}&\int e^{-i\epsilon\gamma_n(\Phi_a M^{ab}\Phi_b)}\,d\Gamma(\gamma_n)\\ &\qquad \equiv \lim_{\zeta\to 0^+}\lim_{L\to\infty}\int_{-L}^{L} e^{-i\epsilon\gamma_n(\Phi_a M^{ab}\Phi_b)}\,\frac{\sin[\epsilon(\delta^2+\zeta)\gamma_n]}{\pi\gamma_n}\,d\gamma_n\\ &\qquad = \mathbb{E}(\epsilon\,\Phi_a M^{ab}\Phi_b \le \epsilon\,\delta^2)\\ &\qquad = \mathbb{E}(\Phi_a M^{ab}\Phi_b \le \delta^2)\,,\end{aligned} \tag{80}$$

which achieves our goal. We note that if the final limit is replaced by $\lim_{\zeta\to 0^-}$, the result becomes $\mathbb{E}(\Phi_\alpha M^{\alpha\beta}\Phi_\beta < \delta^2)$. We normally symbolize the pair of operations by $\int \cdots \mathcal{D}R(\lambda)$, leaving the integral over γ_n implicit.

Remark: For notational simplicity throughout this article, we generally let

$$\begin{aligned}&\int e^{-i\gamma X^2}\frac{\sin(\delta^2\gamma)}{\pi\gamma}\,d\gamma\\ &\qquad \equiv \lim_{\zeta\to 0^+}\lim_{L\to\infty}\int_{-L}^{L} e^{-i\gamma X^2}\,\frac{\sin[(\delta^2+\zeta)\gamma]}{\pi\gamma}\,d\gamma\\ &\qquad = \mathbb{E}(X^2 \le \delta^2)\,.\end{aligned} \tag{81}$$

With (80) we have found a single, universal procedure to create the regularized projection operator $\mathbb{E}$ from the set of constraint operators in a manner that is *completely independent of the nature of the constraints themselves.*

4.4 Basic Second-Class Constraint Example

Consider the two degree of freedom system determined by

$$I = \int [p\dot{q} + r\dot{s} - H(p,q,r,s) - \lambda_1 r - \lambda_2 s]\, dt\ , \tag{82}$$

where we have called the variables of the second degree of freedom r, s, and H is not specified further. The coherent states satisfy $|p,q,r,s\rangle = |p,q\rangle \otimes |r,s\rangle$, which will be useful. We adopt (71) as our formal path integral in the present case, and choose [33]

$$\begin{aligned} \mathbb{E} &= \int e^{-i(\xi_1 R + \xi_2 S)}\, e^{-(\xi_1^2+\xi_2^2)/4}\, d\xi_1 d\xi_2/(2\pi) \\ &= \mathbb{E}(R^2 + S^2 \leq \hbar) \equiv |0_2\rangle\langle 0_2| \end{aligned} \tag{83}$$

which is a projection operator onto the financial vector for the second (constrained) degree of freedom only. With this choice it follows that

$$\begin{aligned} &i\langle p,q,r,s|\mathbb{E}\,(d/dt)\mathbb{E}\,|p,q,r,s\rangle/\langle p,q,r,s|\mathbb{E}\,|p,q,r,s\rangle \\ &\qquad = i\langle p,q|(d/dt)|p,q\rangle - \Im(d/dt)\ln[\langle 0_2|r,s\rangle] \\ &\qquad = p\dot{q} - \Im(d/dt)\ln[\langle 0_2|r,s\rangle]\ , \end{aligned} \tag{84}$$

and

$$\begin{aligned} &\langle p,q,r,s|\mathbb{E}\,\mathcal{H}(P,Q,R,S)\mathbb{E}\,|p,q,r,s\rangle/\langle p,q,r,s|\mathbb{E}\,|p,q,r,s\rangle \\ &\qquad = \langle p,q,0,0|\mathcal{H}(P,Q,R,S)|p,q,0,0\rangle \\ &\qquad = H(p,q,0,0)\ . \end{aligned} \tag{85}$$

Consequently, for this example, (71) becomes

$$\mathcal{M}\int \exp\{i\textstyle\int [p\dot{q} - H(p,q,0,0)]\,dt\}\,\mathcal{D}p\,\mathcal{D}q\ \times\ \langle r'',s''|0_2\rangle\langle 0_2|r',s'\rangle\ , \tag{86}$$

where we have used the fact that at every time slice

$$\int \langle r,s|\mathbb{E}\,|r,s\rangle\, dr\, ds/(2\pi) = \int |\langle 0_2|r,s\rangle|^2\, dr\, ds/(2\pi) = 1\ . \tag{87}$$

Observe, in this path integral quantization, that no variables have been eliminated nor has any domain of integration been reduced; moreover, the operators R and S have remained unchanged. Also observe that the result in (86) is clearly a product of two distinct factors. The first factor describes the true dynamics as if we had solved for the classical constraints and substituted $r = 0$ and $s = 0$ in the classical action from the very beginning, while the second factor characterizes a one-dimensional Hilbert space for the second degree of freedom. Thus we can also drop the second factor completely as well as all the integrations over r and s and still retain the same physics. In this manner we recover the standard result without the use of Dirac brackets or having to initially eliminate the second-class constraints from the theory.

4.5 Conversion Method

One common method to treat second-class constraints is to convert them to first-class constraints and to follow the available procedures for such systems; see, e.g., [12,4,5]. Let us first argue classically, and take as an example a single degree of freedom with canonical variables p and q, a vanishing Hamiltonian, and the second-class constraints $p = 0$ and $q = 0$. This situation may be described by the classical action

$$I = \int [p\dot{q} - \lambda p - \xi q]\, dt \,, \tag{88}$$

where λ and ξ denote Lagrange multipliers. Next, let us introduce a second canonical pair, say r and s, and adopt the classical action

$$I' = \int [p\dot{q} + r\dot{s} - \lambda(p+r) - \xi(q-s)]\, dt \,. \tag{89}$$

Now the two constraints read $p+r=0$ and $q-s=0$ with a Poisson bracket $\{p+r, q-s\} = 0$, characteristic of first-class constraints. We obtain the original problem by imposing the (consistent) *gauge conditions* that $r = 0$ and $s = 0$. Let us look at this example from the projection operator, coherent state approach.

In the first version with one pair of variables, we are led to the reproducing kernel

$$\begin{aligned}
&\langle p'', q''|\mathbb{E}(P^2 + Q^2 \le \hbar)|p', q'\rangle \\
&\quad = \langle p'', q''|0\rangle\langle 0|p', q'\rangle \\
&\quad = e^{-\frac{1}{4}(p''^2 + q''^2 - 2ip''q'')}\, e^{-\frac{1}{4}(p'^2 + q'^2 + 2ip'q')} \,,
\end{aligned} \tag{90}$$

which provides a "bench mark" for this example. As expected the result is a one-dimensional Hilbert space.

In the second version of this problem, we start with the expression

$$\langle p'', q'', r'', s''|\mathbb{E}((P+R)^2 + (Q-S)^2 \le \delta^2)|p', q', r', s'\rangle \tag{91}$$

which involves a constraint with zero in the continuous spectrum. Therefore, following previous examples, we multiply this expression with a suitable factor c_δ and take the limit as $\delta \to 0$. This factor can be chosen so that

$$\begin{aligned}
&\lim_{\delta\to 0} c_\delta\, \langle p'', q'', r'', s''|\mathbb{E}((P+R)^2 + (Q-S)^2 \le \delta^2)|p', q', r', s'\rangle \\
&\quad = e^{-\frac{1}{4}[(p''+r'')^2 + (q''-s'')^2] + \frac{1}{2}i(p''-r'')(q''-s'')} \\
&\qquad \times e^{-\frac{1}{2}i(p'-r')(q'-s') - \frac{1}{4}[(p'+r')^2 + (q'-s')^2]} \,,
\end{aligned} \tag{92}$$

an expression which also describes a one-dimensional Hilbert space. This is a different (but equivalent) representation for the one-dimensional Hilbert space than the one found above. Since it is only one-dimensional we can reduce this reproducing kernel even further, in the fashion illustrated earlier, by choosing a "gauge" where $r'' = s'' = r' = s' = 0$. When this is done the result becomes

$$e^{-\frac{1}{4}(p''^2 + q''^2 - 2ip''q'')} e^{-\frac{1}{4}(p'^2 + q'^2 + 2ip'q')} , \tag{93}$$

which is *identical* to the expression (90) found by quantization of the second-class constraints directly. In this manner we see how the conversion method, in which second-class constraints are turned into first-class constraints by the introduction of auxiliary degrees of freedom, appears within the projection operator, coherent state approach as well. Applications of the conversion method made within the projection operator approach may be found in [31].

4.6 Equivalent Representations

In dealing with quantum mechanics, one may employ many different – yet equivalent – representations of the vectors and operators involved. While, in certain circumstances, some representations may be more convenient than others, the notion that some representations are "better" than others should be resisted.

In the context of coherent-state representations, for example, a change of the financial vector leads to an equivalent representation. If, for a rather general (normalized) financial vector $|\eta\rangle$, we set

$$|p, q; \eta\rangle \equiv e^{-iqP} e^{ipQ} |\eta\rangle , \tag{94}$$

then

$$\psi(p, q; \eta) \equiv \langle p, q; \eta|\psi\rangle \tag{95}$$

defines η-dependent representatives of the abstract vector $|\psi\rangle$. However, all representation dependent aspects disappear when physical questions are asked such as

$$\int |\psi(p, q; \eta)|^2 \, (dp\, dq/2\pi) = \langle\psi|\psi\rangle . \tag{96}$$

More general representation issues may be addressed by using arbitrary unitary operators, say V. Thus if $|p, q\rangle$ denotes elements of one (say) coherent state basis, then $|p, q; V\rangle \equiv V^\dagger |p, q\rangle$ denotes the elements of another basis. Vector and operator representatives, $\psi(p, q; V) \equiv \langle p, q; V|\psi\rangle$ and $A(p', q'; V : p, q; V) \equiv \langle p', q'; V|\mathcal{A}|p, q; V\rangle$, respectively, provide equivalent sets of functional representatives for different V. Evidently the *physics* is unchanged in this transformation; only the intermediate *mathematical representatives* are affected. This formulation is similar to passive coordinate transformations in other disciplines. Another version similar to active coordinate transformations is also possible. In this version the basis vectors, say $|p, q\rangle$, for all relevant (p, q), remain unchanged; instead, the abstract vectors $|\psi\rangle$ and operators $\mathcal{A}$, etc., are transformed: $|\psi\rangle \to V|\psi\rangle$, $\mathcal{A} \to V\mathcal{A}V^\dagger$, etc. It is this form of equivalence that we turn to next.

4.7 Equivalence of Criteria for Second-Class Constraints

Let us return to the simple example of second-class constraints discussed above where, classically, $p = q = 0$. In the associated quantum theory, we chose to express these constraints with the help of the projection operator $\mathbb{E} = \mathbb{E}(P^2+Q^2 \le \hbar) = |0\rangle\langle 0|$, namely, the projection operator onto the ground state of the "Hamiltonian" P^2+Q^2. In turn, this expression led directly to the coherent-state representation of $\mathbb{E}$ given by $\langle p', q'|\mathbb{E}|p, q\rangle = \langle p', q'|0\rangle\langle 0|p, q\rangle$. However, the question arises, what is special about the combination $P^2 + Q^2$? As we shall now argue, any other possible choice leads to an equivalent representation.

As a first example, consider

$$\mathbb{E}(P^2 + \omega^2 Q^2 \le \omega\hbar) = |0;\omega\rangle\langle 0;\omega| = V_\omega^\dagger\,|0\rangle\langle 0|\,V_\omega\,, \tag{97}$$

where V_ω denotes a suitable unitary operator, which establishes the equivalence for any ω, $0 < \omega < \infty$. We emphasize that we do *not* assert the unitary equivalence of P^2+Q^2 and $P^2+\omega^2 Q^2$ for any value of $\omega \ne 1$, only that $|0;\omega\rangle$ and $|0\rangle$ are unitarily related – as are *any* two unit vectors in Hilbert space.

Furthermore, there is nothing sacred about the quadratic combination. For example, for any $0 < \lambda < \infty$, consider $\mathbb{E}(P^2+\lambda Q^4 \le \delta(\hbar)^2) \equiv |0, \lambda\rangle\langle 0, \lambda|$, where we have adjusted $\delta(\hbar)$ to the lowest eigenvalue so as to include only a single eigenvector, $|0, \lambda\rangle$. Since there exists a unitary operator V_λ such that $\langle 0, \lambda| = \langle 0|V_\lambda$, this choice of projection operator leads to an equivalent coherent-state representation as well.

More generally, we are led to reconsider the projection operator

$$\mathbb{E}(\Sigma_a \Phi_a^2 \le \delta(\hbar)^2) = \sum_{j=1}^{J} |j\rangle\langle j|\,, \tag{98}$$

where $\langle j|k\rangle = \delta_{jk}$ and $1 \le J \le \infty$, as determined by the choice of $\delta(\hbar)$. Since all J-dimensional subspaces are unitarily equivalent to each other (with suitable care taken when $J = \infty$), the given prescription is entirely equivalent to any other version, such as

$$\mathbb{E}(\mathcal{F}(\Phi_a) \le \tilde{\delta}(\hbar)^2) = \sum_{\mathsf{j}=1}^{\mathsf{J}} |\mathsf{j}\rangle\langle \mathsf{j}|\,, \tag{99}$$

where $\langle \mathsf{j}|\mathsf{k}\rangle = \delta_{\mathsf{jk}}$, provided that $\tilde{\delta}(\hbar)$ may be – and is – chosen so that $\mathsf{J} = J$. Here $\mathcal{F}(\Phi_a)$ denotes a nonnegative self-adjoint operator that includes all the constraint operators, and for very small $\tilde{\delta}(\hbar)^2$ forces the spectral contribution of each constraint operator to be correspondingly small, just as is the case in (98).

In summary, the general, quadratic criterion we have adopted in (98) has been chosen for simplicity and convenience: any other restriction on the constraint operators leads to an equivalent theory, as in (99), provided that the dimensionality of $\mathbb{E}$ remains the same.

5 Selected Examples of First-Class Constraints

5.1 General Configuration Space Geometry

Although we shall discuss constraints that lead to a general configuration space geometry in this section, we shall for the most part use rather simple illustrative examples. To begin with let us consider the constraint

$$\sum_{j=1}^{J}(q^j)^2 = 1 \,, \tag{100}$$

a condition which puts the classical problem on a (hyper)sphere of unit radius. For convenience in what follows we shall focus as well on the case of a vanishing Hamiltonian so as to isolate clearly the consequences of the constraint independently of any dynamical effects. Adopting a standard vector inner product notation and a different kinematic term, consider the formal path integral

$$\mathcal{M}\int \exp\{i{\textstyle\int}[-q\cdot\dot{p}-\lambda(q^2-1)]\,dt\}\,\mathcal{D}p\,\mathcal{D}q\,\mathcal{D}C(\lambda)\,, \tag{101}$$

the result of which is given by

$$\langle p'',q''|\mathbb{E}\,|p',q'\rangle \tag{102}$$

where

$$\mathbb{E} = \int_{-\infty}^{\infty} e^{-i\lambda(Q^2-1)}\,\frac{\sin(\delta\lambda)}{\pi\lambda}\,d\lambda = \mathbb{E}\,(-\delta \le Q^2-1 \le \delta)\,. \tag{103}$$

In order, ultimately, to obtain a suitable reduction of the reproducing kernel in the present case, we allow for financial vectors other than harmonic oscillator ground states. Thus we let $|\eta\rangle$ denote a general unit vector for the moment; its required properties will emerge from our analysis. In accordance with (101), we choose a phase convention for the coherent states – in particular, in (18) we multiply by $e^{ip\cdot q}$ – so that now the Schrödinger representation of the coherent states reads

$$\langle x|p,q\rangle = e^{ip\cdot x}\,\eta(x-q)\,, \tag{104}$$

which leads immediately to the expression

$$\langle p'',q''|p',q'\rangle = \int \eta^*(x-q'')\,e^{-i(p''-p')\cdot x}\,\eta(x-q')\,d^Jx\,. \tag{105}$$

Consequently, the reproducing kernel that incorporates the projection operator is given, for $0<\delta<1$, by

$$\langle p'',q''|\mathbb{E}\,|p',q'\rangle = \int_{1-\delta\le x^2\le 1+\delta} \eta^*(x-q'')\,e^{-i(p''-p')\cdot x}\,\eta(x-q')\,d^Jx\,. \tag{106}$$

Since $\mathbb{E}$ represents a projection operator, it is evident that this expression defines a reproducing kernel which admits a local integral for its inner product (for any normalized η) with a measure $d^J p\, d^J q/(2\pi)^J$ and an integration domain $\mathbb{R}^{2J}$.

However, if we are willing to restrict our choice of financial vector, we can reduce the number of integration variables and change the domain of integration in a meaningful way. Recall that the group $\mathrm{E}(J)$, the Euclidean group in J-dimensions, consists of rotations that preserve the unit (hyper)sphere in J-dimensions, as well as J translations. As emphasized by Isham [23], this is the natural canonical group for a system confined to the surface of a (hyper)sphere in J dimensions. We can adapt our present coherent states to be coherent states for the group $\mathrm{E}(J)$ without difficulty.

To that end consider the reduction of the reproducing kernel (106) to one for which $q''^2 = q'^2 \equiv 1$. To illustrate the process as clearly as possible let us choose $J = 2$. As a consequence we introduce

$$\langle a'', b'', c'' | a', b', c' \rangle \equiv \langle p'', q'' | \mathbb{E} | p', q' \rangle_{q''^2 = q'^2 = 1} , \tag{107}$$

where $a \equiv p_1$, $b \equiv p_2$, and c arises from the identification $q^1 \equiv \cos(c)$ and $q^2 \equiv \sin(c)$, all relations holding for both end points. Expressed in terms of polar coordinates, r, ϕ, the reduced reproducing kernel becomes

$$\begin{aligned} &\langle a'', b'', c'' | a', b', c' \rangle \\ &= \int_{|r^2-1|\leq\delta} \eta^*(r, \phi - c'')\, e^{-i(a''-a')r\cos\phi - i(b''-b')r\sin\phi}\, \eta(r, \phi - c')\, r\, dr\, d\phi\, . \end{aligned} \tag{108}$$

We next seek to choose η, if at all possible, in such a way that the inner product of this new (reduced) reproducing kernel admits a local integral for its inner product. As a starting point we choose the left-invariant group measure for $\mathrm{E}(2)$ which is given by $M\, da\, db\, dc$, M a constant, with an integration domain $\mathbb{R}^2 \times S^1$. Therefore, we are led to study

$$\begin{aligned} &\int\int_{|r^2-1|<\delta} \eta^*(r, \phi - c'')\, e^{-i(a''-a)r\cos\phi - i(b''-b)r\sin\phi}\, \eta(r, \phi - c)\, r\, dr\, d\phi \\ &\times \int_{|\rho^2-1|<\delta} \eta^*(\rho, \theta - c)\, e^{-i(a-a')\rho\cos\theta - i(b-b')\rho\sin\theta}\, \eta(\rho, \theta - c')\, \rho\, d\rho\, d\theta \\ &\times M\, da\, db\, dc \\ &= (2\pi)^2 M \int \eta^*(r, \phi - c'') e^{-i(a''-a')r\cos\phi - i(b''-b')r\sin\phi}\, \eta(r, \phi - c')\, r\, dr\, d\phi \\ &\times \int |\eta(r, c)|^2\, dc\, , \end{aligned} \tag{109}$$

which leads to the desired result provided (i)

$$\int_0^{2\pi} |\eta(r, c)|^2\, dc = P\, , \qquad 0 < P < \infty\, , \tag{110}$$

is *independent* of r, $|r^2 - 1| < \delta$, and (ii) $M = [(2\pi)^2 P]^{-1}$. Given a general nonvanishing vector $\xi(r, \phi)$, a vector satisfying (110) may always be given by

$$\eta(r,\phi) = \xi(r,\phi)/\sqrt{\int_0^{2\pi} |\xi(r,\theta)|^2 \, d\theta} \tag{111}$$

provided the denominator is positive, and which specifically leads to $P = 1$. In this way we have reproduced the E(2)-coherent states of [24], even including the necessity for a small interval of integration in r, and where financial vectors satisfying (110) were called "surface constant".

Dynamics consistent with the constraint $q^2 = 1$ is obtained in the E(2) case by choosing a Hamiltonian that is a function of the coordinates on the circle, namely $\cos(\theta)$ and $\sin(\theta)$, as well as the rotation generator of E(2), i.e., $-i\partial/\partial\theta$. We refer the reader to [24] for a further discussion of E(2)-coherent states as well as a discussion of the introduction of compatible dynamics. An analogous discussion can be given for the classical constraint $q^2 = 1$ for any value of $J > 2$.

Not only can compact (hyper)spherical configuration spaces be treated in this way, but one may also treat noncompact (hyper)pseudospherical spaces defined by the constraint

$$\Sigma_{i=1}^{I} q^{i\,2} - \Sigma_{j=I+1}^{J} q^{j\,2} = 1\,, \qquad 1 \le I \le J-1\,, \tag{112}$$

appropriate to the Euclidean group $\mathrm{E}(I, J-I)$. Such an analysis would lead to $\mathrm{E}(I, J-I)$-coherent states.

Finally, we comment on the constraint of a general curved configuration space which can be defined by a set of compatible constraints $\phi_a(q) = 0$. Clearly these constraints satisfy $\{\phi_a(q), \phi_b(q)\} = 0$, and define a $(J-A)$-dimensional configuration space in the original Euclidean configuration space $\mathbb{R}^J$. The relevant projection operator $\mathbb{E} = \mathbb{E}(\Sigma_a \Phi_a^2(Q) \le \delta^2)$ is defined in an evident fashion, and the reproducing kernel incorporating the projection operator is defined in analogy with the prior discussion. This reproducing kernel enjoys a local integral representation for its inner product, in fact, this integral is with the same measure and integration domain as without the projection operator. What differs in the present case is that when the reproducing kernel is put on the constraint manifold, the resultant coherent states are generally *not* defined by the action of a group on a fixed financial vector. In short, the relevant coherent states are not group generated, which, in fact, is consistent with their most basic definition; see, e.g., [32] and [26].

5.2 Finite-Dimensional Hilbert Space Examples

Let us consider the case of two degrees of freedom with a "classical" action function given by

$$I = \int [\tfrac{1}{2}(p_1\dot{q}_1 - q_1\dot{p}_1 + p_2\dot{q}_2 - q_2\dot{p}_2) - \lambda(p_1^2 + p_2^2 + q_1^2 + q_2^2 - 4s\hbar)]\, dt\,. \tag{113}$$

For clarity of presentation, we explicitly include $\hbar$ in our classical action, and we continue to make it explicit throughout this section. With the present phase convention for the coherent states, the unconstrained reproducing kernel is given by

$$\begin{aligned}\langle p'', q''|p', q'\rangle &\equiv \langle z''|z'\rangle \\ &= \exp[\Sigma_{j=1}^2(-\tfrac{1}{2}|z_j''|^2 + z''^*{}_j z_j' - \tfrac{1}{2}|z_j'|^2)]\end{aligned} \qquad (114)$$

where $z_j \equiv (q_j + ip_j)/\sqrt{2\hbar}$ for each of the end points.

We next observe that the constraint operator

$$\Phi =: P_1^2 + P_2^2 + Q_1^2 + Q_2^2 : -4s\hbar\mathbb{1} \qquad (115)$$

has discrete eigenvalues, i.e., $2(n_1+n_2-2s)\hbar$, where n_1 and n_2 are nonnegative integers, based on the choice of $|\eta\rangle$ as the ground state for each oscillator. To satisfy $\Phi = 0$ it is necessary that $2s$ be an integer in which case the quantum constraint subspace is $(2s+1)$-dimensional. The projection operator in the present case is defined by

$$\mathbf{E} = \pi^{-1}\int_0^\pi \exp[-i\lambda(: P_1^2 + P_2^2 + Q_1^2 + Q_2^2 : -4s\hbar\mathbb{1})/\hbar]\, d\lambda \qquad (116)$$

which projects onto the appropriate $(2s+1)$-dimensional subspace. It is straightforward to demonstrate that

$$\begin{aligned}\langle z''|\mathbf{E}|z'\rangle &= \exp[-\tfrac{1}{2}\Sigma_{j=1}^2(|z_j''|^2 + |z_j'|^2)][(2s)!]^{-1}(z''^*{}_1 z_1' + z''^*{}_2 z_2')^{2s} \\ &= \exp[-\tfrac{1}{2}\Sigma_{j=1}^2(|z_j''|^2 + |z_j'|^2)]\textstyle\sum_{k=0}^{2s}[k!(2s-k)!]^{-1}(z''^*{}_1 z_1')^k(z''^*{}_2 z_2')^{2s-k} \,.\end{aligned} \qquad (117)$$

The projected reproducing kernel in this case corresponds to a *finite* dimensional Hilbert space; nevertheless, the inner product is given by the same measure and integration domain as in the original, unprojected, infinite dimensional Hilbert space!

Of course, there are other, simpler and more familiar ways to represent a finite-dimensional Hilbert space; but any other representation is evidently equivalent to the one described here.

As the notation suggests the present quantum constraint subspace provides a natural carrier space for an irreducible representation of SU(2) with spin s. We observe that the following three expressions represent generators of the classical rotation group in their action on the constraint hypersurface:

$$\begin{aligned} s_x &= \tfrac{1}{2}(p_1p_2 + q_1q_2)\,, \\ s_y &= \tfrac{1}{2}(q_1p_2 - p_1q_2)\,, \\ s_z &= \tfrac{1}{4}(p_1^2 + q_1^2 - p_2^2 - q_2^2)\,. \end{aligned} \qquad (118)$$

Thus these quantities serve as potential ingredients for a Hamiltonian which is compatible with the constraint.

Although not the subject of this section, we may also observe that an analogous discussion holds in case of the constraint

$$\phi(p,q) = p_1^2 + q_1^2 - p_2^2 - q_2^2 - 2k\hbar = 0 \,, \tag{119}$$

where k is an integer, and the resultant reduced Hilbert space is infinite dimensional for any integral k value. In this case the relevant group is SU(1,1).

5.3 Helix Model

In [13] the authors analyzed the so-called helix model. For details of this model (see also [8,35,47]) and its possible role as a simple analogue of the Gribov problem in non-Abelian gauge models, we refer the reader to their paper. We begin with the classical Hamiltonian for a three-degree of freedom system given by

$$H = \tfrac{1}{2}(p_1^2 + p_2^2 + p_3^2) + U(q_1^2 + q_2^2) + \lambda[g(p_2 q_1 - q_2 p_1) + p_3] \,, \tag{120}$$

where U denotes the potential, which hereafter, following [13], we shall choose as harmonic, namely $U(q_1^2 + q_2^2) = \omega^2(q_1^2 + q_2^2)/2$, because then this special model is fully soluble. Here, $g > 0$ is a coupling constant, and $\lambda = \lambda(t)$ is the Lagrange multiplier which enforces the single first-class constraint

$$\phi(p,q) = g(p_2 q_1 - q_2 p_1) + p_3 = 0 \,. \tag{121}$$

For the first two degrees of freedom we choose coherent states with the phase convention adopted for the previous example, while for the third degree of freedom we return to the original phase convention. This choice means that we consider the formal coherent state path integral given by

$$\begin{aligned}
&\int \exp(i{\textstyle\int}\{\tfrac{1}{2}(p_1\dot q_1 - q_1\dot p_1) + \tfrac{1}{2}(p_2\dot q_2 - q_2\dot p_2) + p_3\dot q_3 \\
&\qquad -\tfrac{1}{2}(p_1^2 + p_2^2 + p_3^2) - \tfrac{1}{2}\omega^2(q_1^2 + q_2^2) \\
&\qquad -\lambda[g(p_2 q_1 - q_2 p_1) + p_3]\}\,dt)\,\mathcal{D}\mu(p,q)\,\mathcal{D}C(\lambda) \\
&= \langle z_1'', z_2'', p_3'', q_3''|\, e^{-i\mathcal{H}T}\,\mathbb{E}\,|z_1', z_2', p_3', q_3'\rangle \,.
\end{aligned} \tag{122}$$

In the present case the relevant projection operator $\mathbb{E}$ is given (for $\hbar = 1$, and $0 < \delta \ll g$) by

$$\mathbb{E} = \mathbb{E}\,((gL_3 + P_3)^2 \le \delta^2) = \sum_{m=-\infty}^{\infty} \mathbb{E}\,((gm + P_3)^2 \le \delta^2)\,\mathbb{E}\,(L_3 = m) \tag{123}$$

where we have used the familiar spectrum for the rotation generator L_3. If $\mathcal{H}_0$ denotes the harmonic oscillator Hamiltonian for the first two degrees of freedom, then it follows that

$$
\begin{aligned}
&\langle z_1'', z_2'', p_3'', q_3''|\, e^{-i\mathcal{H}T}\, \mathbb{E}\, |z_1', z_2', p_3', q_3'\rangle \\
&= \sum_{m=-\infty}^{\infty} \langle z_1'', z_2''| e^{-i\mathcal{H}_0 T} \mathbb{E}\,(L_3 = m)\, |z_1', z_2'\rangle \\
&\times \langle p_3'', q_3''| e^{-iP_3^2 T/2} \mathbb{E}\,(-\delta \le gm + P_3 \le \delta) |p_3', q_3'\rangle \\
&= \exp[-\tfrac{1}{2}(|z_1''|^2 + |z_2''|^2 + |z_1'|^2 + |z_2'|^2)] \\
&\times \sum_{m=-\infty}^{\infty} \left\{ \frac{(z_1''^* + iz_2''^*)(z_1' - iz_2')}{(z''^* - iz_2''^*)(z_1' + iz_2')} \right\}^{m/2} I_m(\sqrt{(z_1''^{*2} + z_2''^{*2})(z_1'^2 + z_2'^2)}\, e^{-i\omega T}) \\
&\times \exp[-\tfrac{1}{2}(gm + p_3'')^2 - \tfrac{1}{2}(gm + p_3')^2 - i\tfrac{1}{2} g^2 m^2 T - igm(q_3'' - q_3')] \\
&\times \frac{2}{\sqrt{\pi}} \frac{\sin[\delta(q_3'' - q_3')]}{(q_3'' - q_3')} + O(\delta^2)\,,
\end{aligned} \tag{124}
$$

where I_m denotes the usual Bessel function.

We observe that the spectrum for the Hamiltonian agrees with the results of [13], and moreover, to leading order in δ, we have obtained gauge-invariant results, i.e., insensitivity to any choice of the Lagrange multiplier function $\lambda(t)$, merely by projecting onto the quantum constraint subspace at $t = 0$. The constrained propagator (124) is composed with the same measure and integration domain as is the unconstrained propagator. We may also divide the constrained propagator by δ and take the limit $\delta \to 0$. The result is a new functional expression for the propagator that fully satisfies the constraint condition, but one that no longer admits an inner product with the same measure and integration domain as before.

5.4 Reparameterization Invariant Dynamics

Let us start with a single degree of freedom ($J = 1$) and the action

$$\int [p\dot{q} - H(p,q)]\, dt\,. \tag{125}$$

We next promote the independent variable t to a dynamical variable, introduce s as its conjugate momentum (often called p_t), enforce the constraint $s + H(p,q) = 0$, and lastly introduce τ as a new independent variable. This modification is realized by means of the classical action

$$\int \{pq^* + st^* - \lambda[s + H(p,q)]\}\, d\tau\,, \tag{126}$$

where $q^* = dq/d\tau$, $t^* = dt/d\tau$, and $\lambda = \lambda(\tau)$ is a Lagrange multiplier. The coherent-state path integral is constructed so that

$$
\begin{aligned}
\mathcal{M} \int \exp(i{\textstyle\int} \{pq^* + st^* - \lambda[s + H(p,q)]\}\, dt)\, \mathcal{D}p\, \mathcal{D}q\, \mathcal{D}s\, \mathcal{D}t\, \mathcal{D}C(\lambda) \\
= \langle p'', q'', s'', t''| \mathbb{E}\, |p', q', s', t'\rangle\,,
\end{aligned} \tag{127}
$$

where

$$\mathbb{E} = \int_{-\infty}^{\infty} e^{-i\xi[S+\mathcal{H}(P,Q)]} \, \frac{\sin(\delta\xi)}{\pi\xi} \, d\xi$$
$$= \mathbb{E}\,(-\delta \leq S + \mathcal{H}(P,Q) \leq \delta)\,. \tag{128}$$

The result in (127) and (128) represents as far as we can go without choosing $\mathcal{H}(P,Q)$.

To gain further insight into such expressions, we specialize to the case of the nonrelativistic free particle, $\mathcal{H} = P^2/2$. Then it follows that

$$\begin{aligned}\langle p'', q'', s'', t''|\mathbb{E}\,|p', q', s', t'\rangle \\ = \pi^{-1} \int_{-\infty}^{\infty} & \exp[-\tfrac{1}{2}(k-p'')^2 - \tfrac{1}{2}(\tfrac{1}{2}k^2 + s'')^2 \\ & + ik(q''-q') - i\tfrac{1}{2}k^2(t''-t') \\ & - \tfrac{1}{2}(k-p')^2 - \tfrac{1}{2}(\tfrac{1}{2}k^2 + s')^2]\, dk \\ & \times \frac{2\sin[\delta(t''-t')]}{(t''-t')} + O(\delta^2)\,. \end{aligned} \tag{129}$$

For any δ such that $0 < \delta \ll 1$, we observe that this expression represents a reproducing kernel which in turn defines an associated reproducing kernel Hilbert space composed, as usual, of bounded, continuous functions given, for arbitrary complex numbers $\{\alpha_k\}$, phase-space points $\{p_k, q_k, s_k, t_k\}$, and $K < \infty$, by

$$\psi(p, q, s, t) \equiv \sum_{k=0}^{K} \alpha_k \langle p, q, s, t|\mathbb{E}\,|p_k, q_k, s_k, t_k\rangle\,, \tag{130}$$

or as the limit of Catchy sequences of such functions in the norm defined by means of the inner product given by

$$(\psi, \psi) = \textstyle\int |\psi(p, q, s, t)|^2 \, dp\, dq\, ds\, dt/(2\pi)^2 \tag{131}$$

integrated over $\mathbb{R}^4$.

Let us next consider the reduction of the reproducing kernel given by

$$\begin{aligned}\langle p'', q'', t''|p', q', t'\rangle \\ \equiv \lim_{\delta\to 0} \frac{1}{4\sqrt{\pi}\,\delta} & \int \langle p'', q'', s'', t''|\mathbb{E}\,|p', q', s', t'\rangle \, ds''\, ds' \\ = \pi^{-1/2} \int & \exp[-\tfrac{1}{2}(k-p'')^2 - \tfrac{1}{2}(k-p')^2 \\ & + ik(q''-q') - i\tfrac{1}{2}k^2(t''-t')]\, dk\,, \end{aligned} \tag{132}$$

which in turn generates a *new* reproducing kernel in the indicated variables. For the resultant kernel it is straightforward to demonstrate, for any t, that

$$\int \langle p'', q'', t''|p, q, t\rangle\langle p, q, t|p', q', t'\rangle \, dp\, dq/(2\pi) = \langle p'', q'', t''|p', q', t'\rangle\,. \tag{133}$$

This relation implies that the span of the vectors $\{|p,q\rangle \equiv |p,q,0\rangle\}$ is identical with the span of the vectors $\{|p,q,t\rangle\}$, meaning further that the states $\{|p,q,t\rangle\}$ form a set of *extended coherent states*, which are "extended" with respect to t in the sense of [34]. Observe how the time variable has become distinguished by the criterion (133). Consequently, we may properly interpret

$$\langle p'',q'',t''|p',q',t'\rangle \equiv \langle p'',q''|e^{-i(P^2/2)(t''-t')}|p',q'\rangle\,, \tag{134}$$

namely, as the conventional, single degree of freedom, coherent-state matrix element of the evolution operator appropriate to the free particle.

To further demonstrate this interpretation as the dynamics of the free particle, we may pass to sharp q matrix elements with the observation that

$$\begin{aligned}\langle q''|e^{-i(P^2/2)(t''-t')}|q'\rangle \\ &\equiv \frac{\pi^{1/2}}{(2\pi)^2}\int \langle p'',q''|e^{-i(P^2/2)(t''-t')}|p',q'\rangle\,dp''\,dp' \\ &= \frac{1}{2\pi}\int \exp[\,ik(q''-q')-i\tfrac{1}{2}k^2(t''-t')]\,dk \\ &= \frac{e^{i(q''-q')^2/2(t''-t')}}{\sqrt{2\pi i(t''-t')}}\,,\end{aligned} \tag{135}$$

which is clearly the usual result.

5.5 Elevating the Lagrange Multiplier to an Additional Dynamical Variable

Sometimes it is useful to consider an alternative formulation of a system with constraints in which the initial Lagrange multipliers are regarded as dynamical variables, complete with their own conjugate variables, and to introduce new constraints as needed. For example, let us start with a single degree of freedom system with a single first-class constraint specified by the action functional

$$\textstyle\int [p\dot q - H(p,q) - \lambda\phi(p,q)]\,dt\,, \tag{136}$$

where $\phi(p,q)$ represents the constraint and λ the Lagrange multiplier. Instead, let us replace this action functional by

$$\textstyle\int [p\dot q + \pi\dot\lambda - H(p,q) - \sigma\pi - \theta\phi(p,q)]\,dt\,. \tag{137}$$

In this expression we have introduced π as the canonical conjugate to λ, the Lagrange multiplier σ to enforce the constraint $\pi = 0$, and the Lagrange multiplier θ to enforce the original constraint $\phi = 0$. Observe that

$\{\pi, \phi(p,q)\} = 0$, and therefore the constraints remain first class in the new form. The path integral expression for the extended form reads

$$\mathcal{M} \int \exp\{i \textstyle\int [p\dot{q} + \pi\dot{\lambda} - H(p,q) - \sigma\pi - \theta\phi(p,q)]\, dt\}\, \mathcal{D}p\, \mathcal{D}q\, \mathcal{D}\pi\, \mathcal{D}\lambda\, \mathcal{D}C(\sigma,\theta)$$
$$= \langle p'', q'', \pi'', \lambda'' | e^{-i\mathcal{H}T} \mathbf{E}\, | p', q', \pi', \lambda' \rangle \; . \tag{138}$$

In this expression, we may choose

$$\mathbf{E} = \mathbf{E}\,(\Phi(P,Q)^2 \leq \delta^2)\, \mathbf{E}\,(\Pi^2 \leq \delta'^2) \tag{139}$$

involving two possibly distinct regularization parameters. Consequently, the complete propagator factors into two terms,

$$\langle p'', q'', \pi'', \lambda'' | e^{-i\mathcal{H}T} \mathbf{E}\, | p', q', \pi', \lambda' \rangle$$
$$= \langle p'', q'' | e^{-i\mathcal{H}T} \mathbf{E}\,(\Phi(P,Q)^2 \leq \delta^2) | p', q' \rangle \langle \pi'', \lambda'' | \mathbf{E}\,(\Pi^2 \leq \delta'^2) | \pi', \lambda' \rangle \; . \tag{140}$$

The first factor is exactly what would be found by the appropriate path integral of the original classical system with only the single constraint $\phi(p,q) = 0$ and the single Lagrange multiplier λ. The second factor represents the modification introduced by considering the extended system. Note, however, that with a suitable δ'-limit the second factor reduces to a product of terms, one depending on the " $''$ " arguments, the other depending on the " $'$ " arguments, just as was the case previously. This result for the second factor implies that it has become the reproducing kernel for a *one- dimensional* Hilbert space, and when multiplied by the first factor it may be ignored entirely. In this way it is found that the quantization of the original and extended systems leads to identical results.

6 Special Applications

6.1 Algebraically Inequivalent Constraints

The following example is suggested by Problem 5.1 in [21]. Consider the two-degree of freedom system with vanishing Hamiltonian described by the classical action

$$I = \textstyle\int (p_1\dot{q}_1 + p_2\dot{q}_2 - \lambda_1 p_1 - \lambda_2 p_2)\, dt \; . \tag{141}$$

The equations of motion become

$$\dot{q}_j = \lambda_j \; , \qquad \dot{p}_j = 0 \; , \qquad p_j = 0 \; , \qquad j = 1, 2 \; . \tag{142}$$

Evidently the Poisson bracket $\{p_1, p_2\} = 0$.

As a second version of the same dynamics, consider the classical action

$$I = \int (p_1\dot{q}_1 + p_2\dot{q}_2 - \lambda_1 p_1 - \lambda_2 e^{cq_1} p_2)\, dt\,, \tag{143}$$

which leads to the equations of motion

$$\dot{q}_1 = \lambda_1\,,\;\; \dot{q}_2 = \lambda_2 e^{cq_1}\,,\;\; \dot{p}_1 = -c\lambda_2 e^{cq_1} p_2\,,\;\; \dot{p}_2 = 0\,,\;\; p_1 = e^{cq_1} p_2 = 0\,. \tag{144}$$

Since $e^{cq_1}p_2 = 0$ implies that $p_2 = 0$, it follows that the two formulations are equivalent despite the fact that in the second case $\{p_1, e^{cq_1}p_2\} = -c\,e^{cq_1}p_2$, which has a fundamentally different algebraic structure when $c \neq 0$ as compared to $c = 0$.

Let us discuss these two examples from the point of view of a coherent state, projection operator quantization. For the first version we consider

$$\mathcal{M}\int \exp[i\textstyle\int(p_1\dot{q}_1 + p_2\dot{q}_2 - \lambda_1 p_1 - \lambda_2 p_2)\,dt]\,\mathcal{D}p\,\mathcal{D}q\,\mathcal{D}C(\lambda)\,, \tag{145}$$

defined in a fashion to yield

$$\langle p'', q''|\mathbb{E}|p', q'\rangle \tag{146}$$

where, for ease of evaluation, we may choose

$$\mathbb{E} = \mathbb{E}(P_1^2 \le \delta^2)\mathbb{E}(P_2^2 \le \delta^2)\,. \tag{147}$$

In particular this choice leads to the fact that

$$\langle p'', q''|\mathbb{E}|p', q'\rangle$$
$$= \pi^{-1}\prod_{l=1}^{2}\int_{-\delta}^{\delta} \exp[-\tfrac{1}{2}(k_l - p_l'')^2 + ik_l(q_l'' - q_l') - \tfrac{1}{2}(k_l - p_l')^2]\,dk_l\,. \tag{148}$$

Let us reduce this reproducing kernel, in particular, by multiplying this expression by $\pi/(2\delta)^2$ and passing to the limit $\delta \to 0$. The result is the reduced reproducing kernel given by

$$\exp[-\tfrac{1}{2}(p_1''^2 + p_2''^2)]\,\exp[-\tfrac{1}{2}(p_1'^2 + p_2'^2)]\,, \tag{149}$$

which clearly characterizes a particular representation of a one-dimensional Hilbert space in which every vector is proportional to $\exp[-\frac{1}{2}(p_1^2 + p_2^2)]$. This example, of course, is related to the reduction examples given earlier. Moreover, we can introduce an integral representation over the remaining p variables for the inner product if we so desire.

Let us now turn attention to the second formulation of the problem by focusing [for a different $C(\lambda)$] on

$$\mathcal{M}\int \exp[i\textstyle\int(p_1\dot{q}_1 + p_2\dot{q}_2 - \lambda_1 p_1 - \lambda_2 e^{cq_1} p_2)\,dt]\,\mathcal{D}p\,\mathcal{D}q\,\mathcal{D}C(\lambda)\,. \tag{150}$$

This expression again leads (for a different $\mathbb{E}$) to

$$\langle p'', q''|\mathbb{E}|p', q'\rangle\ , \tag{151}$$

where in the present case the fully *reduced* form of this expression is proportional to

$$\begin{aligned}\int \exp[-\tfrac{1}{2}(k_2 - p_2'')^2 + ik_2(q_2'' - q_2') - \tfrac{1}{2}(k_2 - p_2')^2]\\ \times\exp[-\tfrac{1}{2}(k_1 - p_1'')^2 + ik_1q_1'' - \tfrac{1}{2}i\lambda_1 k_1]\\ \times\exp[-ixk_1 - i\lambda_2 e^{cx}k_2 + ix\kappa_1]\\ \times\exp[-\tfrac{1}{2}i\lambda_1\kappa_1 - i\kappa_1 q_1' - \tfrac{1}{2}(\kappa_1 - p_1')^2]\\ \times dk_2\, dk_1\, dx\, d\kappa_1\, d\lambda_1\, d\lambda_2\ .\end{aligned} \tag{152}$$

When normalized appropriately, this expression is evaluated as

$$\exp[-\tfrac{1}{2}(p_1''^2 + p_2''^2 + icp_1'')]\ \exp[-\tfrac{1}{2}(p_1'^2 + p_2'^2 - icp_1')]\ , \tag{153}$$

which once again represents a one-dimensional Hilbert space although it has a different representation than in the case $c = 0$.

Thus we have obtained a c-dependent family of distinct but equivalent quantum representations for the same Hilbert space, reflecting the c-dependent family of equivalent classical solutions.

6.2 Irregular Constraints

In discussing constraints one often pays considerable attention to the regularity of the expressions involved. Consider, once again, the simple example of a single constraint $p = 0$ as illustrated by the classical action

$$I = \int(p\dot{q} - \lambda p)\, dt\ . \tag{154}$$

The equations of motion read $\dot{q} = \lambda$, $\dot{p} = 0$, and $p = 0$. On the other hand, one may ask about imposing the constraint $p^3 = 0$ or possibly $p^{1/3} = 0$, etc., instead of $p = 0$. Let us incorporate several such examples by studying the classical action

$$\int(p\dot{q} - \lambda p|p|^{\gamma})\, dt\ , \qquad \gamma > -1\ . \tag{155}$$

Here the equations of motion include $\dot{q} = \lambda(\gamma + 1)\,|p|^{\gamma}$ which, along with the constraint $p|p|^{\gamma} = 0$, may cause some difficulty in seeking a classical solution of the equations of motion. When $\gamma \neq 0$, such constraints are said to be *irregular* [21]. It is clear from (9) that irregular constraints lead to considerable difficulty in conventional phase-space path integral approaches.

Let us examine the question of irregular constraints from the point of view of a coherent state, projection operator, phase-space path integral quantization. We first observe that the operator $P|P|^\gamma$ is well defined by means of its spectral decomposition. Moreover, for any $\gamma > -1$, it follows that

$$\begin{aligned} &\int e^{-i\xi P|P|^\gamma} \frac{\sin(\delta^{\gamma+1}\xi)}{\pi\xi}\, d\xi \\ &\quad = \mathbb{E}(-\delta^{\gamma+1} \leq P|P|^\gamma \leq \delta^{\gamma+1}) \\ &\quad = \mathbb{E}(-\delta \leq P \leq \delta)\,. \end{aligned} \tag{156}$$

Thus, from the operator point of view, it is possible to consider the constraint operator $P|P|^\gamma$ just as easily as P itself. In particular, it follows that

$$\langle p'', q''|\mathbb{E}|p', q'\rangle = \mathcal{M} \int \exp[i{\textstyle\int}(p\dot{q} - \lambda p|p|^\gamma)\, dt]\, \mathcal{D}p\, \mathcal{D}q\, \mathcal{D}C_\gamma(\lambda)\,, \tag{157}$$

where we have appended γ to the measure for the Lagrange multiplier λ to emphasize the dependence of that measure on γ. The reduction of the reproducing kernel proceeds as with the cases discussed earlier, and we determine for all γ that

$$\lim_{\delta\to 0} \frac{\sqrt{\pi}}{(2\delta)} \langle p'', q''|\mathbb{E}|p', q'\rangle = e^{-\frac{1}{2}(p''^2 + p'^2)}\,, \tag{158}$$

representative of a one-dimensional Hilbert space. Note that, like the classical theory, the ultimate form of the quantum theory is independent of γ.

It is natural to ask how one is to understand this acceptable behavior for the quantum theory for irregular constraints and the difficulties they seem to present to the classical theory. Just like the classical and quantum Hamiltonians, the connection between the classical and quantum constraints is given by

$$\phi(p, q) \equiv \langle p, q|\Phi(P, Q)|p, q\rangle = \langle 0|\Phi(P + p, Q + q)|0\rangle\,. \tag{159}$$

With this rule we typically find that $\phi(p, q) \neq \Phi(p, q)$ due to the fact that $\hbar \neq 0$, but the difference between these expressions is generally qualitatively unimportant. In certain circumstances, however, that difference is qualitatively significant even though it is quantitatively very small. Since that difference is $O(\hbar)$, let us explicitly exhibit the appropriate $\hbar$-dependence hereafter.

First consider the case of $\gamma = 2$. In that case

$$\langle p, q|P^3|p, q\rangle = \langle 0|(P + p)^3|0\rangle = p^3 + 3\langle P^2\rangle p\,, \tag{160}$$

where we have introduced the shorthand $\langle(\cdot)\rangle \equiv \langle 0|(\cdot)|0\rangle$. Since $\langle P^2\rangle = \hbar/2$ it follows that for the quantum constraint P^3, the corresponding classical constraint function is given by $p^3 + (3\hbar/2)p$. For $|p| \gg \sqrt{\hbar}$, this constraint is adequately given by p^3. However, when $|p| \ll \sqrt{\hbar}$ – *as must eventually be the case in order to actually satisfy the classical constraint* – then the functional form of the constraint is effectively $(3\hbar/2)p$. In short, if the quantum constraint operator is P^3, then the classical constraint function is in fact *regular* when the constraint vanishes.

A similar analysis holds for a general value of γ. The classical constraint is given by

$$\begin{aligned}\phi_\gamma(p) &= (\pi\hbar)^{-1/2}\int (k+p)|k+p|^\gamma e^{-k^2/\hbar}\,dk \\ &= (\pi\hbar)^{-1/2}\int k|k|^\gamma e^{-(k-p)^2/\hbar}\,dk\,. \end{aligned} \tag{161}$$

For $|p| \gg \sqrt{\hbar}$ this expression effectively yields $\phi_\gamma(p) \simeq p|p|^\gamma$. On the other hand, for $p \approx 0$, and more especially for $|p| \ll \sqrt{\hbar}$, this expression shows that the constraint function vanishes *linearly*, specifically as $\phi_\gamma(p) \simeq \kappa\, p$, where

$$\kappa \equiv 2(\hbar^\gamma/\pi)^{1/2}\int y^2|y|^\gamma e^{-y^2}\,dy = 2(\hbar^\gamma/\pi)^{1/2}\Gamma((\gamma+3)/2) \equiv \hbar^{\gamma/2}\kappa_o\,. \tag{162}$$

A rough, but qualitatively correct expression for this behavior is given by

$$\phi_\gamma(p) \simeq \kappa_o\, p(\hbar + p^2\kappa_o^{-2/\gamma})^{\gamma/2}\,. \tag{163}$$

Thus, from the present point of view, irregular constraints do not arise from consistent quantum constraints; instead, irregular constraints arise as limiting expressions of consistent, regular classical constraints as $\hbar \to 0$.

7 Some Other Applications of the Projection Operator Approach

There have been several cases in which the projection operator has been used to study constrained systems. In [43], as well as in [45] and [17], the projection operator formalism has been applied to a simple $0+1$ model of a gauge theory. Govaerts [16] applied the projection operator scheme to study the relativistic particle in a reparameterization invariant form. In [30] the authors have studied first-class constraints, while in [31] they studied second-class constraint situations from the point of view of projection operator quantization. In addition, they have discussed in a general way the application of projection operator techniques to gauge theory [45]. Fermion systems have been treated, e.g., in [25]. Shabanov [44] has incorporated the projection operator into his Physics Reports review of gauge theories, and developed an algorithm for how the projection operator approach may be incorporated into lattice gauge theory calculations. Shabanov has also shown how the projection operator approach may be especially useful in ensuring that constraints are satisfied in an ion-surface interaction [6]. In addition, Klauder [28] has applied the projection operator method in a study of quantum gravity. Finally, a U(1) Chern-Simons model has been studied and solved with the projection operator method using coherent states in [18].

Projection operators have also been used previously in the study of constrained system quantization. For example, as noted earlier, some aspects of a coherent state quantization procedure that emphasized projection operators for systems with closed first-class constraints have been presented in [42]. In addition, we thank M. Henneaux for his thoughtful comments as this approach was being developed, as well as for pointing out that projection operators for closed first-class constraints also appear in [21]. Please note that this very short list does not pretend to be complete regarding prior considerations of projection operator investigations in connection with constrained systems.

Acknowledgements

The present paper represents a summary of some of the author's principal contributions to the projection operator approach for the quantization of systems with constraints for the four years from 1996 through 1999. It is a pleasure to thank my coauthors J. Govaerts and S. Shabanov who have shared in this general project. In addition to these two individuals, thanks are also extended to M. Henneaux and B. Whiting for many discussions regarding constraints and their quantization.

References

1. A. Ashtekar, G.T. Horowitz: Phys. Rev. D **26**, 3342 (1986)
2. N. Aronszajn: Proc. Cambridge Phil. Soc. **39**, 133 (1943)
3. N. Aronszajn: Trans. Amer. Math. Soc. **68**, 337 (1950)
4. I.A. Batalin, E.S. Fradkin: Phys. Lett. B **180**, 157 (1987)
5. I.A. Batalin, E.S. Fradkin: Nucl. Phys. B **279**, 524 (1987)
6. A. Borisov, S.V. Shabanov: in preparation
7. N.H. Christ, T.D. Lee: Phys. Rev. D **22**, 939 (1980)
8. B.S. DeWitt: In: *General Relativity: An Einstein Centenary Survey*, ed. by S.W. Hawking and W. Israel (Cambridge University Press, Cambridge 1979)
9. P.A.M. Dirac: *The Principles of Quantum Mechanics*, 3rd edn. (Oxford University Press, Oxford 1956)
10. P.A.M. Dirac: *Lectures on Quantum Mechanics* (Belfer Graduate School of Science, Yeshiva University, New York 1964)
11. L.D. Faddeev: Theor. Math. Phys. **1**, 1 (1970)
12. L.D. Faddeev, S.L. Shatashvili: Phys. Lett. B **167**, 225 (1986)
13. R. Friedberg, T.D. Lee, Y. Pang, H.C. Ren: Ann. Phys. **246**, 381 (1996)
14. D.M. Gitman, I.V. Tyutin: *Quantization of Fields with Constraints* (Springer-Verlag, Berlin 1990)
15. J. Govaerts: 'Hamiltonian Quantisation and Constrained Dynamics'. In: *Leuven Notes in Mathematical and Theoretical Physics, Vol. 4, Series B: Theoretical Particle Physics* (Leuven University Press, 1991)

16. J. Govaerts: J. Phys. A **30**, 603 (1997)
17. J. Govaerts, J.R. Klauder: Ann. Phys. **274**, 251 (1999)
18. J. Govaerts, B. Deschepper: J. Phys. A **33**, 1031 (2000)
19. H. Grundling, C.A. Hurst: Comm. Math. Phys. **98**, 369 (1985)
20. P. Hájiĉek: J. Math. Phys. **27**, 1800 (1986)
21. M. Henneaux, C. Teitelboim: *Quantization of Gauge Systems* (Princeton University Press, Princeton, New York 1992)
22. A. Higuchi: Class. Quantum Grav. **8**, 2023 (1991)
23. C.J. Isham: In: *Relativity, Groups, and Topology, II, Les Houches 1983*, ed. by B.S. DeWitt and R. Stora, (North Holland, Amsterdam 1984)
24. C.J. Isham, J.R. Klauder: J. Math. Phys. **32**, 607 (1991)
25. G. Junker, J.R. Klauder: Eur. Phys. J. C **4**, 173 (1998)
26. J.R. Klauder: Modern Phys. Lett. A **8**, 1735 (1993)
27. J.R. Klauder: Ann. Phys. **254**, 419 (1997)
28. J.R. Klauder: J. Math. Phys. **40**, 5860 (1999)
29. J.R. Klauder: Nucl. Phys. B **547**, 397 (1999)
30. J.R.Klauder, S.V. Shabanov: Phys. Lett. B **398**, 116 (1997)
31. J.R. Klauder, S.V. Shabanov: Nucl. Phys. B **511**, 713 (1998)
32. J.R. Klauder, B.-S. Skagerstam: *Coherent States: Applications to Physics and Mathematical Physics* (World Scientific, Singapore 1985)
33. J.R. Klauder, E.C.G. Sudarshan: *Fundamentals of Quantum Optics*, (W.A. Benjamin, Inc., New York 1986)
34. J.R. Klauder, B.F. Whiting: J. Phys. A **26**, 1697 (1993)
35. K. Kuchar: Phys. Rev. D **34**, 3031 (1986)
36. N.P. Landsman: J. Geom. Phys. **15**, 285 (1995)
37. J. McKenna, J.R. Klauder: J. Math. Phys. **5**, 878 (1964)
38. H. Meschkowski: *Hilbertsche Raum mit Kernfunktion* (Springer, Berlin 1962)
39. L.V. Prokhorov: Sov. J. Nucl. Phys. **35**, 129 (1982)
40. L.V. Prokhorov, S.V. Shabanov: Sov. Phys. Uspekhi **34**, 108 (1991)
41. F. Riesz, B. Sz.-Nagy: *Functional Analysis*, (Dover Publications, Inc., New York 1990)
42. S.V. Shabanov: 'Path Integral in Holomophic Representation without Gauge Fixation', Dubna preprint E2-89-687 (unpublished); 'The Role of Gauge Invariants in Path Integral Construction', Dubna preprint E2-89-688 (unpublished); 'Phase Space Structure in Gauge Theories' (in Russian), JINR Lecture Notes **54** (1989)
43. S.V. Shabanov: 'Path Integral in Holomorphic Representation without Gauge Fixation'. In: *Path Integrals: Dubna '96* (Publishing Department, Joint Institute for Nuclear Research 1997); quant-ph/9608018
44. S.V. Shabanov: Phys. Rep. **326**, 1 (2000); hep-th/0002043
45. S.V. Shabanov, J.R. Klauder: Phys. Lett. B **456**, 38 (1999)
46. C. Teitelboim: J. Math. Phys. **25**, 1093 (1984)
47. V.M. Villanueva, J. Govaerts, J.-L. Lucio-Martinez: 'Quantisation without Gauge Fixing: Avoiding Gribov Problems through the Physical Propagator'; hep-th/9909033 (to appear)

Algebraic Methods of Renormalization

Klaus Sibold

Universität Leipzig, Institut für Theoretische Physik,
D-04109 Leipzig, Augustusplatz 10/11, Germany

Abstract. After some remarks to subtraction schemes yielding finite Green functions in perturbation theory the action principle will be presented. It provides the backbone for the quantization of gauge theories independent from the scheme chosen for actual calculations. As examples supersymmetric theories will be treated as well as the electroweak standard model. The emphasis will be on the structural aspects of the renormalization problem.

1 Generalities

1.1 Renormalization Schemes

The perturbative definition of quantum field theories is usually based on the Gell-Mann Low formula

$$\begin{aligned} G(x_1,\ldots,x_n) &\equiv \langle T(\varphi(x_1)\ldots\varphi(x_n))\rangle \\ &= \frac{\langle T\varphi^{(o)}(x_1)\ldots\varphi^{(o)}(x_n)e^{i\int\mathcal{L}^{(o)}_{\text{int}}}\rangle_{(o)}}{\langle Te^{i\int\mathcal{L}^{(o)}_{\text{int}}}\rangle_{(o)}}\,, \end{aligned} \tag{1}$$

which expresses the vacuum expectation value of time ordered products of interacting fields $\varphi(x)$ as a (formal) power series of those of free fields involving the interaction Lagrangian $\mathcal{L}^{(o)}_{\text{int}}$. The actual evaluation of (1) proceeds via Wicks theorem and yields the Feynman rules of the model in question. The fields $\varphi^{(o)}(x)$ obey the free field equations

$$\begin{aligned} (\Box+m^2)\varphi &= 0 \quad (\text{spin } 0)\,, \\ (i\,\partial\!\!\!/ - m)\psi &= 0 \quad (\text{spin } 1/2)\,, \\ \Box A_\mu &= 0 \quad (\text{spin 1; Feynman gauge})\,, \end{aligned} \tag{2}$$

whereas $\mathcal{L}^{(o)}_{\text{int}}$ consists of field monomials and derivatives such that its dimension does not exceed four. The reason for this latter restriction becomes evident when one performs power counting in a one-particle-irreducible (1PI) Feynman diagram γ with N_a external lines of type a, m independent closed loops, I internal lines and V vertices of dimension d_i. The superficial degree of divergence in the ultraviolet region (large internal momenta) turns out to be

$$\begin{aligned} d(\gamma) &= 4 - \sum_a N_a d_a + \sum_i^V (d_i - 4) \,, \\ m &= I - V + 1 \,. \end{aligned} \tag{3}$$

Here, d_a denotes the UV-dimension of fields ($d_a = 1$ for $a =$ spin 0, spin 1; $d_a = 3/2$ for $a =$ spin 1/2). As free propagators we have assumed the usual ones:

$$\begin{aligned} \langle T\varphi^{(0)}\varphi^{(0)}\rangle\rangle^{\text{F.T.}} &= \frac{i}{k^2 - m^2 + i\varepsilon} \,, \\ \langle T\psi^{(0)}\psi^{(0)}\rangle^{\text{F.T.}} &= \frac{\not{k} + m}{k^2 - m^2 + i\varepsilon} \,, \\ \langle TA_\mu^{(0)}A_\nu^{(0)}\rangle^{\text{F.T.}} &= \eta_{\mu\nu}\frac{-i}{k^2 + i\varepsilon} \text{ (Feynman gauge).} \end{aligned} \tag{4}$$

It is obvious from (3) that every diagram with N external lines can be made divergent if there is a vertex with $d_i > 4$. If all vertices of $\mathcal{L}_{\text{int}}$ satisfy $d_i \leq 4$ then the respective theory is called power counting renormalizable: only finitely many classes of graphs are divergent (at most those with $0, 1, 2, 3, 4$ external lines).

Giving meaning to (1) thus is the same as rendering at least the individual terms of the series finite. The difficulty in this problem is, of course, not to obtain just finiteness. This could always be achieved by a trivial assignment, e.g. replacing every term by its tree approximation. The aim rather is to give a proper mathematical definition which maintains the axioms to be obeyed by the Green functions or at least the S-matrix. We want to maintain Lorentz covariance and the S-matrix should be unitary and causal. If (1) is defined completely ad libitum then one can certainly not hope to satisfy any axioms.

Some popular methods to proceed are the following ones.

Pauli-Villars Regularization. The free propagators $\Delta_{\text{F}}(k; m)$ in (4) are replaced by sums

$$\Delta_{\text{reg}} = \Delta_{\text{F}}(k; m) + \sum_l c_l \Delta_{\text{F}}(k; M_l) \,, \tag{5}$$

with coefficients c_l determined in such a way that all diagrams become finite. One then adds to $\mathcal{L}_{\text{int}}$ counterterms depending on M_l such that in the limit $M_l \to \infty$ all diagrams still stay finite. Whereas the regularized model violates at least unitarity, the renormalized theory (with the limit $M_l \to \infty$ performed) may satisfy the axioms. Abelian gauge invariance can be maintained in the course of regularization, whereas non-abelian invariance is broken. (For original reference and convenient application one may consult [14].)

Analytical Regularization. The propagator denominators are modified into $(k^2 - m^2 + i\varepsilon)^\lambda$ with complex λ and the dimension of the integration measure $d^4k_1 \cdots d^4k_m$ also is continued to a complex number. Then divergences are represented as poles in these complex numbers that can be removed by adding appropriate counterterms leading to the same poles. Abelian gauge invariance can be maintained, non-abelian not. (The original reference is [19].)

Dimensional Regularization. Here, the dimension of the integration measure is continued from four to n. The γ and momentum algebra is modified accordingly. Divergences exhibit themselves as poles in $(n-4)$ that can be removed. This method is widely used in practice since it also maintains non-abelian gauge invariance as long as no γ_5 or $\epsilon_{\mu\nu\rho\sigma}$ is present in the theory. In the case of chiral fermions it requires a priori non-invariant counterterms, in particular it does not maintain supersymmetry. (As an original reference one may read [23], as a useful modern version [4]; for the never ending discussion on the γ_5-problem one should consult [7].)

Momentum Space Subtractions. Since the propagators (4) are rational functions of the momenta one can render all one-loop diagrams with $d(\gamma) \geq 0$ finite by subtracting the Taylor expansion around vanishing external momenta up to and including the term of order $d(\gamma)$. Multiloop diagrams are recursively treated with the help of the forest formula [25] which disentangles all overlapping divergences. Vanishing masses require additional care because subtraction at zero momentum can introduce spurious infrared divergences. One adds an auxiliary mass $M(s-1)$, where s is treated like an external momentum: it participates in subtractions and is put equal to 1 only at the very end of the calculation [11]. This renormalization scheme is not very convenient for explicit calculations, but essentially all relevant theorems in renormalization theory have been proven in this context, hence we shall refer to it whenever an explicit use of a specific scheme is required. (A very readable account is provided in [9].)

One important feature is common to all these approaches. Whenever one has removed an infinity, one has defined at the same time the finite part of the diagrams in question. These finite parts can be modified order by order through local counterterms in the action (with dimensions less than or equal to four). It is the most fundamental theorem in renormalization theory that whenever one has defined Green functions satisfying the axioms in one scheme then the Green functions in any other scheme are related to them by adding finite counterterms [5]. One can formulate this statement also as follows: by applying any specific scheme one has implicitly specified normalization conditions. Going over to another scheme but maintaining the specific normalization conditions is always possible by adding finite counterterms.[1]

[1] One caveat is in order: If the normalization conditions chosen are singular (e.g. involve infinite limits like in MS or $\overline{\text{MS}}$) this transition might not be possible.

1.2 The Action Principle

Let a set of Green functions $G_n(x_1, \dots, x_n)$ be given. It depends on the masses and the couplings as parameters and involves relations amongst different Green functions as consequences of equations of motion. The action principle answers the question: How can one express the variation of parameters and fields in terms of Green functions and possibly, additional vertices? The most concise way of formulating the answer is in terms of generating functionals.

$$Z \equiv Z(J) = \langle T e^{i \int dx \varphi(x) J(x)} \rangle \tag{6}$$

denotes the generating functional for (general) Green functions:

$$G(x_1 \dots x_n) = \frac{\delta}{i\delta J(x_1)} \cdots \frac{\delta}{i\delta J(x_n)} Z(J)\Big|_{J=0} \; . \tag{7}$$

If free fields $\varphi(x) = \varphi^{(0)}(x)$ appear in (6) we deal with

$$Z_0 \equiv Z_0(J) = \langle T e^{i \int dx \varphi^{(0)}(x) J(x)} \rangle \, , \tag{8}$$

the generating functional for free Green functions. It consists of products of free propagators with no points coinciding, e.g. for one scalar field:

$$Z_0(J) = e^{\frac{1}{2} \int dx_1 dx_2 iJ(x_1) \Delta_F(x_1 - x_2) iJ(x_2)} \, . \tag{9}$$

Not worrying about divergences but proceeding formally one can derive the following *renormalizable* formula

$$Z(J) = \frac{e^{i \int \mathcal{L}_{\text{int}}(\frac{\delta}{i\delta J})} Z_0(J)}{e^{i \int \mathcal{L}_{\text{int}}(\frac{\delta}{i\delta J})} Z_0(J)\Big|_{J=0}} \; . \tag{10}$$

Here, e.g. for a φ^4-theory

$$\mathcal{L}_{\text{int}} = -\frac{\lambda}{4!} \varphi^4 \, , \tag{11}$$

which links in a very definite manner the free propagators of (4). For convergent diagrams the derivation of (10) is certainly rigorous, hence it has enormous heuristic value: it governs the correct combinatorics compatible with the axioms. Enlarging (11) to

$$\mathcal{L}_{\text{int}} = z \frac{1}{2} \partial\varphi \partial\varphi - \frac{1}{2} m^2 a \varphi^2 - \frac{\lambda + c}{4!} \varphi^4 \, , \tag{12}$$

i.e. permitting counterterms, does not violate combinatorial relations and points to an even more remarkable formula:

$$Z(J) = e^{i(z\Delta_1 - a\Delta_2 - (\lambda + c)\Delta_4)} Z_0(J) \, . \tag{13}$$

Here, in addition to the manipulations of (10), we assign $\Delta_1 \equiv \left[\int \frac{1}{2}\partial\varphi\partial\varphi\right]_4$, $\Delta_2 \equiv \left[\int \frac{1}{2}m^2\varphi^2\right]_4$ and $\Delta_4 \equiv \left[\int \frac{1}{4!}\varphi^4\right]_4$, via the rule that they be considered as vertices of UV-dimension four in the power counting formula and in the evaluation of the diagrams according to Zimmermann. I.e. diagrams are worked out as prescribed in (10) with (12) inserted. Then they are rendered convergent by the momentum subtractions where the power counting formula (3) is being used and $\Delta_1, \Delta_2, \Delta_4$ represent vertices of dimension four.

The validity of (13) assumed, it is clear what the response to variation of λ (the physical coupling is):

$$\frac{\partial Z}{\partial \lambda} = \left(\frac{\partial z}{\partial \lambda}\Delta_1 - \frac{\partial a}{\partial \lambda}\Delta_2 - (1 + \frac{\partial c}{\partial \lambda})\Delta_2\right) Z(J) \,. \tag{14}$$

That is, the diagrams of $Z(J)$ acquire an additional "insertion" of the vertex

$$\frac{\partial z}{\partial \lambda}\Delta_1 - \frac{\partial a}{\partial \lambda}\Delta_2 - (1 + \frac{\partial c}{\partial \lambda})\Delta_2 \,. \tag{15}$$

As far as combinatorics is concerned they are to be treated as before. As far as power counting is concerned the additional vertex has assigned degree four. By introducing the suggestive notion of Γ_{eff} of Zimmermann

$$\begin{aligned} \Gamma_{\text{eff}} &\equiv \Gamma_{\text{cl}} + \Gamma_{\text{int}} \\ &= (1+z)\Delta_1 - (1+a)\Delta_2 - (\lambda + c)\Delta_4 \,, \end{aligned} \tag{16}$$

we can write

$$\frac{\partial Z}{\partial \lambda} = \left[\frac{\partial \Gamma_{\text{eff}}}{\partial \lambda}\right]_4 \cdot Z(J) \tag{17}$$

instead of (14).

The derivative $\partial/\partial m^2$ operates on Z_0 also, hence is slightly more involved than $\partial/\partial\lambda$. But the analysis yields a result analogous to (17)

$$m^2\frac{\partial Z}{\partial m^2} = \left[m^2\frac{\partial \Gamma_{\text{eff}}}{\partial m^2}\right]_4 \cdot Z(J) \,. \tag{18}$$

In fact, if ∇ denotes the derivative w.r.t. any parameter in a model, then one can show that

$$\nabla Z = [\nabla \Gamma_{\text{eff}}]_4 \cdot Z(J) \,. \tag{19}$$

Here, as in (16), Γ_{eff} denotes the classical action plus all counterterms.

The variation w.r.t. to a field can also be expressed on Z, but most suggestively on another functional: That of 1PI Green functions Γ. It is obtained from $Z(J)$ by first defining the functional $Z_c(J)$ of connected Green functions via

$$Z(J) = e^{iZ_c(J)} \,. \tag{20}$$

Then Γ is constructed as Legendre transform of $Z_c(J)$

$$\Gamma(\varphi) = Z_c(Z(\varphi)) - \int dx J(\varphi)\varphi \,, \tag{21}$$

$$\varphi \quad = \frac{\delta Z_c}{\delta J} \,. \tag{22}$$

Explicitly, (22) is solved for $J = J(\varphi)$ and this result inserted in (21). In terms of Γ the desired relation reads

$$\frac{\delta \Gamma}{\delta \varphi} = \text{``}\left[\frac{\delta \Gamma_{\text{eff}}}{\delta \varphi}\right]_{4-d(\varphi)}\text{''} \cdot \Gamma \,. \tag{23}$$

The quotation marks indicate that linear terms in $\delta\Gamma_{\text{eff}}/\delta\varphi$ are not considered as vertices for Γ, whereas non-linear terms form the additional vertex in question carrying the power counting $4 - d(\varphi)$. As an example for the φ^4-theory

$$\frac{\delta \Gamma}{\delta \varphi} = -(1+z)\Box\varphi - m^2(1+a)\varphi - \frac{\lambda + c}{3!}\left[\varphi^3\right]_3 \cdot \Gamma \,. \tag{24}$$

Transformed back to $Z(J)$, one obtains

$$-JZ(J) = \left[\frac{\delta \Gamma_{\text{eff}}}{\delta \varphi}\right]_3 \cdot Z(J) \,. \tag{25}$$

If a field φ_1 transforms linearly into a field φ_2 one also has a very suggestive result

$$\varphi_2 \frac{\delta \Gamma}{\delta \varphi_1} = \left[\varphi_2 \frac{\delta \Gamma_{\text{eff}}}{\delta \varphi_1}\right]_\delta \cdot \Gamma \,, \tag{26}$$

with $\delta = 4 - d(\varphi_1) + d(\varphi_2)$. It reads on Z

$$-J_1 \frac{\delta \Gamma}{\delta J_2} = \left[\varphi_2 \frac{\delta \Gamma_{\text{eff}}}{\delta \varphi_1}\right]_\delta \cdot Z \,. \tag{27}$$

The simplicity of these relations in unfortunately lost when considering non-linear field transformations. Looked at on the level of diagrams, it is obvious that non-linear variations may introduce additional loops, i.e. additional divergences which are not necessarily dealt with in a naive fashion. If classically

$$\delta\varphi = Q(\varphi) \tag{28}$$

with Q non-linear (dim $Q = \delta$), then one introduces an external field q, changes Γ_{eff} into

$$\Gamma^q_{\text{eff}} = \Gamma_{\text{eff}} + [\int q\, Q]_4 + \text{corrections} \,, \tag{29}$$

"corrections" denoting q-dependent counterterms, and finds

$$iJ(x)\frac{\delta Z^q}{\delta q} = [\hat{Q}]_{3+\delta} \cdot Z \,, \tag{30}$$

$$\frac{\delta \Gamma}{\delta \varphi(x)} \frac{\delta \Gamma^q}{\delta q}\Big|_{q=0} = [\hat{Q}]_{3+\delta} \cdot \Gamma \,, \tag{31}$$

for the variation. Here

$$\hat{Q} = \frac{\delta \Gamma^q_{\text{eff}}}{\delta q} \frac{\delta \Gamma_{\text{eff}}}{\delta \varphi}\Big|_{q=0} + \text{corrections} \,. \tag{32}$$

Important for applications is not the explicit knowledge of $\hat{Q}$ but rather that the r.h.s of (30) and (31) form local insertions of well-defined power counting. Only this property will be used in the general analysis.

1.3 Green Functions and Operators

The Gell-Mann-Low formula complemented by a subtraction prescription and normalization conditions yields unique Green functions. These are not measurable and contain in general more information than available experimentally: they constitute continuations off the physical mass shell. Relevant, however, are only values on the physical mass shell. The most important quantity in practice is certainly the S-Matrix. Its elements are given by

$$\begin{aligned} &{}_{\text{in}}\langle p_1 \dots p_n | S | q_1 \dots q_l \rangle_{\text{in}} = \\ &\left(\frac{-i}{\sqrt{z}}\right)^{n+l} \lim \prod_{k=1}^{n} (p_k^2 - m^2) \prod_{j=1}^{l} (q_j^2 - m^2) \tilde{G}(-p_1 \dots -p_n, q_1 \dots q_l) \,. \end{aligned} \tag{33}$$

Here, l particles are incoming (with momenta $q_1 \dots q_l$), n particles are outgoing (with momenta $p_1 \dots p_n$) and lim indicates the on shell transition $p_i^2 \to m^2, q_j^2 \to m^2$ $(p_i^0 > 0, q_j^0 > 0)$. $\tilde{G}$ denotes the Fouriertransform of the $(n+l)$-point Green function $G(y_1 \dots y_n, x_1 \dots x_l)$. z is the wave function renormalization defined as the residue of the 2-point-function at the pole of the physical mass:

$$G(x,y) = \frac{i}{(2\pi)^4} \int dp e^{ip(x-y)} \frac{z}{p^2 - m^2 + i\varepsilon} (1 + \mathcal{O}(p^2 - m^2)) \,. \tag{34}$$

Analogously the matrix elements of an operator $\mathcal{O}$ in Hilbert space can be obtained from its Green functions with elementary fields

$$\begin{aligned} &{}_{\text{out}}\langle p_1 \dots p_n | \mathcal{O} | q_1 \dots q_l \rangle_{\text{in}} = \\ &\left(\frac{-i}{\sqrt{z}}\right)^{n+l} \lim \prod_{k=1}^{n} (p_k^2 - m^2) \prod_{j=1}^{l} (q_j^2 - m^2) \tilde{G}_{\mathcal{O}(x)}(-p_1 \dots -p_n, q_1 \dots q_l) \,, \end{aligned} \tag{35}$$

where as before lim denotes the on-shell transition and $\tilde{G}_{\mathcal{O}(x)}$ is the Fourier transform of

$$G_{\mathcal{O}(x)}(y_1 \dots y_n, x_1 \dots x_l) = \langle T\mathcal{O}(x)\varphi(y_1)\cdots\varphi(x_l)\rangle \ . \tag{36}$$

Eqns. (33) and (35) are formulated for one scalar field, but the generalization to other fields should be obvious: The factors $(p^2 - m^2)$ are to be replaced by the inverse of the respective propagator and associated spin factors.

2 The Quantization of Gauge Theories

2.1 The Abelian Case

If one aims at a construction of field theories as independent as possible from the renormalization scheme employed, Ward identities (WI) turn out to be the most convenient tool. E.g. the classical action

$$\Gamma_{\text{inv}} = \int -\frac{1}{4}F^{\mu\nu}F_{\mu\nu} + \bar{\psi}(i\,\partial\!\!\!/ - m + e\,A\!\!\!/)\psi \ , \tag{37}$$

where $F_{\mu\nu} = \partial_\mu A_\nu - \partial_\nu A_\mu$, is invariant under the local gauge transformation

$$\delta A_\mu = \partial_\mu \omega(x) \ , \tag{38}$$
$$\delta\psi \ = ie\omega\psi \ , \tag{39}$$
$$\delta\bar{\psi} \ = -ie\omega\bar{\psi} \ , \tag{40}$$

and this is expressed in functional form as

$$w\Gamma_{\text{inv}} \equiv -\partial\frac{\delta\Gamma_{\text{inv}}}{\delta A} - ie\bar{\psi}\frac{\overrightarrow{\delta}}{\delta\bar{\psi}}\Gamma_{\text{inv}} + ie\Gamma_{\text{inv}}\frac{\overleftarrow{\delta}}{\delta\psi}\psi = 0 \ . \tag{41}$$

In the perturbative treatment of field theories the classical action can be identified with the lowest order approximation of Γ, the generating functional of vertex functions:

$$\Gamma = \Gamma^{(0)} + \hbar\Gamma^{(1)} + \hbar^2\Gamma^{(2)} + \dots \ , \tag{42}$$
$$\Gamma^{(0)} = \Gamma_{\text{cl}} \ .$$

It is well-known, that one has to fix the gauge if one wants to describe propagation. Eq. (37) is generalized to

$$\Gamma_{\text{cl}} \ = \Gamma_{\text{inv}} + \Gamma_{\text{g.f.}} \ , \tag{43}$$
$$\Gamma_{\text{g.f.}} \equiv \int -\frac{1}{2\xi}(\partial A)^2 + \frac{1}{2}M^2A^2$$

and (41) to

$$w\Gamma = -\frac{1}{\xi}(\Box + \xi M^2)\partial A \ . \tag{44}$$

(We added also a mass term for the vector field.) For Z, the generating functional of general Green functions, (44) corresponds to

$$\begin{aligned} wZ &\equiv -i\,\partial J \cdot Z + ie\bar{\eta}\frac{\overrightarrow{\delta}}{\delta\bar{\eta}}Z - ieZ\frac{\overleftarrow{\delta}}{\delta\eta}\eta \\ &= -\frac{1}{\xi}(\Box + \xi M^2)\,\partial\frac{\delta Z}{\delta J}\,. \end{aligned} \tag{45}$$

In order to establish (44) or (45) to all orders of perturbation theory one may e.g. employ an invariant regularization scheme (like dimensional or Pauli-Villars). Then the counterterms have the same form as those of $\Gamma_{\rm cl}$ in (43) and their coefficients are fixed by normalization conditions . It is well-known, that the parameter e in (41) can indeed be identified with the physical charge and no vertex function is required to fix its value (Thirring's theorem; Thomson limit of the WI). Like the classical theory, the renormalized one also maintains parity and charge conjugation invariance.

If one does not want to rely on an invariant scheme one proceeds as follows. One admits at first *all* possible counterterms to $\Gamma_{\rm cl}$ compatible with power counting renormalizability (i.e. with dimension ≤ 4) and then tries to fix their coefficients consistently order by order in perturbation theory. The most important aid originates from the observation that the WI operator w in (41) satisfies the algebraic relation:

$$[w(x), w(y)] = 0\,. \tag{46}$$

Now we use the action principle to deduce

$$w(x)\Gamma + \frac{1}{\xi}(\Box + \xi M^2)\partial A = [P(x)]_4 \cdot \Gamma\,, \tag{47}$$

where $P(x)$ can be expressed in a basis of field monomials of dimension four

$$P(x) = \sum_i a_i P_i(x)\,, \tag{48}$$

We used the fact that terms linear in the quantized fields can never be renormalization parts. Imposing parity invariance as a defining symmetry we can restrict P and hence P_i to parity invariant terms. Since the WI (44) holds in the classical approximation the insertion $[P(x)] \cdot \Gamma$ is necessarily of one loop order, hence

$$[P(x)]_4 \cdot \Gamma = P(x) + \mathcal{O}(\hbar) \tag{49}$$

Applying (46) to Γ and using (47), (49) we find

$$w(x)P(y) - w(y)P(x) = 0\,, \tag{50}$$

implying restrictions for the terms P_i which can possibly contribute to P. A short explicit calculation shows that the solution of (50) is given by

$$P(x) = w(x)\int dz P_{\rm var}(z)\,, \tag{51}$$

$$P_{\text{var}} = a_1(\partial A)^2 + a_2 A^2 + a_3(A^2)^2 + a_4\bar{\psi} i \partial\!\!\!/ \psi \,. \tag{52}$$

Adding P_{var} to Γ_{cl} as counterterm one can therefore establish the WI (44) by suitably adjusting the values of $a_1, \ldots, a_4$. The coefficients of the counterterms P_{inv} are fixed by the normalization conditions. Obviously the values of these counterterm coefficients are scheme-dependent, whereas the Green functions are not: after imposing the symmetries ((44), parity) and the normalization conditions they are uniquely fixed.

Up to now the WI served the mathematical purpose to define the theory. But it also implies immediate physical consequences. Applying the LSZ reduction technique to (45) one obtains the operator equation

$$(\Box + \xi M^2)\partial A^{\text{Op}} \stackrel{*}{=} 0 \tag{53}$$

($\stackrel{*}{=}$ means "on the physical mass shell"). Hence the ghost field ∂A, which causes the indefinite metric of the Fock space, is a free field and the construction of the physical Hilbert space for the free theory survives interaction! A graphical representation of this situation (for $\xi = 1$) is obtained by multiplying (45) with the inverse of $(\Box + M^2)$ and then displaying the terms as follows:

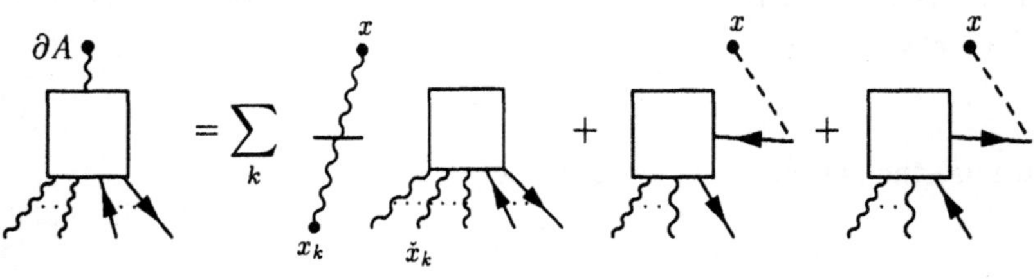

(54)

The broken line •----• represents the inverse of $(\Box + M^2)$, i.e. a scalar propagator. Crossing off ˇ means that this leg is to be omitted. The dashed line •~|~• indicates that that this line is replaced by $\partial_\mu \delta(x - x_k)$. Going on shell implies that the leg starting with ∂A just propagates freely.

2.2 BRS Transformations

For non-abelian gauge transformations

$$\delta A^a_\mu = \partial_\mu \omega^a + f^{abc} \omega^b A^c_\mu \tag{55}$$

(f^{abc} structure constants of a simple gauge group) and the associated invariant action

$$\Gamma_{\text{inv}} = -\frac{1}{4g^2} \operatorname{Tr} \int F^{\mu\nu} F_{\mu\nu} \,, \tag{56}$$

$$F_{\mu\nu} = \tau^a F^a_{\mu\nu} \,, \tag{57}$$

$$\left[\tau^a, \tau^b\right] = i f^{abc} \tau^c \tag{58}$$

(matter suppressed for a moment) one might be tempted to closely follow the abelian case and write down the WI

$$w^a \Gamma_{\rm cl} \equiv -\left(\partial \frac{\delta}{\delta A^a} + f^{abc} A^c \frac{\delta}{\delta A^a}\right) \Gamma_{\rm cl}$$
$$= -\frac{1}{\xi}(\delta^{ac}\Box + f^{abc} A^b_\mu \partial^\mu)\partial^\nu A^c_\nu \equiv -\frac{1}{\xi}\left(D_{\rm g.f.}\partial A\right)^a \qquad (59)$$

for

$$\Gamma_{\rm cl} = \Gamma_{\rm inv} + \Gamma_{\rm g.f.}\,, \qquad (60)$$
$$\Gamma_{\rm g.f.} \equiv {\rm Tr}\int -\frac{1}{2\xi}(\partial A)^2\,.$$

But then the second term of the r.h.s of (59) indicates that the field ∂A^a is *not* free, hence the longitudinal vector field interacts. The WI will neither define the theory nor yield unitarity! In explicit calculations Feynman found out, that in one-loop diagrams the unitarity violating contributions of ∂A^a could be compensated by the exchange of an additional multiplet of scalars in the adjoint representation, yet quantized as if they were fermions. Faddeev and Popov formulated their effect in functional form on Z, whereas Becchi-Rouet-Stora (BRS) [2] rewrote this identity and discovered that it expressed a *symmetry* of Z (and hence also of the action) if one interpreted the terms accordingly.

It is instructive to repeat their reasoning on the classical action [16]. We add a $\phi\pi$ term to $\Gamma_{\rm cl}$

$$\Gamma_{\phi\pi} = \frac{1}{\xi}{\rm Tr}\int \bar{c} D_{\phi\pi} c\,, \qquad (61)$$

with a differential operator $D_{\phi\pi}$ which originates from the gauge transformation of ∂A:

$$\delta(\partial A^a(x)) = (D_{\phi\pi})^a(x) = \Box\omega^a + f^{abc}\partial^\mu(A^b_\mu \omega^c)\,, \qquad (62)$$

it is just the adjoint to $D_{\rm g.f.}$

$${\rm Tr}\int \omega'(x)(D_{\phi\pi}\omega(x)) = {\rm Tr}\int (D_{\rm g.f.}\omega'(x))\omega(x)\,. \qquad (63)$$

The local WI is changed into

$$w^a \Gamma_{\rm cl} = -\frac{1}{\xi}\left((D_{\rm g.f}\partial A)^a + f^{abc}\partial^\mu(\partial_\mu \bar{c}^b\, c^c) - f^{abc'} f^{c'a'c}\partial^\mu \bar{c} A^b_\mu c^c\right)\,. \qquad (64)$$

We now multiply by $c^a(x)$, integrate over x, use (63) and integration by parts and arrive at

$${\rm Tr}\int dx\, c(x) w(x)\Gamma_{\rm cl} \equiv {\rm Tr}\int dx\,(\partial_\mu c + i[c, A_\mu])\frac{\delta \Gamma_{\rm cl}}{\delta A_\mu}$$
$$= \frac{1}{\xi}{\rm Tr}\int dx\,(-\partial A\, D_{\phi\pi} c + icc\, D_{\rm g.f.}\bar{c})\,. \qquad (65)$$

Now the r.h.s contains precisely equation of motion terms

$$\frac{\delta\Gamma}{\delta\bar{c}} = \frac{1}{\xi}D_{\phi\pi}c\,, \tag{66}$$

$$\frac{\delta\Gamma}{\delta c} = \frac{1}{\xi}D_{\mathrm{g.f.}}\bar{c}\,. \tag{67}$$

Hence we may insert them and rewrite equation (65) as a *homogeneous* identity

$$\mathrm{Tr}\int dx\left(sA_\mu\frac{\delta}{\delta A_\mu} + s\bar{c}\frac{\delta}{\delta\bar{c}} + sc\frac{\delta}{\delta c}\right)\Gamma_{\mathrm{cl}} = 0\,, \tag{68}$$

where

$$sA_\mu = \partial_\mu c + i[c, A_\mu]\,, \tag{69}$$

$$sc \;= icc\,, \quad sc^a = -\frac{1}{2}f^{abc}c^b c^c\,, \tag{70}$$

$$s\bar{c} \;= \partial A\,. \tag{71}$$

These transformations (BRS) thus constitute a symmetry of Γ_{cl}. It is clear that one may add gauge invariant matter terms and achieve for them BRS invariance as well by the prescription

$$s\phi = ic^a T^a\phi\,. \tag{72}$$

Here, ϕ is a multiplet of complex scalar fields. As on A_μ the BRS transformation is just the gauge transformation with field c as transformation "parameter".

2.3 The Slavnov–Taylor Identity

Since indeed the hope that BRS invariance will define the theory and yield unitarity materializes as fact it is important to formulate this invariance in a way which survives all orders and is independent from the renormalization scheme. It is convenient to introduce a Lagrange multiplier field B for fixing the gauge

$$\Gamma_{\mathrm{g.f.}} \equiv \mathrm{Tr}\int(\frac{\xi}{2}B^2 + B\partial A)\,. \tag{73}$$

B transforms under rigid gauge transformations in the adjoint representation

$$\delta_\omega B = i\,[\omega, B] \tag{74}$$

and forms together with $\bar{c}$ a doublet under BRS transformations

$$s\bar{c} \;= B\,, \tag{75}$$

$$sB = 0\,. \tag{76}$$

This presentation renders BRS transformations nilpotent on all fields

$$s^2\varphi = 0\ , \quad \varphi = A_\mu, c, \bar{c}, \text{matter}\ . \tag{77}$$

The associated $\phi\pi$-terms combine with $\Gamma_{\text{g.f.}}$ to a BRS variation

$$\Gamma_{\text{g.f.}} + \Gamma_{\phi\pi} = s\text{Tr}\int (\frac{\xi}{2}\bar{c}B + \bar{c}\partial A)\ . \tag{78}$$

As pointed out in Chap. 1 non-linear field transformations are best dealt with via external fields.

$$\Gamma_{\text{ext.f.}} = \int \left(\text{Tr}(\rho^\mu\, sA_\mu + \sigma\, sc) + Y\, s\phi + \bar{Y}\, s\bar{\phi}\right)\ . \tag{79}$$

The external fields ρ^μ, σ transform under the adjoint, the fields $Y, \bar{Y}$ under the contragradient representation of ϕ w.r.t. rigid transformations, hence rigid invariance can be maintained. Expressed on Z, the generating functional for general Green functions, BRS invariance reads as follows

$$\mathbf{s}\,Z \equiv \int \left(\text{Tr}\left(\frac{\delta Z}{\delta\rho^\mu}J^\mu + \frac{\delta Z}{\delta j_B}j_{\bar{c}} + \frac{\delta Z}{\delta\sigma}j_c\right) + \frac{\delta Z}{\delta Y}j_\phi + \frac{\delta Z}{\delta\bar{Y}}j_{\bar{\phi}}\right) = 0\ . \tag{80}$$

Going over to connected Green functions Z_c via (20) and by Legendre transformation (21) to the 1PI Green functions Γ one arrives at the Γ-bilinear form

$$\mathbf{s}(\Gamma) \equiv \int \left(\text{Tr}\left(\frac{\delta\Gamma}{\delta\rho^\mu}\frac{\delta\Gamma}{\delta A_\mu} + B\frac{\delta\Gamma}{\delta\bar{c}} + \frac{\delta\Gamma}{\delta\sigma}\frac{\delta\Gamma}{\delta c}\right) + \frac{\delta\Gamma}{\delta Y}\frac{\delta\Gamma}{\delta\phi} + \frac{\delta\Gamma}{\delta\bar{Y}}\frac{\delta\Gamma}{\delta\bar{\phi}}\right) = 0\ . \tag{81}$$

It is clear that

$$\Gamma_{\text{cl}} = \Gamma_{\text{inv}} + \Gamma_{\phi\pi} + \Gamma_{\text{ext.f.}} \tag{82}$$

satisfies the ST identities (81), the gauge condition

$$\frac{\delta\Gamma}{\delta B} \equiv \xi B + \partial A\ , \tag{83}$$

and the rigid WI

$$\mathcal{W}_\omega\Gamma \equiv -i\int \delta_\omega\,\underline{\varphi}\frac{\delta\Gamma}{\delta\underline{\varphi}} = 0\ . \tag{84}$$

The claim is now that (81), (83) and (84) together with normalization conditions uniquely fix Γ to all orders if the matter representation satisfies

$$r^{(1)} = k\frac{d^{abc}}{d^2}\text{Tr}T^aT^bT^c\ , \tag{85}$$

i.e. the Adler - Bardeen anomaly is absent (k is a numerical factor).

In order to proof this claim one proceeds as follows. One first observes that $\Gamma_{\text{g.f.}}$ is the solution of (83) due to linearity in the fields. For establishing (84) one uses the method presented in Sect. 2.1:

$$\mathcal{W}^a \Gamma = \Delta^a \cdot \Gamma = \Delta^a + \mathcal{O}(\hbar) \tag{86}$$

is the most general deviation from rigid invariance (action principle). The algebra of the WI operators

$$[\mathcal{W}^a, \mathcal{W}^b] = i f^{abc} \mathcal{W}^c \tag{87}$$

implies consistency conditions for the insertions Δ^a

$$\mathcal{W}^a \Delta^b - \mathcal{W}^b \Delta^a = i f^{abc} \Delta^c \,. \tag{88}$$

It has been shown [2] that their most general solution is

$$\Delta^a = \mathcal{W}^a \hat{\Delta} \,. \tag{89}$$

Hence one can establish the rigid WI to all orders by absorbing $\hat{\Delta}$ as counterterm.

For the proof of (81) one also starts with the action principle

$$\mathbf{s}(\Gamma) = \Delta \cdot \Gamma = \Delta + \mathcal{O}(\hbar) \,. \tag{90}$$

A useful consistency condition is found by observing that the operator

$$\mathbf{s}_\Gamma \equiv \int \left(\frac{\delta \Gamma}{\delta A} \frac{\delta}{\delta \rho} + \frac{\delta \Gamma}{\delta \rho} \frac{\delta}{\delta A} + \frac{\delta \Gamma}{\delta \sigma} \frac{\delta}{\delta c} + \frac{\delta \Gamma}{\delta c} \frac{\delta}{\delta \sigma} + \right. \\ \left. \frac{\delta \Gamma}{\delta \bar{Y}} \frac{\delta}{\delta \bar{\phi}} + \frac{\delta \Gamma}{\delta \bar{\phi}} \frac{\delta}{\delta \bar{Y}} + \frac{\delta \Gamma}{\delta Y} \frac{\delta}{\delta \phi} + \frac{\delta \Gamma}{\delta \phi} \frac{\delta}{\delta Y} + B \frac{\delta}{\delta \bar{c}} \right) \tag{91}$$

satisfies

$$\begin{aligned} \mathbf{s}_\gamma \, \mathbf{s}(\gamma) &= 0 \quad \forall \gamma \,, \\ \mathbf{s}_\gamma \mathbf{s}_\gamma &= 0 \quad \forall \gamma \text{ with } \mathbf{s}(\gamma) = 0 \,. \end{aligned} \tag{92}$$

Applied to (90) this means

$$\mathbf{s}_{\Gamma_{\text{cl}}} \Delta = 0 \,. \tag{93}$$

with $\mathbf{s}_{\Gamma_{\text{cl}}} \mathbf{s}_{\Gamma_{\text{cl}}} = 0$, since Γ_{cl} satisfies the ST identity. The analysis of (93) is easy for all terms containing external fields but tricky for the terms made up by A_μ and c only [2]. The result is

$$\Delta = \mathbf{s}_\Gamma \hat{\Delta} + r \mathcal{A} \,, \tag{94}$$

$$\mathcal{A} = \operatorname{Tr} \int c \, \partial^\mu A^\nu \partial^\rho A^\sigma \epsilon_{\mu\nu\rho\sigma} \,. \tag{95}$$

The one-loop value for r is given by (85). A non-renormalization theorem [1] then says: if $r^{(n)} \neq 0$ for some n, then $n = 1$. I.e. if one arranges the multiplets T^a such that $r^{(1)} = 0$ there will never arise any anomaly in the ST identity.

3 Applications

3.1 The Electroweak Standard Model

In practice the most important example is the electroweak standard model (SM).

The Problems. If one attempts to renormalize it to all orders of perturbation theory one is faced with the observation that no obvious invariant regularization is known. Dimensional regularization has to cope with the γ_5-problem, whereas BPHZ or analytic regularization spoil BRS invariance. Hence an all order treatment can only be based on the algebraic method which we exemplified above. It has to deal with the following peculiarities specific to the SM:

1. The gauge group $SU(2) \times U(1)$ is not semisimple and the position of the unbroken $U(1)$ subgroup has to be determined and fixed in the course of renormalization.
2. The photon has to be kept massless. Its mixing with Z_μ has to be controlled such that a particle interpretation is possible. Off-shell IR problems have to be avoided.
3. $W_\mu^\pm$, Z_μ are unstable: the definition of their mass is non-trivial and gauge parameter dependence is a crucial issue.

A solution of problem 1) has been constructed by establishing
- the Slavnov–Taylor identity,
- (deformed) rigid Ward identities,
- an abelian local Ward identity.

The solution of problem 2) requires
- careful IR power counting,
- suitable normalization conditions.

For problem 3) a complete solution to all orders is not yet known, but the ST identity and reasonable normalization conditions guarantee unitarity and permit the LSZ asymptotic limit at least in a formal sense.

Under the simplifying assumption that CP is maintained and families are not mixed problems 1) and 2) have been solved by E. Kraus [8]. The remarks that follow are based on this paper.

The Abelian Subgroup. It turns out that fixing the abelian subgroup is equivalent to finding equations whose solutions together with normalization conditions characterize uniquely the model. In order not to miss parameters or representations one *sharpens* the algebraic method: one does not give the WI-operators beforehand but prescribes only type (scalar, vector, spinor) and number of fields and the *algebra* of the WI-operators. For the rigid transformations one requires

$$[W_\alpha, W_\beta] = i\hat{\epsilon}_{\alpha\beta\gamma}\tilde{I}_{\gamma\gamma'}W_{\gamma'} \tag{96}$$

for consistency with the ST identity

$$W_\alpha \mathbf{s}(\Gamma) - \mathbf{s}_\Gamma W_\alpha \Gamma = 0 \qquad \forall \Gamma . \tag{97}$$

$$\hat{\epsilon}_{\alpha\beta\gamma} : \begin{cases} \hat{\epsilon}_{+-3} = i \\ \hat{\epsilon}_{+-4} = 0 \end{cases} \text{ is totally antisymmetric.}$$

$$\tilde{I} = \begin{pmatrix} 0\,1\,0\,0 \\ 1\,0\,0\,0 \\ 0\,0\,0\,0 \\ 0\,0\,0\,0 \end{pmatrix} \text{ correlates } +,- \text{ of the electric charge.}$$

We present a sample for the respective ansatz (contribution of some vector fields)

$$W_\alpha = \tilde{I}_{\alpha\dot{\alpha}} \int \mathrm{d}^4x \cdots V_b^\mu \hat{a}^V_{bc,\alpha'} \tilde{I}_{cc'} \frac{\delta}{\delta V^\mu_{c'}} + \cdots \tag{98}$$

$$\begin{aligned} \mathbf{s}(\Gamma) = & \int z_4 (\sin\theta_3^g \partial_\mu c_Z + \cos\theta_3^g \partial_\mu c_A) \times \\ & \times \left(\sin\theta_4^V \frac{\delta\Gamma}{\delta Z_\mu} + \cos\theta_4^V \frac{\delta\Gamma}{\delta A_\mu} \right) + \cdots \\ & + \frac{\delta\Gamma}{\delta\rho_3^\mu} z_g \left(\cos\theta_3^V \frac{\delta\Gamma}{\delta Z_\mu} - \sin\theta_3^V \frac{\delta\Gamma}{\delta A_\mu} \right) + \cdots \end{aligned} \tag{99}$$

(Here the first two lines stand for linearly transforming vector pieces in ST, the third line for non-linearly transforming ones. ρ_3^μ is an external field coupled to a part of the BRS transformations of A_μ.)

The parameters $\hat{a}^V_{\dots}$, θ_3^g, $\theta_{3,4}^V$, z_4, z_g are to be chosen such that (96) and (97) is satisfied. In order to make this ansatz conceivable we give their tree approximation values in the conventional parameterization.

$$\hat{a}^V_\alpha = \mathcal{O}^T(\theta_w)\hat{\epsilon}_\alpha \mathcal{O}(\theta_w) , \tag{100}$$

$$\mathcal{O}(\theta_w) = \begin{pmatrix} 1 & 0 & 0 & 0 \\ 0 & 1 & 0 & 0 \\ 0 & 0 & \cos\theta_w & -\sin\theta_w \\ 0 & 0 & \sin\theta_w & \cos\theta_w \end{pmatrix} , \tag{101}$$

$$\begin{aligned} \theta_3^g &= \theta_3^V = \theta_4^V = \theta_w , \\ z_4 &= z_g = 1 . \end{aligned} \tag{102}$$

It is to be noted that electric charge and Faddeev-Popov-charge neutrality is naively maintained by the ansatz. Similarly one chooses the parameter values such that

$$W_+^* = W_- , \qquad W_{\substack{\pm\\3\\4}} \xrightarrow{\mathrm{CP}} -W_{\substack{\mp\\3\\4}} . \tag{103}$$

This work has to be performed for all sectors (vectors, scalars, fermions, ghosts); then the eqns. (96), (97) have to be solved and the solutions have to be parameterized such that the free parameters can either be fixed naively or via normalization conditions. It is most remarkable that a *non-diagonal* transformation

$$s\bar{c}_a = \hat{g}_{ab} B_b \tag{104}$$

is compatible with the algebra. If one has succeeded with this first step, namely solving the algebra, one can go on and find now in a second step the most general classical solution of the rigid WI

$$W_\alpha \Gamma_{\rm cl} = 0 \tag{105}$$

and the ST identity

$$\mathbf{s}(\Gamma_{\rm cl}) = 0\,. \tag{106}$$

As experience tells one and as one confirms in the present case too, this yields all possible renormalizations in the form of possible redefinitions of fields and parameters. With the help of the action principle and the consistency conditions as inferred from (96), (97) one performs in an analogous manner the search for the solutions of the WI's (105), (106) to all orders. Here contact is made with the work of BBBC [1] because it turns out that this analysis can equivalently be performed in terms of unphysical fields (V^a_μ, $a = 1, 2, 3, 4$). The absence of unitarity ruining anomalies follows from the structure of the standard multiplets.

In a third and last step one can now indeed proceed to the identification of the abelian subgroup. First of all one has to note that the naive electromagnetic WI operator

$$W_{\rm em} = \int {\rm d}^4x\, w_{\rm em} = i \int {\rm d}^4x \sum_a Q^{\rm em}_a \phi_a \frac{\delta}{\delta \phi_a} \tag{107}$$

is *not* abelian. Furthermore

$$w_{\rm em}\Gamma = \Box B_{\rm em} + Q_{\rm em} \cdot \Gamma \tag{108}$$

with $Q_{\rm em}$ a non-trivial insertion. On the non-integrated level it is not the electromagnetic direction which is abelian in the sense of having a trivial right hand side (which could then naively be constructed to all orders). It rather turns out that $w^Q_4 := w_{\rm em} - w_3$ is a good starting point leading eventually to

$$\hat{w}^Q_4 = g_1 w^Q_4 - \frac{1}{r^V_Z} \sin\theta^V \partial_\mu \frac{\delta}{\delta Z_\mu} - \frac{1}{r^V_A} \cos\theta^V \partial_\mu \frac{\delta}{\delta A_\mu}\,. \tag{109}$$

This operator is singled out by

$$[\hat{w}^Q_4, W_\alpha] \qquad = 0\,, \tag{110}$$

$$\mathbf{s}_\Gamma \hat{w}^Q_4 \Gamma - \hat{w}^Q_4 \mathbf{s}(\Gamma) = 0 \qquad \forall \Gamma\,. \tag{111}$$

It satisfies a local WI

$$\hat{w}_4^Q \Gamma = \frac{\sin\theta^V}{r_Z^V} \Box B_Z + \frac{\cos\theta^V}{r_A^V} \Box B_A \tag{112}$$

which can be postulated and established to all orders of perturbation theory because the right hand side is linear in propagating fields. This WI, which can only be required – due to its characterization by (110) and (111) – *after* the rigid WI and the ST identity have been established, fixes eventually the instabilities of the abelian subgroup.

The parameter

$$g_1 = \frac{e}{\cos\theta_w} + o(\hbar) \tag{113}$$

is in QED-like normalization conditions fixed on this local WI. The parameters θ^V, $r_{Z,A}^V$ are determined in W_{∂_μ}.

As far as interpretation is concerned one has to note that algebraically no distinction is possible between gauging the electromagnetic current or lepton- and quark number currents; hence this local WI is needed as additional requirement.

Photon/Z Mixing. In order to keep the photon massless and also control the mixing one imposes as normalization conditions

$$\Gamma_{AA}^T(p^2=0) \quad = 0\,, \tag{114}$$
$$\Gamma_{ZA}^T(p^2=0) \quad = 0\,, \tag{115}$$
$$\mathrm{Re}\ \Gamma_{ZZ}^T(p^2=M_Z^2) = 0\,. \tag{116}$$

Eqns. (114) and (115) are automatic if one uses the BPHZL scheme [25,10,11]. The crucial point is now to check that one has indeed enough parameters at ones disposal to satisfy these normalization conditions after having arranged all WI's. Since IR dangerous terms like $\int \bar{c}_A c_Z$ are to be avoided as counterterms this task is non-trivial. It turns out that one can avoid off-shell IR danger by using the freedom left in the transformation law (104) of the antighost fields. This introduces a *ghost angle* θ_G as an important parameter into the theory.

All other masses are similarly introduced via two-point-functions in order to ensure poles. This in turn requires to have non-trivial parameter dependence in the rigid WI operators: they become deformed in higher orders. Likewise follows the antighost equation as a consequence of local WI and ST. Had one fixed and a priori prescribed WI operators and the antighost equation as a postulate one could not have fixed all masses as physical ones – as poles of propagators.

As an overall consistency check one derives the Callan-Symanzik equation because it controls the motion of all parameters of the theory under renormalization. The outcome is as follows:

$$C\Gamma = \mathrm{soft} \cdot \Gamma \tag{117}$$

- exists IR-wise,
- contains β-functions for mass ratios,
- shows that θ_w and θ_G are independently renormalized.

This yields a consistent picture.

We collect the result:

- the algebra of WI operators (96), (97)
- rigid WI + ST + local WI
- on-shell normalization conditions

determine Γ uniquely to all orders. The rigid WI operators are deformed; as a new parameter enters the angle θ_G in the ghost sector. It goes along with non-diagonal transformations (104) of the antighost in higher orders.

3.2 Supersymmetry in Non-linear Realization

If supersymmetry is realized on fields of canonical dimension (spin 0, dim 1; spin 1/2, dim 3/2; spin 1, dim 1) then the transformations are non-linear.

The Wess-Zumino Model For a chiral multiplet the transformations read:

$$\begin{aligned} \delta_\alpha A &= \psi_\alpha \,, & \bar\delta_{\dot\alpha} A &= 0 \,, \\ \delta_\alpha \psi^\beta &= 2\,\delta_\alpha{}^\beta (m\bar A + g\bar A^2) \,, & \bar\delta_{\dot\alpha}\psi_\beta &= 2i\,\partial\!\!\!/_{\beta\dot\alpha} A \,, \\ \delta_\alpha \bar A &= 0 \,, & \bar\delta_{\dot\alpha}\bar A &= \bar\psi_{\dot\alpha} \,, \\ \delta_\alpha \bar\psi_{\dot\beta} &= 2i\,\partial\!\!\!/_{\alpha\dot\beta}\bar A \,, & \bar\delta_{\dot\alpha}\bar\psi^{\dot\beta} &= -2\,\delta_{\dot\alpha}{}^{\dot\beta}(mA + gA^2) \,. \end{aligned} \tag{118}$$

The algebra of these transformations closes only on-shell, i.e. by use of the equations of motion (for the spinor fields). The non-linear transformations can be dealt with by coupling them to external fields

$$\Gamma_{\text{ext}} = \int dx \left(2u(gA^2 + mA) + 2\bar u(g\bar A^2 + m\bar A)\right) \,, \tag{119}$$

resulting in Γ-bilinear WI's for SUSY [15]

$$\begin{aligned} \mathcal{W}_\alpha(\Gamma) &\equiv -i\int dx \left(\psi_\alpha \frac{\delta\Gamma}{\delta A} + \frac{\delta\Gamma}{\delta\bar u}\frac{\delta\Gamma}{\delta\psi^\alpha} - 2i\,\partial\!\!\!/_{\alpha\dot\beta}\bar A\frac{\delta\Gamma}{\delta\bar\psi_{\dot\beta}} + 2iu\,\partial\!\!\!/_{\alpha\dot\beta}\bar\psi^{\dot\beta}\right) = 0 \,, \\ \bar{\mathcal{W}}_{\dot\alpha}(\Gamma) &\equiv -i\int dx \left(\bar\psi_{\dot\alpha}\frac{\delta\Gamma}{\delta\bar A} - \frac{\delta\Gamma}{\delta u}\frac{\delta\Gamma}{\delta\bar\psi^{\dot\alpha}} - 2i\frac{\delta\Gamma}{\delta\psi_\beta}\,\partial\!\!\!/_{\beta\dot\alpha}A - 2i\bar u\,\partial\!\!\!/_{\beta\dot\alpha}\psi^\beta\right) = 0 \,. \end{aligned} \tag{120}$$

They give rise to linearized functional operators which govern the transformations of insertions

$$\mathcal{W}_\alpha^\Gamma \equiv -i \int dx \left(\psi_\alpha \frac{\delta}{\delta A} + \frac{\delta \Gamma}{\delta \bar{u}} \frac{\delta}{\delta \psi^\alpha} + \frac{\delta \Gamma}{\delta \psi^\alpha} \frac{\delta}{\delta \bar{u}} - 2i \, \partial\!\!\!/_{\alpha\dot{\beta}} \frac{\delta}{\delta \bar{\psi}_{\dot{\beta}}} \right) , \quad (121)$$

(analogously for $\bar{\mathcal{W}}^\Gamma \dot{\alpha}$) that satisfy the identities

$$\mathcal{W}_\alpha^\Gamma \mathcal{W}_\beta(\Gamma) + \mathcal{W}_\beta^\Gamma \mathcal{W}_\alpha(\Gamma) = 0 \,, \quad (122)$$
$$\bar{\mathcal{W}}_{\dot{\alpha}}^\Gamma \bar{\mathcal{W}}_{\dot{\beta}}(\Gamma) + \bar{\mathcal{W}}_{\dot{\beta}}^\Gamma \bar{\mathcal{W}}_{\dot{\alpha}}(\Gamma) = 0 \,, \quad (123)$$
$$\mathcal{W}_\alpha^\Gamma \bar{\mathcal{W}}_{\dot{\beta}}(\Gamma) + \bar{\mathcal{W}}_{\dot{\beta}}^\Gamma \mathcal{W}_\alpha(\Gamma) = 2\sigma^\mu_{\alpha\dot{\beta}} \mathcal{W}_\mu \Gamma \,. \quad (124)$$

Here, $\mathcal{W}_\mu \equiv -i \int \sum_\varphi \partial_\mu \varphi \frac{\delta}{\delta \varphi}$ is the functional generator for translations. These identities serve in the course of renormalization as consistency conditions that constrain possible deviations from symmetry. The problem of on-shell closure is solved by admitting in Γ_{eff} a term $a \int u\bar{u}$, i.e. *bilinear* in the external fields. Its effect can be easily traced in the classical approximation. According to (121) it contributes to the transformation of ψ_α, via $\delta\Gamma/\delta\bar{u} = au + \ldots$. But acting on the term au by a second transformation one has to use the laws given by (121) also (or more precisely by its conjugate): $\delta\Gamma/\delta\bar{\psi}$ – this is an equation of motion! Hence it is clear that on this functional level, where the external fields also transform, the functional transformations do indeed close – the on-shell closure problem is solved. The result for the Wess-Zumino model is [15] that a unique Γ can be found satisfying the WI's (120). It is deducible by solving the consistency conditions, which admit no anomaly.

3.3 SUSY Gauge Theories

A further obstacle is provided in gauge theories. Here every SUSY transformation has to be accompanied by a gauge transformation and thus a gauge fixing term can never be naively invariant. This problem is in its renormalization aspect solved by "ghostifying" the transformations and again by admitting in Γ_{eff} terms that are bilinear in the external fields. This has first been shown in [24] and then worked out in detail with a slightly changed method of proof in [12]. We shall present the example of SQED [6]. Apart from ghosts for gauge transformations one introduces ghosts for SUSY and translations as well. Those of the gauge transformations (c) are the usual $\phi\pi$-ghosts and they propagate. Those of SUSY ($\epsilon, \bar{\epsilon}$) and translations (ω^ν) are constants and do not propagate. Their Grassmann character is always opposite to the one of the naive transformation. For the field of the vector multiplet A_μ (photon);

$\lambda, \bar{\lambda}$ (photino) they read

$$sA_\mu = \partial_\mu c + i(\epsilon\sigma_\mu\bar{\lambda} - \lambda\sigma_\mu\bar{\epsilon}) - i\omega^\nu\partial_\nu A_\mu \,, \tag{125}$$
$$s\lambda^\alpha = \frac{i}{2}(\epsilon\sigma^{\rho\sigma})^\alpha F_{\rho\sigma} - i\epsilon^\alpha\, eQ_L(|\phi_L|^2 - |\phi_R|^2) - i\omega^\nu\partial_\nu\lambda^\alpha \,, \tag{126}$$
$$s\bar{\lambda}_{\dot{\alpha}} = \frac{-i}{2}(\bar{\epsilon}\bar{\sigma}^{\rho\sigma})_{\dot{\alpha}} F_{\rho\sigma} - i\bar{\epsilon}_{\dot{\alpha}}\, eQ_L(|\phi_L|^2 - |\phi_R|^2) - i\omega^\nu\partial_\nu\bar{\lambda}_{\dot{\alpha}} \,, \tag{127}$$

For (charged) matter multiplets ϕ, ψ they have the form

$$s\phi_L = -ieQ_L c\phi_L + \sqrt{2}\,\epsilon\psi_L - i\omega^\nu\partial_\nu\phi_L \,, \tag{128}$$
$$s\phi_L^\dagger = +ieQ_L c\phi_L^\dagger + \sqrt{2}\,\bar{\psi}_L\bar{\epsilon} - i\omega^\nu\partial_\nu\phi_L^\dagger \,, \tag{129}$$
$$s\psi_L^\alpha = -ieQ_L c\psi_L^\alpha - \sqrt{2}\,\epsilon^\alpha\, m\phi_R^\dagger - \sqrt{2}\,i(\bar{\epsilon}\bar{\sigma}^\mu)^\alpha D_\mu\phi_L - i\omega^\nu\partial_\nu\psi_L^\alpha \,, \tag{130}$$
$$s\bar{\psi}_{L\dot{\alpha}} = +ieQ_L c\bar{\psi}_{L\dot{\alpha}} + \sqrt{2}\,\bar{\epsilon}_{\dot{\alpha}}\, m\phi_R + \sqrt{2}\,i(\epsilon\sigma^\mu)_{\dot{\alpha}}(D_\mu\phi_L)^\dagger - i\omega^\nu\partial_\nu\bar{\psi}_{L\dot{\alpha}} \,. \tag{131}$$

In order that the algebra closes up to equations of motion

$$s^2\phi = 0 \qquad \text{mod(equation of motion)} \,, \tag{132}$$

one has to incorporate field dependent gauge transformations and also transformations of the constant ghosts

$$sc = 2i\epsilon\sigma^\nu\bar{\epsilon}A_\nu - i\omega^\nu\partial_\nu c \,, \tag{133}$$
$$s\epsilon^\alpha = s\bar{\epsilon}^{\dot{\alpha}} = 0 \,, \tag{134}$$
$$s\omega^\nu = 2\epsilon\sigma^\nu\bar{\epsilon} \,, \tag{135}$$
$$s\bar{c} = B - i\omega^\nu\partial_\nu\bar{c} \,, \tag{136}$$
$$sB = 2i\epsilon\sigma^\nu\bar{\epsilon}\partial_\nu\bar{c} - i\omega^\nu\partial_\nu B \,. \tag{137}$$

The gauge fixing term can now be introduced as generalized BRS variation

$$\begin{aligned} \Gamma_{\text{g.f.}} &= s\int dx\bar{c}\left(\partial^\mu A_\mu + \frac{\xi}{2}B\right) \\ &= \int dx\Big(B\partial^\mu A_\mu + \frac{\xi}{2}B^2 - \bar{c}\Box c \\ &\qquad - \bar{c}\partial^\mu(i\epsilon\sigma_\mu\bar{\lambda} - i\lambda\sigma_\mu\bar{\epsilon}) + \xi i\epsilon\sigma^\nu\bar{\epsilon}(\partial_\nu\bar{c})\bar{c}\Big) \,. \end{aligned} \tag{138}$$

The complete action

$$\Gamma = \Gamma_{\mathrm{inv}} + \Gamma_{\mathrm{g.f.}} + \Gamma_{\mathrm{ext.f.}} + \Gamma_{\mathrm{bil}} \,, \tag{139}$$

$$\begin{aligned}\Gamma_{\mathrm{inv}} = \int dx \Big\{ & -\frac{1}{4} F_{\mu\nu}F^{\mu\nu} + \frac{1}{2}\bar{\tilde{\gamma}} i\gamma^\mu \partial_\mu \tilde{\gamma} + |D_\mu \phi_L|^2 + |D_\mu \phi_R^\dagger|^2 + \bar{\Psi} i\gamma^\mu D_\mu \Psi \\ & - \sqrt{2} e Q_L \left(\bar{\Psi} P_R \tilde{\gamma} \phi_L - \bar{\Psi} P_L \tilde{\gamma} \phi_R^\dagger + \phi_L^\dagger \bar{\tilde{\gamma}} P_L \Psi - \phi_R \bar{\tilde{\gamma}} P_R \Psi \right) \\ & - \frac{1}{2} \left(eQ_L |\phi_L|^2 + eQ_R |\phi_R|^2 \right)^2 - m\bar{\Psi}\Psi - m^2 (|\phi_L|^2 + |\phi_R|^2) \Big\} \,, \end{aligned} \tag{140}$$

$$\begin{aligned}\Gamma_{\mathrm{ext.f.}} = \int dx \Big(& Y_\lambda^\alpha s\lambda_\alpha + Y_{\bar{\lambda}\dot{\alpha}} s\bar{\lambda}^{\dot{\alpha}} \\ & + Y_{\phi_L} s\phi_L + Y_{\phi_L^\dagger} s\phi_L^\dagger + Y_{\psi_L}^\alpha s\psi_{L\alpha} + Y_{\bar{\psi}_L \dot{\alpha}} s\bar{\psi}_L^{\dot{\alpha}} + (L \to R) \Big) \,, \end{aligned} \tag{141}$$

$$\Gamma_{\mathrm{bil}} = -(Y_\lambda \epsilon)(\bar{\epsilon} Y_{\bar{\lambda}}) - 2(Y_{\psi_L}\epsilon)(\bar{\epsilon} Y_{\bar{\psi}_L}) - 2(Y_{\psi_R}\epsilon)(\bar{\epsilon} Y_{\bar{\psi}_R}) \,, \tag{142}$$

satisfies

$$\mathbf{s}(\Gamma) = 0 \,, \tag{143}$$

the generalized ST identity. The linearized ST operator $\mathbf{s}_{\mathcal{F}}$

$$\mathbf{s}_{\mathcal{F}} = \int \left(s\varphi_i' \frac{\delta}{\delta\varphi_i'} + \frac{\delta\mathcal{F}}{\delta Y_i}\frac{\delta}{\delta\varphi_i} + \frac{\delta\mathcal{F}}{\delta\varphi_i}\frac{\delta}{\delta Y_i} \right) . \tag{144}$$

(φ_i' are all linear and φ_i all non-linear transforming fields) is nilpotent

$$\mathbf{s}_{\mathcal{F}}^2 = 0 \,, \tag{145}$$

provided $\mathcal{F}$ satisfies (143) and the linear identity

$$i\epsilon\sigma^\mu \frac{\delta\mathcal{F}}{\delta Y_{\bar{\lambda}}} - i\frac{\delta\mathcal{F}}{\delta Y_\lambda}\sigma^\mu\bar{\epsilon} + i\omega^\nu\partial_\nu(i\epsilon\sigma^\mu\bar{\lambda} - i\lambda\sigma^\mu\bar{\epsilon}) - 2i\epsilon\sigma_\nu\bar{\epsilon}F^{\nu\mu} = 0 \,. \tag{146}$$

(This condition is equivalent to $\mathbf{s}_{\mathcal{F}}^2 A_\mu = 0$ which is true for $\mathcal{F} = \Gamma_{\mathrm{cl}}$.) It satisfies the consistency condition

$$\mathbf{s}_{\mathcal{F}}\mathbf{s}(\mathcal{F}) = 0 \tag{147}$$

for every functional $\mathcal{F}$ which fulfills (146).

It can be shown [6] that gauge condition, ST identity (143) and (146) can be solved, i.e. a Γ can be found satisfying these equations to all orders.

On the functional level the SUSY transformation law of the fields is defined by $\partial_\epsilon \mathbf{s}_\Gamma$ and $\partial_{\bar{\epsilon}} \mathbf{s}_\Gamma$. One finds e.g. in one-loop order

$$\partial_\epsilon \mathbf{s}_\Gamma \psi = \int dy \phi^\dagger(y) \Gamma_{\phi^\dagger \epsilon Y_\psi}(y, x) \,, \tag{148}$$

with $\Gamma_{\phi^\dagger \epsilon Y_\psi}$ truly being a non-local kernel [6]. Hence for the interacting fields the transformations become non-local, the origin being the gauge fixing term which is not supersymmetric.

In order to clarify on which level, i.e. on which quantities supersymmetry is realized one studies the effect of the respective charges as operators in Fock or Hilbert space [18]. If one defines on the asymptotic "in" fields $\phi^{\rm in}$ a SUSY charge $Q^{\rm in}_\alpha$ which generates the linear transformations of the free theory, then $Q^{\rm out}_\alpha$ is given by the time evolution governed by the S-Matrix

$$Q^{\rm out}_\alpha = S Q^{\rm in}_\alpha S^\dagger \, . \tag{149}$$

It turns out that

$$Q^{\rm out}_\alpha = Q^{\rm in}_\alpha - i \left[Q^{\rm BRS}, \partial_{\epsilon^\alpha} \Gamma^{\rm Op}_{\rm eff} \right] \, . \tag{150}$$

Here, $Q^{\rm BRS}$ is the charge operator generating the ordinary BRS transformations and all operators are constructed by LSZ reduction from the respective insertions into the functional of general Green functions. This result shows that SUSY is a symmetry between physical states only.

It is remarkable that for an interacting field $\phi^{\rm Op}$ one can still derive

$$i \left[Q_\alpha(x^0), \phi(x) \right]^{\rm Op} = - \partial_{\epsilon^\alpha} \frac{\delta \Gamma^{\rm Op}_{\rm eff}}{\delta Y(x)} \, . \tag{151}$$

Here $Q_\alpha(x^0)$ denotes the time dependent SUSY charge (s. [18]). Summarizing one can say that the non-linear realization of supersymmetry can be mastered in perturbation theory and its symmetry effects show up only between physical states.

Acknowledgments I thank the organizers for the invitation to such a nice school with lively atmosphere, my collaborators M. Roth, Ch. Rupp and R. Scharf for very efficient help in preparing the notes.

References

1. G. Bandelloni, C. Becchi, A. Blasi, R. Collina: Ann. Inst. Henri Poincaré **28**, 225, 255 (1978)
2. C. Becchi, A. Rouet, R. Stora: Comm. Math. Phys. **42**, 127 (1975)
3. C. Becchi, A. Rouet, R. Stora: Ann. Phys. (N.Y.) **98**, 287 (1976)
4. J. C. Collins: *Renormalization* (Cambridge University Press, 1984)
5. K. Hepp: 'Renormalization Theory'. In: *Statistical Mechanics and Quantum Field Theory*, ed. by C. DeWitt, R. Stora (Gordon & Breach, 1971)
6. W. Hollik, E. Kraus, D. Stöckinger: Eur. Phys. J. C **11**, 365 (1999)
7. F. Jegerlehner: 'Facts of life with γ_5', preprint DESY 00-075 (2000)
8. E. Kraus: Ann. Phys. (NY) **262**, 155 (1998)
9. J.H. Lowenstein: 'BPHZ-Renormalization'. In: *Renormalization Theory, Erice, 1975*, ed. by A.S. Wightman, G. Velo (Reidel, Dordrecht 1976)
10. J.H. Lowenstein, W. Zimmermann: Comm. Math. Phys. **44**, 73 (1975)
11. J.H. Lowenstein: Comm. Math. Phys. **47**, 53 (1976)

12. N. Maggiore: Int. J. Mod. Phys. A **10**, 3781 (1995)
13. N. Maggiore, O. Piguet, S. Wolf: Nucl. Phys. B **458**, 403 (1996); B **476**, 329 (1996)
14. W. Pauli, F. Villars: Rev. Mod. Phys. **21**, 434 (1949)
15. O. Piguet, K. Sibold: Nucl. Phys. B **253**, 269 (1985)
16. O. Piguet: 'Renormalisation en Théorie Quantique des Champs', Troisième Cycle de la Physique, Genève, Semestre d'hiver 1982-1983
17. O. Piguet: 'Renormalisation des Théories de Jauge', Troisième Cycle de la Physique, Genève, Semestre d'été 1983
18. C. Rupp, R. Scharf, K. Sibold: in preparation
19. E. Speer: *Generalized Feynman Amplitudes* (Princeton University Press, 1969)
20. E. Speer: J. Math. Phys. **9**, 1404 (1968)
21. E. Speer: 'Analytical Renormalization'. In: *Renormalization Theory, Erice, 1975*, ed. by A.S. Wightman, G. Velo (Reidel, Dordrecht 1976)
22. K. Symanzik: 'Lectures on Lagrangian Quantum Field Theory', Islamabad 1968, DESY internal report
23. G. 't Hooft, M. Veltman: Nucl. Phys. B **44**, 189 (1972)
24. P. L. White: Class. Quant. Grav. **9**, 413, 1663 (1992)
25. W. Zimmermann: Comm. Math. Phys. **11**, 1 (1968)
26. W. Zimmermann: Comm. Math. Phys. **15**, 208 (1969)

Functional Integrals for Quantum Theory

Ludwig Streit[1,2]

[1] Universität Bielefeld, BiBoS, D-33615 Bielefeld, Germany
[2] Universidade da Madeira, CCM, P-9000 Funchal, Portugal

1 Introduction

35 years exactly lie between my first Schladming lecture and this - presumably the last one I give here. In 1965 I presented "An introduction to theories of integration over function spaces" [24]. Functional integrals in physics at that time primarily meant Feynman (and maybe Wiener) integrals; with the Euclidean, functional integral approach to QFT still waiting in the wings.

But I do recall from those times a twofold fascination:

- The discrepancy between the physical neatness and the mathematical messiness of Feynman's quantization. Integration, for the mathematician, is with respect to a measure, and there is no real or even complex valued measure on path space that would produce the Feynman "integral".
- The mysterious, and again, mathematically completely unjustified power of the reasoning in the 1960 paper of F. Coester and R. Haag [7] where the authors extract quantum field theoretical dynamics from a non-existent vacuum density functional. (In fact the existence of such densities would essentially contradict Haag's own theorem on the inequivalence of representations for free and interacting fields!)

In the years since that time I have tried to understand these miraculous workings a little better. Some mathematics had to be created to do so [10], and then some things became clearer:

- The Feynman integral might be like those "integrals" of physics in which generalized functions or "distributions" appear.
- If we admit generalized functions for densities, then there will be no contradiction with Haag's theorem.

Hence our program for these lectures will be

- to establish a mathematically acceptable functional calculus which includes generalized functionals, in other words to extend distribution theory to infinitely many variables,
- and then put this theory to work in quantum field theory and for a better understanding of Feynman integrals.

By what should we replace Lebesgue integration in infinite dimension? Or – to state this in another way – what is a good set of independent coordinates?

White noise analysis [10], [19], [22] attempts an answer to these questions[1]. Hence, in the following sections we shall introduce nonlinear functionals of white noise, and – in analogy to the usual test functions and distributions – we shall introduce spaces of smooth and of generalized functionals of white noise. Then examples of applications in quantum physics will be discussed.

2 White Noise Analysis

Gaussian white noise appears naturally wherever one tries to model random events occurring independently at different points in time (and/or space), the prime example being thermal noise in electric circuits: you can hear it if you tune your radio to where there is **no** station and turn it on loud enough.

Of course this strict independence, expressed in the expectation

$$E\left(\omega(s)\omega(t)\right) = \delta\left(s-t\right)$$

is an idealization, but it is one of those which physics invokes frequently to facilitate a simple and transparent analysis, an analysis that in many instances would become much more cumbersome for models with "colored noise". In particular white noise ω appears as the velocity of Brownian motion modeled by the Wiener process $B(t)$

$$\omega(t) = \frac{d}{dt}B(t).$$

With its independence and invariance properties Gaussian white noise ω is ideally suited as a universal "coordinate system" for stochastic and infinite dimensional analysis. Hence we shall develop white noise calculus in what follows, noting however that most results have a straightforward extension to more general Gaussian [16] and even non-Gaussian [17] settings.

Brownian motion is (almost surely) non-differentiable in the usual sense, i.e. white noise sample "functions" are distributions:

$$\omega \in S^*(R) \tag{1}$$

Hence a certain amount of care is needed if we want to profit from the advantages of white noise modeling in a nonlinear setting; a well-known example is stochastic integration where e.g. the distinction between the integrals in the sense of Ito and of Stratonovich essentially results from carefully defining them as nonlinear functionals of white noise.

[1] It appears in the 2000 Mathematics Subject Classification under "60H40 White Noise Theory"

From the fact that we consider white noise as a δ-correlated mean zero process we are immediately led to its characteristic function

$$C(f) = E\left(e^{i\int \omega(t)f(t)dt}\right) = \int_{S^*(R)} e^{i\int \omega(t)f(t)dt} d\mu(\omega) = e^{-\frac{1}{2}\int f^2(t)dt} .$$

For nonlinear functionals

$$\varphi \in L^2(d\mu)$$

of white noise ω we shall make the ansatz

$$\varphi(\omega) = \sum_{n=0}^{\infty} \int d^n t F_n(t_1, \ldots, t_n) : \omega(t_1) \ldots \omega(t_n) : . \qquad (2)$$

Here

$$: \omega^{\otimes n}(t) :=: \omega(t_1) \ldots \omega(t_n) :$$

is the well-known "normal ordered" or "Wick" or "Hermite product". It amounts to an orthogonalization

$$E(: \omega^{\otimes n}(s) :: \omega^{\otimes m}(t) :) = \delta_{m,n} n! \prod_{i=1}^{n} \delta(s_i - t_i)$$

(assuming the s_i, t_i ordered) and can e.g. be defined recursively

$$\begin{aligned} &: \omega(t_1) \ldots \omega(t_n) := \\ &: \omega(t_1) \ldots \omega(t_{n-1}) : \omega(t_n) - \sum_{i=1}^{n-1} \delta(t_n - t_i) : \prod_{k \neq i}^{n-1} \omega(t_k) : . \end{aligned}$$

In this fashion all the terms in the normal ordered expansion are orthogonal to each other, and we find immediately the L^2 norm of φ :

$$\|\varphi\|^2_{L^2(d\mu)} = E(\varphi^* \varphi) = \sum_{n=0}^{\infty} n! \int d^n t \, |F_n(t_1, \ldots, t_n)|^2 .$$

On the right-hand side we recognize a Fock space norm:

$$\varphi \in L^2(d\mu) \leftrightarrow \{F_n\} \in H_{Fock}.$$

This is the Gelfand-Ito-Segal isomorphism between square integrable white noise functionals and bosonic Fock space vectors.

2.1 Smooth and Generalized Functionals

Smooth and generalized functionals are constructed in analogy to the usual test functions and distributions; we recall that test functions in Schwartz space are characterized by a sequence of norms

$$f \in S(R^n) \text{ iff } \ |f|_p < \infty \text{ for all } p$$

where we may choose for the norms e.g.

$$|f|_p = |H^p_{osc} f|_{L^2}. \tag{3}$$

Likewise we now introduce for white noise functionals the norms

$$\|\varphi\|^2_{p,q} = \sum_{n=0}^{\infty} (n!)^{1+\beta}\, 2^{nq}\, |F_n|^2_p \tag{4}$$

for $p, q > 0$. Putting $\beta = 0$ for the moment we define the space (S) of smooth functionals by

$$(S) = \left\{\varphi : \|\varphi\|_{p,q} < \infty \text{ for all } p, q\right\}.$$

These functionals then have kernels which are Schwartz test functions and whose norms $|F_n|^2_p$ decrease rapidly as n becomes large.

The space $(S)^*$ of generalized functionals ("Hida distributions") is then defined as the dual of (S):

$$(S) \subset L^2(d\mu) \subset (S)^*.$$

Similarly, for $\beta = 1$, we have the "Kondratiev spaces" [15] $(S)^{\pm 1}$, with

$$(S)^1 \subset (S) \subset L^2(d\mu) \subset (S)^* \subset (S)^{-1}.$$

2.2 Characterization of Generalized Functionals $\Phi \in (S)^*$

In many physics applications Φ will be given in terms of a "source functional", such as

$$T\Phi(f) = E\left(\Phi(\omega) e^{i\int \omega(t) f(t) dt}\right)$$

or

$$S\Phi(f) = E\left(\Phi(\omega) : e^{\int \omega(t) f(t) dt} :\right)$$

Fortunately, any of these expressions provides a complete characterization of generalized white noise functionals Φ [16].

Theorem 1. *A functional $G(f)$, $f \in S(R)$, is the T-transform of a unique generalized white noise functional $\Phi \in (S)^*$ iff for all $f_i \in S(R)$, $G(zf_1 + f_2)$ is analytic in the whole complex z-plane and of second order exponential growth*

$$|G(zf)| < a e^{b|zf|^2_p}$$

for some $p > 0$.

Since there are many instances where physics gives us (some control of) the source functionals, it is often not too hard to see whether they fall into our framework.

Often, also, we are given a sequence of approximations and would like to ensure their convergence. The following corollary gives sufficient conditions for the existence of a limiting (generalized) functional.

Corollary 1. *Let $\{\Phi_n\}_{n\in N}$ denote a sequence of generalized white noise functionals with the following properties:*

1. *For all $f \in S(R)$, the T-transforms $\{T\Phi_n(f)\}_{n\in N}$ are Cauchy sequences.*
2. *There exist C_i, p such that the bound*

$$|T\Phi_n(zf)| \leq C_1 \exp(C_2 |z|^2 |f|_p^2)$$

holds uniformly in n.

Then there is a unique

$$\lim \Phi_n = \Phi \in (S)^* \ . \tag{5}$$

Similarly we can control the integration of families of Hida distributions with respect to a parameter as follows:

Corollary 2. *Let (Ω, B, m) be a measure space, and Φ_λ in $(S)^*$ for $\lambda \in \Omega$. We assume that the T-transform of Φ_λ*

1. *is an m-measurable function of λ for any test function f,*
2. *obeys an estimate*

$$| (T\Phi_\lambda)(zf) | \leq C_1(\lambda) \exp(C_2(\lambda) \mid z \mid^2 |f|_p^2)$$

for some fixed p and for $C_1 \in L^1(m), C_2 \in L^\infty(m)$.

Then Φ_λ is (Bochner-) integrable

$$\int_\Omega dm(\lambda)\Phi_\lambda \in (S)^* \tag{6}$$

and we may interchange T-transform and integration:

$$T(\int_\Omega dm(\lambda)\Phi_\lambda)(f) = \int_\Omega dm(\lambda)(T(\Phi_\lambda)(f)). \tag{7}$$

Example 1. A Fourier representation for Donsker's δ-function:

$$\delta(<\omega, g> - a) \equiv \frac{1}{2\pi} \int d\lambda e^{i\lambda(<\omega,g>-a)} \ . \tag{8}$$

Remarks:

1. Analogous statements are true for the S-transform.
2. Generalized functionals in the Kondratiev space are characterized by local analyticity and local boundedness of the source functionals.
3. Hence, in either case, linear combinations, but also products of source functionals are again admissible source functionals. This induces an algebraic structure on the space of generalized functionals, via

$$S(\Phi)S(\Psi) = S(\Phi \diamond \Psi);$$

As it turns out this product is simply the Wick product:

$$: \omega^{\otimes n}(s) : \diamond : \omega^{\otimes m}(t) :=: \omega^{\otimes n}(s)\omega^{\otimes m}(t) : \; .$$

4. For the Kondratiev space of generalized white noise functionals even more can be said [15]. By Remark 2, analytic functions g of source functionals are again admissible, and this induces an analytic "Wick calculus" on distribution space

$$g\left(S(\Phi)\right) = S\left(g^{\diamond}(\Phi)\right)$$

with

$$g^{\diamond}(\Phi) \equiv \sum_n a_n \Phi^{\diamond n} \; for \; g(z) = \sum_n a_n z^n. \tag{9}$$

2.3 Calculus

Test functionals $\varphi \in (S)$ admit directional ("Gateaux") derivatives:

$$D_h\varphi(\omega) = \lim_{\varepsilon \to 0} \frac{\varphi(\omega + \varepsilon h) - \varphi(\omega)}{\varepsilon}$$

for even for *generalized* functions $h \in S^*(R)$. Hence, for any such h, the adjoint

$$D_h^* = -D_h - < \omega, h >$$

acts continuously on generalized functions $\Phi \in (S)^{-1}$of white noise.

In particular we may put $h = \delta_t$, writing in this case

$$D_h \equiv \partial_t$$

which provides us with the natural notion of a gradient ("Frechet derivative")

$$\nabla\varphi = \{\partial_t\varphi : t \in R\} \; .$$

The "carré du champ" functional is a test functional

$$|\nabla\varphi|^2 = \int dt \, |\partial_t\varphi|^2 \in (S) \text{ for all } \varphi \in (S).$$

For much more detail on white noise calculus, see [10].

3 Quantum Field Theory

3.1 The Vacuum Density

The analogue of a ground state distribution of particle configurations x

$$d\nu(x) = \rho(x)\, d^n x \tag{10}$$

would be

$$d\nu(\varphi) = \rho(\varphi)\, d^\infty \varphi, \tag{11}$$

a probability measure on field configurations, with vacuum density ρ. Unfortunately, $d^\infty\varphi$ refuses to exist; but even a more modest attempt

$$d\nu(\varphi) = \rho(\varphi)\, d\nu_0(\varphi) \tag{12}$$

– where now ρ would be the density of the physical vacuum with respect to the (well defined) free vacuum measure ν_0 – will not work: Haag's theorem about inequivalence of free and interacting fields excludes the existence of such a density. In fact the saga of constructive quantum field theory was in essence the quest for *inequivalent* measures ν

$$\langle \Omega | \phi(x_1) \dots \phi(x_n) | \Omega \rangle \overset{!}{=} \int d\nu(\phi) \phi(x_1) \dots \phi(x_n) \tag{13}$$

on infinite dimensional spaces.

What however if, guided by singular measures arising from generalized functions such as e.g. Dirac's δ–function, one would look for measures arising from positive *generalized* functions in infinite dimensional analysis? White Noise Analysis [10] provides such a framework with Gaussian white noise as independent coordinates. Gaussian white noise gives of course rise to a Gaussian measure μ and we might begin our quest for vacuum densities in the corresponding L^2 space

$$L^2(d\mu) \equiv (L^2)\,.$$

In the light of the above however, this will not be rich enough, we shall need to go beyond (L^2) to a suitable space of generalized functions of white noise such as:

$$(S) \subset L^2(d\mu) \subset (S)^*.$$

Fortunately, by a theorem of Kondratiev and of Yokoi[2], and just as in finite dimensional distribution theory, all positive generalized functions turn out to be measures, hence candidates for vacuum densities. Do they provide an appropriate framework for QFT:

[2] A very general version of this theorem, valid even in non-Gaussian settings, can be found in [17], together with references to earlier work.

$$d\nu = \rho d\mu \text{ with } \rho \in (S)^* \,? \tag{14}$$

Indeed, physical vacuum densities (Euclidean and Minkowski, $P(\varphi)_2$, Sine-Gordon, etc.) are all Hida distributions:

$$\langle 0 | F(\varphi) | 0 \rangle = < \rho, F >$$

for suitable positive $\rho \in (S)^*$, cf. [23]. Proofs use the "Froehlich bounds" on moments

$$\langle 0 | \varphi^n(f) | 0 \rangle = O((n!)^{1/2} |f|_p^n) .$$

In particular such an estimate holds for canonical Bose fields if they obey a ϕ-bound

$$\pm\varphi(f) \leq aH + b|f|_p^2 + c, \text{ with } a, b, c \geq 0. \tag{15}$$

3.2 Dynamics in Terms of the Vacuum

Quantum field dynamics in terms of the ground state goes back to the work of Coester and Haag [7], and of Araki [6], in 1960. It has a counterpart in non-relativistic quantum mechanics: the "ground state representation", in contradistinction to the usual Schrödinger representation has been extensively studied since the seventies. Its strength is in handling extremely singular interactions such as e.g. the so-called pseudo-potentials or zero-range interactions. So let us take a quick glimpse at quantum mechanics in terms of the ground state.

Quantum Mechanics in Terms of the Ground State

	Schrödinger Representation	*Ground State Repn.*
State Space $\mathfrak{H}$	$\mathfrak{H} = L^2(R^n, d^n x)$	$\widetilde{\mathfrak{H}} = L^2(R^n, d\nu)$
Ground State Ω	$\Omega(x) = < x\|\Omega >$	$\widetilde{\Omega}(x) = 1$
States ψ	$\psi(x) = < x\|\psi >$	$\widetilde{\psi}(x) = \frac{< x\|\psi >}{< x\|\Omega >}$
Hamiltonian H	$H = \nabla_x^* \nabla_x + V(x)$	$\widetilde{H} = \nabla_x^* \nabla_x$
Energy Form $\varepsilon(\psi)$	$\varepsilon(\psi) = (\nabla\psi, \nabla\psi) + (\psi, V\psi)$	$\varepsilon(\psi) = \int \left(\nabla\widetilde{\psi}\right)^2 d\nu$
= Dirichlet Form for:	Brownian m. with killing	distorted Brownian m.

With the choice of the "ground state measure"

$$d\nu(x) = < x|\Omega >^2 d^n x \tag{16}$$

the two representations, whenever both exist, are unitarily equivalent.

On the other hand one may start by choosing a measure ν for the ground state representation

The energy form

$$\varepsilon(\psi) = \int \left(\nabla\widetilde{\psi}(x)\right)^2 d\nu(x) = <\psi|H|\psi> \tag{17}$$

then serves as definition of the Hamiltonian in terms of the ground state measure ν. Not all measures qualify. To obtain a unique self-adjoint Hamiltonian from the quadratic form we need that the form be closable, technically: that the gradient operator ∇ has a densely defined adjoint in $L^2(R^n, d\nu)$, a condition that in practice is not too hard to check.

Note the universal form of the Hamiltonian

$$\widetilde{H} = \nabla^* \nabla \tag{18}$$

for any ground state measure ν; it is the latter which contains exclusively the dynamical information.

The energy form ε is a Dirichlet form and gives rise to a diffusion process which solves the stochastic differential equation proposed by Ezawa, Klauder, and Shepp;

$$dy = \beta(y)dt + dB,$$

where the drift β is given in terms of the ground state wave function:

$$\beta(x) = \nabla_x \log \Omega^2(x) \tag{19}$$

This is quantum dynamics in terms of the ground state. What is its scope? It turns out that we are confronted with a vast extension of Schrödinger theory. While formally we can recuperate the potential from the ground state through the eigenvalue relation $H\Omega = 0$, which gives

$$V(x) = \frac{\Delta_x \Omega(x)}{\Omega(x)}, \tag{20}$$

the ground state representation extends to $\Omega(x)$, and hence to perfectly well-defined dynamics, for which V will not be a valid perturbation of the free Hamiltonian, or will not even exist! Ground states are always smoother than the corresponding potentials, so that the former may survive in limiting cases where the latter fail to exist. All this is treated in detail in [4], [5]; here we only mention zero range, multiparticle "pseudopotentials" as an example which fits nicely into this scheme.

Quantum field theory Our goal is, as in the quantum mechanical case, to define the Hamiltonian by its energy form. Where do we get the measures? Here the Kondratiev-Yokoi theorem is helpful since it states that all positive Hida distributions $\rho > 0$ are indeed measures, in the sense that

$$< \rho, \Psi > = \int \Psi(\omega)\, d\nu(\omega) \tag{21}$$

for smooth white noise functionals $\Psi \in (S)$. Hence, in terms of a generalized, positive "vacuum density" $\rho \in (S)^*$, we shall make the ansatz

$$\varepsilon(\Psi) = \langle \Psi | H | \Psi \rangle = < \rho, |\nabla \Psi|^2 > , \qquad \Psi \in (S) ,$$

since it generalizes

$$\varepsilon(\Psi) = \int |\nabla \Psi|^2(\omega)\, \rho(\omega)\, d\mu(\omega) = \int |\nabla \Psi|^2(\omega)\, d\nu(\omega)$$

to cases where

$$\rho = \frac{d\nu}{d\mu} \tag{22}$$

fails to be integrable, i.e. is a *generalized* density.

We consider ρ admissible if the resulting form is closable. Of course the first candidate that one will want to check is ρ with the source functional ("T-transform")

$$T\rho(f) = < \rho, e^{i<\omega,f>} > = e^{-\frac{1}{2}\left(f,(-\Delta+m^2)^{-1/2} f\right)} , \tag{23}$$

i.e. the vacuum density of a relativistic free field. This was done in [11].

For the vacuum densities of constructive quantum field theory closability was established in [2], [3], including the realization of the fields as infinite dimensional diffusions.

What about the field theory counterpart of the stochastic evolution equation? This comes up in the stochastic quantization program with all its difficulties, but a Gaussian "toy model" is quite tractable [12].

4 Feynman Integrals

There is indeed a quantum jump from the classical stochastic scenario with its statistical averaging over random effects on the one hand, and quantum mechanics with its superposition of amplitudes, capable of interfering constructively or destructively with each other. This quantum scenario is maybe most explicit in Feynman's formulation of quantum mechanics with its "sum over histories", often written as

$$N \int d^\infty x(\tau) e^{\frac{i}{\hbar} S[x]} \tag{24}$$

which as if by magic uses the purely classical action functional S to obtain a prescription for quantization. This feat alone is sufficient to explain why this expression has become one of physics' favorite concepts since its invention by Feynman in the forties.

It is all the more remarkable how, over half a century, mathematicians have had a rather hard time making sense of formulas such as (24).

The point is that mathematicians rightly say that there is no such thing as Feynman's infinite dimensional integration measure:

$$N \int d^{\infty}x(\tau) = ??? \tag{25}$$

Many remedies were proposed over the years, such as
– well if infinite dimension is a problem, let's look at finite dimensional approximations,
or:
– let's go "Euclidean", i.e. to imaginary time, where we have the Feynman-Kac formula

$$\langle F \rangle_E = N \int d^{\infty}x(\tau) e^{-\frac{1}{\hbar} S_E[x]} F[x] = \int d\mu F. \tag{26}$$

None of this is needed. All you need to recognize is that the Feynman average

$$\langle F \rangle \quad \text{“}=\text{”} \quad N \int d^{\infty}x(\tau) e^{\frac{i}{\hbar} S[x]} F[x] \tag{27}$$

while not the action of a measure on F, is indeed the action of a generalized white noise functional

$$\langle F \rangle = \langle I, F \rangle \text{ with } I \in (S)^*. \tag{28}$$

This in itself is remarkable since the Feynman integral becomes well-defined and manageable on a mathematical level, but the result goes far beyond an abstract existence theorem: the proof is constructive, and the construction is intuitive.

The basic ideas are

- Brownian paths

$$x(t_0 + \tau) = x_0 + \left(\frac{\hbar}{m}\right)^{1/2} B(\tau). \tag{29}$$

(Note that the "velocity" of Brownian motion and hence of these paths x is white noise ω: $B(\tau) = \int_0^{\tau} \omega(s) ds$.)

- The free Feynman integrand should then be

$$I_0(x, t \mid x_0, t_0) \tag{30}$$

$$= N \exp\left(\frac{i+1}{2} \int_{\mathbf{R}} \omega^2(\tau)\, d\tau\right) \delta(x(t) - x), \tag{31}$$

where the exponential contains the $\frac{i}{\hbar}$ times the free action, while the δ-function serves to pin down the final position of the paths; the extra seemingly spurious real term in the exponent actually is crucial to compensate the Gaussian fall-off of the white noise measure.

- To tackle this expression it is helpful to calculate its "T -transform" defined for general(ized) white noise functionals Φ as follows

$$T\Phi(\Phi) = E\left(\Phi(\omega)e^{i\int \Phi(t)\omega(t)dt}\right),$$

i.e. we should calculate an (infinite dimensional) Gauss-Fourier transform.

This can be done rather straightforwardly for the free case and gives, with $m = \hbar = 1$,

$$TI_0(\Phi) = \frac{1}{(2\pi i\,|t-t_0|)^{\frac{d}{2}}} \exp\left[-\frac{i}{2}\int_{\mathbf{R}} \xi^2(\tau)\,\mathrm{d}\tau \right. \tag{32}$$

$$\left. -\frac{1}{2i\,|t-t_0|}\left(\int_{t_0}^{t} \xi(\tau)\,\mathrm{d}\tau + x - x_0\right)^2\right].$$

In fact

$$TI_0(\xi) \equiv K_0^{(\xi)}(x,t|x_0,t_0) \tag{33}$$

has a physical meaning, it obeys a Schrödinger equation

$$\left(i\partial_t + \frac{1}{2}\Delta_d - \dot{\xi}(t)\cdot x\right) K_0^{(\xi)}(x,t|x_0,t_0) = 0 \tag{34}$$

with the initial condition

$$\lim_{t\searrow t_0} K_0^{(\xi)}(x,t|x_0,t_0) = \delta(x-x_0).$$

4.1 The Interactions

The big question is: Which potentials may be included in this framework? Over the recent years various classes of admissible potentials have been identified [8], [9], [13], [14], [18], such as potentials which are superpositions of delta functions, or Fourier transforms of measures. The most recent, and rather surprising, class is given in the form

$$V(x) = \int_{\mathbf{R}^d} e^{\alpha\cdot x}\mathrm{d}m(\alpha). \tag{35}$$

where m is any complex measure with

$$\int_{\mathbf{R}^d} e^{C|\alpha|}\,\mathrm{d}|m|(\alpha) < \infty, \quad \forall C > 0. \tag{36}$$

Example 2. Dirac measure $m(\alpha) := g\,\delta_a(\alpha)$, $g \in \mathbf{R}$., *i.e.* $V(x) = g\,e^{ax}$. *Likewise, potentials proportional to* $\sinh(ax)$, $\cosh(ax)$.

Example 3. The *Morse potential* $V(x) := g(e^{-2ax} - 2\gamma e^{-ax})$ *with* $g, a, x \in \mathbf{R}$ *and* $\gamma > 0$.

Example 4. A Gaussian measure m *gives* $V(x) = ge^{bx^2}$ *with* $b \in \mathbf{R}$.

Example 5. Entire functions of arbitrary high order of growth are in this class. $m(\alpha) := \Theta(\alpha)\exp\left(-k\alpha^{1+1/n}\right)$ *with* n, $k > 0 \Rightarrow V$ *is entire of order* $1 + n$.

For all of these, the construction of the Feynman integrand I is perturbative

$$I = I_0 \cdot \exp\left(-i\int_{t_0}^{t} V(x(\tau))\,\mathrm{d}\tau\right) \tag{37}$$

$$= \sum_{n=0}^{\infty} \frac{(-i)^n}{n!} \int_{[t_0,t]^n} \int_{\mathbf{R}^{dn}} I_0 \cdot \prod_{j=1}^{n} e^{\alpha_j \cdot x(\tau_j)} \prod_{j=1}^{n} \mathrm{d}m(\alpha_j)\,\mathrm{d}^n\tau \tag{38}$$

Theorem 2. *Let* V *be as above. Then*

$$I := \sum_{n=0}^{\infty} \frac{(-i)^n}{n!} \int_{[t_0,t]^n} \int_{\mathbf{R}^{dn}} I_0 \cdot \prod_{j=1}^{n} e^{\alpha_j \cdot x(\tau_j)} \prod_{j=1}^{n} \mathrm{d}m(\alpha_j)\,\mathrm{d}^n\tau \tag{39}$$

exists as a generalized white noise functional.

The proof can be found in [18] and is reasonably straightforward: the principal object we need to control is

$$I_0 \cdot \prod_{j=1}^{n} e^{\alpha_j \cdot x(\tau_j)}. \tag{40}$$

Its T-transform is an exercise in (generalized) Gaussian integration. It can be done in closed form and the Characterization Theorem 1 then ensures that we are dealing with a bona fide generalized white noise functional. The rest are integrations and limits which we control by the two corollaries of Theorem 1.

The main issue from the physics point of view is of course whether the corresponding Feynman integral solves the Schrödinger equation as in the free case (34):

Theorem 3. *Let V be as above. Then the T-transform of I solves the Schrödinger equation for all $x, x_0, t_0 < t$*

$$\left(i\frac{\partial}{\partial t}+\frac{1}{2}\Delta_d-gV\left(x\right)-x\cdot\overset{\bullet}{\xi}\left(t\right)\right)K^{(\xi)}\left(x,t\mid x_0,t_0\right)=0. \tag{41}$$

with initial condition

$$\lim_{t\searrow t_0}K^{(\xi)}(x,t\mid x_0,t_0)=\delta(x-x_0)\ . \tag{42}$$

The proof is by verification, and in fact the result extends to time dependent potentials. We shall not reproduce here the explicit, but longish form of the propagator for the general case. Instead, because of its interest in applications we present a particular example in more detail.

4.2 The Morse Potential

Its Hamilton operator is

$$H:=-\frac{1}{2}\Delta+g\left(e^{-2ax}-2\gamma e^{-ax}\right)\ . \tag{43}$$

Remark 1. H is essentially self-adjoint for $g\geq 0$ and it is not essentially self-adjoint for $g<0$.

The Green function, the eigenvectors and the discrete eigenvalues are not analytic in g.

The propagator, with $\xi\equiv 0$, is in this case

$$\begin{aligned} K(x,t\mid x_0,t_0) &= \\ & K_0\left(x,t\mid x_0,t_0\right)\sum_{n=0}^{\infty}\frac{(-ig)^n}{n!}\left(t-t_0\right)^n\sum_{j_1,\dots,j_n=1}^{2}\left(-2\gamma\right)^{2n-\sum_{k=1}^{n}j_k} \\ & \times\int_{[0,1]^n}\exp\left\{-a\sum_{l=1}^{n}j_l\left(\sigma_l x+\left(1-\sigma_l\right)x_0\right)\right\} \\ & \times\exp\left\{-\frac{i}{2}\left(t-t_0\right)a^2\sum_{l=1}^{n}\sum_{k=1}^{n}j_kj_l\left[\sigma_j\sigma_k-\sigma_j\wedge\sigma_k\right]\right\}\mathrm{d}^n\sigma. \end{aligned} \tag{44}$$

It is not hard to verify that, in spite of the above remark, this is a convergent series! In fact it is implicit in our construction (39) that the propagators for all the potentials (35) admit convergent perturbation series *for their propagators.* It is an easy exercise to verify that this is not the case for the corresponding "Euclidean" heat equations.

References

1. S. Albeverio, T. Hida, J. Potthoff, L. Streit: Phys. Lett B **217**, 511 (1989)
2. S. Albeverio, T. Hida, J. Potthoff, M. Roeckner, L. Streit: Rev. Math. Phys. **1**, 291 (1990)
3. S. Albeverio, T. Hida, J. Potthoff, M. Roeckner, L. Streit: Rev. Math. Phys. **1**, 313 (1990)
4. S. Albeverio, R. Hoegh-Krohn, L. Streit: J. Math. Phys. **18**, 907 (1977)
5. S. Albeverio, R. Hoegh-Krohn, L. Streit: J. Math. Phys. **21**, 1636 (1980)
6. H. Araki: J. Math. Phys. **1**, 492 (1960)
7. F. Coester, R. Haag: Phys. Rev. **117**, 1137 (1960)
8. M. de Faria, J. Potthoff, L. Streit: J. Math. Phys. **32**, 2123 (1991)
9. M. Grothaus, D.C. Khandekar, J.L. Silva, L. Streit: J. Math. Phys. **38**, 3278 (1997)
10. T. Hida, H.H. Kuo, J. Potthoff, L. Streit: *White Noise. An Infinite Dimensional Calculus* (Kluwer Academic, 1993)
11. T. Hida, J. Potthoff, L. Streit: Comm. Math. Phys **116**, 235 (1988)
12. T. Hida, L. Streit: Nagoya Math. J. **68**, 21 (1977)
13. T. Hida, L. Streit: Stoch. Proc. Appl. **16**, 55 (1983)
14. D.C. Khandekar, L. Streit: Annalen d. Physik **1**, 49 (1992)
15. Y.G. Kondratiev, P. Leukert, L. Streit: Acta Appl. Math. **44**, 269 (1996)
16. Y.G. Kondratiev, P. Leukert, J. Potthoff, L. Streit, W. Westerkamp: J. Funct. Analysis **141**, 301 (1996)
17. Yu. Kondratiev, L. Streit, W. Westerkamp, Y.A. Yan: Hiroshima Math. J. **28**, 213 (1998)
18. T. Kuna, L. Streit, W. Westerkamp: J. Math. Phys. **39**, 4476 (1998)
19. H.H. Kuo: *White Noise Distribution Theory* (CRC Press, 1996)
20. A Lascheck, P. Leukert, L. Streit, W. Westerkamp: Rep. Math. Phys. **33**, 221 (1993)
21. A. Lascheck, P. Leukert, L. Streit, W. Westerkamp: Soochow Math. J. **20**, 401 (1994)
22. N. Obata: *White Noise Calculus and Fock Space* (Springer LNM 1577, 1994)
23. J. Potthoff, L. Streit: J. Funct. Anal. **111**, 295 (1993)
24. L. Streit: Acta Physica Austriaca, Suppl. II, 2 (1966)

Seminars

In addition to the lectures several participants gave interesting seminar talks fitting to the general theme of the School. In what follows, we give a complete list of the seminar speakers (alphabetically ordered) and the titles of their contributions.

A. Kling (TU Wien)

BRST-cohomology of super-D-strings

W. H. Klink (Univ. Graz)

Quantization and point-form relativistic quantum mechanics

N. Ilieva (Univ. Wien)

2-dimensional anyons and the temperature dependence of commutator anomalies

L. Martinovic (Acad. Sci. Bratislava)

Symmetries and vacuum structure on the light front

A. Nefediev (ITEP Moscow)

Relativistic constrained systems in the einbein field formalism

H. Nikolic (Rudjer Boskovic Inst. Zagreb)

Classical relativistic effects in non-inertial frames treated by Fermi coordinates

D. Nogradi (Univ. Budapest)

Geometric quantization of global Liouville mechanics

R. Pullirsch (TU Wien)

Quantization of chaotic field theories

A. Ruffing (TU München)

Quantization and special functions

L. Snobl (TU Prague)

Construction of quantum doubles from solutions of the Yang-Baxter equation

D. Sorokin (INFN Padova)

Superbranes in the superembedding approach

T. Strobl (Univ. Jena)

Group theoretical and projection quantization

T. Sykora (Univ. Prague)

Schwinger terms in the fully quantized 1+1 dimensional model – exact solution

L. Theussl (ISN Grenoble)

From the Bethe-Salpeter equation to non-relativistic approaches with effective two-body interactions

S. Vernov (State Univ. Moscow)

Quantization close to non-stationary classical fields in terms of Bogoliubov group variables

M. Volkov (Univ. Jena)

Euclidean Freedman-Schwarz model

Lecture Notes in Physics

For information about Vols. 1–536 please contact your bookseller or Springer-Verlag

Vol. 537: S. Cotsakis, G. W. Gibbons (Eds.), Mathematical and Quantum Aspects of Relativity and Cosmology. Proceedings, 1998. XII, 251 pages. 1999.

Vol. 538: Ph. Blanchard, D. Giulini, E. Joos, C. Kiefer, I.-O. Stamatescu (Eds.), Decoherence: Theoretical, Experimental, and Conceptual Problems. Proceedings, 1998. XII, 345 pages. 2000.

Vol. 539: A. Borowiec, W. Cegła, B. Jancewicz, W. Karwowski (Eds.), Theoretical Physics. Fin de Siècle. Proceedings, 1998. XX, 319 pages. 2000.

Vol. 540: B. G. Schmidt (Ed.), Einstein's Field Equations and Their Physical Implications. Selected Essays. 1999. XIII, 429 pages. 2000.

Vol. 541: J. Kowalski-Glikman (Ed.), Towards Quantum Gravity. Proceedings, 1999. XII, 376 pages. 2000.

Vol. 542: P. L. Christiansen, M. P. Sørensen, A. C. Scott (Eds.), Nonlinear Science at the Dawn of the 21st Century. Proceedings, 1998. XXVI, 458 pages. 2000.

Vol. 543: H. Gausterer, H. Grosse, L. Pittner (Eds.), Geometry and Quantum Physics. Proceedings, 1999. VIII, 408 pages. 2000.

Vol. 544: T. Brandes (Ed.), Low-Dimensional Systems. Interactions and Transport Properties. Proceedings, 1999. VIII, 219 pages. 2000

Vol. 545: J. Klamut, B. W. Veal, B. M. Dabrowski, P. W. Klamut, M. Kazimierski (Eds.), New Developments in High-Temperature Superconductivity. Proceedings, 1998. VIII, 275 pages. 2000.

Vol. 546: G. Grindhammer, B. A. Kniehl, G. Kramer (Eds.), New Trends in HERA Physics 1999. Proceedings, 1999. XIV, 460 pages. 2000.

Vol. 547: D. Reguera, G. Platero, L. L. Bonilla, J. M. Rubí(Eds.), Statistical and Dynamical Aspects of Mesoscopic Systems. Proceedings, 1999. XII, 357 pages. 2000.

Vol. 548: D. Lemke, M. Stickel, K. Wilke (Eds.), ISO Surveys of a Dusty Universe. Proceedings, 1999. XIV, 432 pages. 2000.

Vol. 549: C. Egbers, G. Pfister (Eds.), Physics of Rotating Fluids. Selected Topics, 1999. XVIII, 437 pages. 2000.

Vol. 550: M. Planat (Ed.), Noise, Oscillators and Algebraic Randomness. Proceedings, 1999. VIII, 417 pages. 2000.

Vol. 551: B. Brogliato (Ed.), Impacts in Mechanical Systems. Analysis and Modelling. Lectures, 1999. IX, 273 pages. 2000.

Vol. 552: Z. Chen, R. E. Ewing, Z.-C. Shi (Eds.), Numerical Treatment of Multiphase Flows in Porous Media. Proceedings, 1999. XXI, 445 pages. 2000.

Vol. 553: J.-P. Rozelot, L. Klein, J.-C. Vial Eds.), Transport of Energy Conversion in the Heliosphere. Proceedings, 1998. IX, 214 pages. 2000.

Vol. 554: K. R. Mecke, D. Stoyan (Eds.), Statistical Physics and Spatial Statistics. The Art of Analyzing and Modeling Spatial Structures and Pattern Formation. Proceedings, 1999. XII, 415 pages. 2000.

Vol. 555: A. Maurel, P. Petitjeans (Eds.), Vortex Structure and Dynamics. Proceedings, 1999. XII, 319 pages. 2000.

Vol. 556: D. Page, J. G. Hirsch (Eds.), From the Sun to the Great Attractor. X, 330 pages. 2000.

Vol. 557: J. A. Freund, T. Pöschel (Eds.), Stochastic Processes in Physics, Chemistry, and Biology. X, 330 pages. 2000.

Vol. 558: P. Breitenlohner, D. Maison (Eds.), Quantum Field Theory. Proceedings, 1998. VIII, 323 pages. 2000

Vol. 559: H.-P. Breuer, F. Petruccione (Eds.), Relativistic Quantum Measurement and Decoherence. Proceedings, 1999. X, 140 pages. 2000.

Vol. 560: S. Abe, Y. Okamoto (Eds.), Nonextensive Statistical Mechanics and Its Applications. IX, 272 pages. 2001.

Vol. 561: H. J. Carmichael, R. J. Glauber, M. O. Scully (Eds.), Directions in Quantum Optics. XVII, 369 pages. 2001.

Vol. 562: C. Lämmerzahl, C. W. F. Everitt, F. W. Hehl (Eds.), Gyros, Clocks, Interferometers...: Testing Relativistic Gravity in Space. XVII,507 pages. 2001.

Vol. 563: F. C. Lázaro, M. J. Arévalo (Eds.), Binary Stars. Selected Topics on Observations and Physical Processes. 1999. X, 332 pages. 2001.

Vol. 564: T. Pöschel, S. Luding (Eds.), Granular Gases. VIII, 457 pages. 2001.

Vol. 565: E. Beaurepaire, F. Scheurer, G. Krill, J.-P. Kappler (Eds.), Magnetism and Synchrotron Radiation. XIV, 396 pages. 2001.

Vol. 566: J. L. Lumley (Ed.), Fluid Mechanics and the Environment: Dynamical Approaches. VIII, 412 pages. 2001.

Vol. 567: D. Reguera, L. L. Bonilla, J. M. Rubí (Eds.), Coherent Structures in Complex Systems. IX, 465 pages. 2001.

Vol. 568: P. A. Vermeer, S. Diebels, W. Ehlers, H. J. Herrmann, S. Luding, E. Ramm (Eds.), Continuous and Discontinuous Modelling of Cohesive-Frictional Materials. XIV, 307 pages. 2001.

Vol. 569: M. Ziese, M. J. Thornton (Eds.), Spin Electronics. XVII, 493 pages. 2001.

Vol. 570: S. G. Karshenboim, F. S. Pavone, F. Bassani, M. Inguscio, T. W. Hänsch (Eds.), The Hydrogen Atom: Precision Physics of Simple Atomic Systems. XXIII, 293 pages. 2001.

Vol. 572: H. Latal, W. Schweiger (Eds.), Methods of Quantization. XI, 224 pages. 2001.

Monographs

For information about Vols. 1–25
please contact your bookseller or Springer-Verlag

Vol. m 26: M. Dineykhan, G. V. Efimov, G. Ganbold, S. N. Nedelko, Oscillator Representation in Quantum Physics. IX, 279 pages. 1995.

Vol. m 27: J. T. Ottesen, Infinite Dimensional Groups and Algebras in Quantum Physics. IX, 218 pages. 1995.

Vol. m 28: O. Piguet, S. P. Sorella, Algebraic Renormalization. IX, 134 pages. 1995.

Vol. m 29: C. Bendjaballah, Introduction to Photon Communication. VII, 193 pages. 1995.

Vol. m 30: A. J. Greer, W. J. Kossler, Low Magnetic Fields in Anisotropic Superconductors. VII, 161 pages. 1995.

Vol. m 31 (Corr. Second Printing): P. Busch, M. Grabowski, P.J. Lahti, Operational Quantum Physics. XII, 230 pages. 1997.

Vol. m 32: L. de Broglie, Diverses questions de mécanique et de thermodynamique classiques et relativistes. XII, 198 pages. 1995.

Vol. m 33: R. Alkofer, H. Reinhardt, Chiral Quark Dynamics. VIII, 115 pages. 1995.

Vol. m 34: R. Jost, Das Märchen vom Elfenbeinernen Turm. VIII, 286 pages. 1995.

Vol. m 35: E. Elizalde, Ten Physical Applications of Spectral Zeta Functions. XIV, 224 pages. 1995.

Vol. m 36: G. Dunne, Self-Dual Chern-Simons Theories. X, 217 pages. 1995.

Vol. m 37: S. Childress, A.D. Gilbert, Stretch, Twist, Fold: The Fast Dynamo. XI, 406 pages. 1995.

Vol. m 38: J. González, M. A. Martín-Delgado, G. Sierra, A. H. Vozmediano, Quantum Electron Liquids and High-Tc Superconductivity. X, 299 pages. 1995.

Vol. m 39: L. Pittner, Algebraic Foundations of Non-Com-mutative Differential Geometry and Quantum Groups. XII, 469 pages. 1996.

Vol. m 40: H.-J. Borchers, Translation Group and Particle Representations in Quantum Field Theory. VII, 131 pages. 1996.

Vol. m 41: B. K. Chakrabarti, A. Dutta, P. Sen, Quantum Ising Phases and Transitions in Transverse Ising Models. X, 204 pages. 1996.

Vol. m 42: P. Bouwknegt, J. McCarthy, K. Pilch, The W3 Algebra. Modules, Semi-infinite Cohomology and BV Algebras. XI, 204 pages. 1996.

Vol. m 43: M. Schottenloher, A Mathematical Introduction to Conformal Field Theory. VIII, 142 pages. 1997.

Vol. m 44: A. Bach, Indistinguishable Classical Particles. VIII, 157 pages. 1997.

Vol. m 45: M. Ferrari, V. T. Granik, A. Imam, J. C. Nadeau (Eds.), Advances in Doublet Mechanics. XVI, 214 pages. 1997.

Vol. m 46: M. Camenzind, Les noyaux actifs de galaxies. XVIII, 218 pages. 1997.

Vol. m 47: L. M. Zubov, Nonlinear Theory of Dislocations and Disclinations in Elastic Body. VI, 205 pages. 1997.

Vol. m 48: P. Kopietz, Bosonization of Interacting Fermions in Arbitrary Dimensions. XII, 259 pages. 1997.

Vol. m 49: M. Zak, J. B. Zbilut, R. E. Meyers, From Instability to Intelligence. Complexity and Predictability in Nonlinear Dynamics. XIV, 552 pages. 1997.

Vol. m 50: J. Ambjørn, M. Carfora, A. Marzuoli, The Geometry of Dynamical Triangulations. VI, 197 pages. 1997.

Vol. m 51: G. Landi, An Introduction to Noncommutative Spaces and Their Geometries. XI, 200 pages. 1997.

Vol. m 52: M. Hénon, Generating Families in the Restricted Three-Body Problem. XI, 278 pages. 1997.

Vol. m 53: M. Gad-el-Hak, A. Pollard, J.-P. Bonnet (Eds.), Flow Control. Fundamentals and Practices. XII, 527 pages. 1998.

Vol. m 54: Y. Suzuki, K. Varga, Stochastic Variational Approach to Quantum-Mechanical Few-Body Problems. XIV, 324 pages. 1998.

Vol. m 55: F. Busse, S. C. Müller, Evolution of Spontaneous Structures in Dissipative Continuous Systems. X, 559 pages. 1998.

Vol. m 56: R. Haussmann, Self-consistent Quantum Field Theory and Bosonization for Strongly Correlated Electron Systems. VIII, 173 pages. 1999.

Vol. m 57: G. Cicogna, G. Gaeta, Symmetry and Perturbation Theory in Nonlinear Dynamics. XI, 208 pages. 1999.

Vol. m 58: J. Daillant, A. Gibaud (Eds.), X-Ray and Neutron Reflectivity: Principles and Applications. XVIII, 331 pages. 1999.

Vol. m 59: M. Kriele, Spacetime. Foundations of General Relativity and Differential Geometry. XV, 432 pages. 1999.

Vol. m 60: J. T. Londergan, J. P. Carini, D. P. Murdock, Binding and Scattering in Two-Dimensional Systems. Applications to Quantum Wires, Waveguides and Photonic Crystals. X, 222 pages. 1999.

Vol. m 61: V. Perlick, Ray Optics, Fermat's Principle, and Applications to General Relativity. X, 220 pages. 2000.

Vol. m 62: J. Berger, J. Rubinstein, Connectivity and Superconductivity. XI, 246 pages. 2000.

Vol. m 63: R. J. Szabo, Ray Optics, Equivariant Cohomology and Localization of Path Integrals. XII, 315 pages. 2000.

Vol. m 64: I. G. Avramidi, Heat Kernel and Quantum Gravity. X, 143 pages. 2000.

Vol. m 65: M. Hénon, Generating Families in the Restricted Three-Body Problem. Quantitative Study of Bifurcations. XII, 301 pages. 2001.

Vol. m 66: F. Calogero, Classical Many-Body Problems Amenable to Exact Treatments. XIX, 749 pages. 2001.

Vol. m 67: A. S. Holevo, Statistical Structure of Quantum Theory. IX, 159 pages. 2001.

GPSR Compliance

The European Union's (EU) General Product Safety Regulation (GPSR) is a set of rules that requires consumer products to be safe and our obligations to ensure this.

If you have any concerns about our products, you can contact us on ProductSafety@springernature.com

In case Publisher is established outside the EU, the EU authorized representative is:

Springer Nature Customer Service Center GmbH
Europaplatz 3
69115 Heidelberg, Germany

Batch number: 10584448

Printed by Printforce, the Netherlands